FOOD COMPOSITION AND ANALYSIS

Methods and Strategies

FOOD COMPOSITION AND ANALYSIS

Methods and Strategies

Edited by
A. K. Haghi, PhD, and Elizabeth Carvajal-Millan, PhD

Apple Academic Press

TORONTO   NEW JERSEY

Apple Academic Press Inc. | Apple Academic Press Inc.
3333 Mistwell Crescent | 9 Spinnaker Way
Oakville, ON L6L 0A2 | Waretown, NJ 08758
Canada | USA

©2014 by Apple Academic Press, Inc.

First issued in paperback 2021

Exclusive worldwide distribution by CRC Press, a member of Taylor & Francis Group
No claim to original U.S. Government works

ISBN 13: 978-1-77463-315-1 (pbk)
ISBN 13: 978-1-926895-85-7 (hbk)

Library of Congress Control Number: 2014934523

Library and Archives Canada Cataloguing in Publication

Food composition and analysis: methods and strategies/edited by A.K. Haghi, PhD, and Elizabeth Carvajal-Millan, PhD.

Includes bibliographical references and index.
ISBN 978-1-926895-85-7 (bound)
1. Food--Composition. 2. Food--Composition--Methodology. 3. Food--Analysis. I. Haghi, A. K., editor of compilation II. Carvajal-Millan, Elizabeth, editor of compilation

TX541.F65 2014   664'.07   C2014-901332-9

Apple Academic Press also publishes its books in a variety of electronic formats. Some content that appears in print may not be available in electronic format. For information about Apple Academic Press products, visit our website at **www.appleacademicpress.com** and the CRC Press website at **www.crcpress.com**

ABOUT THE EDITORS

A. K. Haghi, PhD

A. K. Haghi, PhD, holds a BSc in urban and environmental engineering from the University of North Carolina (USA); a MSc in mechanical engineering from North Carolina A&T State University (USA); a DEA in applied mechanics, acoustics and materials from Université de Technologie de Compiègne (France); and a PhD in engineering sciences from Université de Franche-Comté (France). He is the author and editor of 65 books as well as 1000 published papers in various journals and conference proceedings. Dr. Haghi has received several grants, consulted for a number of major corporations, and is a frequent speaker to national and international audiences. Since 1983, he served as a professor at several universities. He is currently Editor-in-Chief of the *International Journal of Chemoinformatics and Chemical Engineering* and *Polymers Research Journal* and on the editorial boards of many international journals. He is a member of the Canadian Research and Development Center of Sciences and Cultures (CRDCSC), Montreal, Quebec, Canada.

Elizabeth Carvajal-Millan, PhD

Elizabeth Carvajal-Millan, PhD, has been working as a Research Scientist at the Research Center for Food and Development (CIAD) in Hermosillo, Mexico, since 2005. She obtained her PhD in France at Ecole Nationale Supérieure Agronomique de Montpellier (ENSAM), her MSc degree at CIAD, and undergraduate degree at the University of Sonora, Sonora, Mexico. Her research interests are focused on biopolymers, particularly in the extraction and characterization of polysaccharides of high value added co-products, recovered from the food industry, especially ferulated arabinoxylans. She has published more than 40 refereed papers, 15 chapters in books, over 60 conference presentations, and one patent.

CONTENTS

LIST OF CONTRIBUTORS

M. Anachkov
Institute of Catalysis, Bulgarian Academy of Sciences, Sofia 1113, Bulgaria.

E. A. Averjanova
Sochi Branch of the Russian Geographical Society, 354024, Sochi, Kurortny pr., 113.
E-mail: geo@opensochi.org
Tel. +7 (8622) 619857

R. Badonia
Veraval Research Centre, Central Institute of Fisheries Technology, 362 029, Gujarat.

Vasudha Bansal
Agrionics Division (DU-1), Central Scientific Instruments Organisation (CSIO), CSIR
Chandigarh, India.

Mário C. J. Bier
Bioprocess Engineering and Biotechnology Department, Federal University of Paraná (UFPR), CEP 81531-990, Curitiba–PR, Brazil.

V.V. Borodai
Eco biotechnology and biodiversity department, Biotechnology faculty, National University of Life and Environmental Sciences of Ukraine, Kyiv 15, Geroiv Oborony str., Kyiv, Ukraine.
E-mail: veraboro@gmail.com

C. Bueno-Ferrer
Analytical Chemistry, Nutrition & Food Sciences Department, University of Alicante, P.O. Box 99, E-03080, Alicante, Spain.
Telephone: +34-965909660
E-mail: alfjimenez@ua.es

N. Burgos
Analytical Chemistry, Nutrition & Food Sciences Department, University of Alicante, P.O. Box 99, E-03080, Alicante, Spain.
Telephone: +34-965909660
E-mail: alfjimenez@ua.es

A. V. Bychkova
Federal state budgetary institution of science Emanuel Institute of Biochemical Physics of Russian Academy of Sciences, Kosygina str., 4, Moscow, 119334, Russia.
E-mail: annb0005@yandex.ru

Elizabeth Carvajal-Millán
Laboratory of Biopolymers, CTAOA. Research Center for Food and Development, CIAD, AC., Hermosillo, Sonora 83000, Mexico.
E-mail address: arascon@.ciad.mx

Cesar de Morais Coutinho
Department of Food Research - Campus of Frederico Westphalen – CAFW, Federal University of Santa Maria – RS.

Himani Dhanze
Research Scholar, Department of Veterinary Public Health, College of Veterinary & Animal Sciences, CSK-HPKV, Palampur-176062, Himachal Pradesh, India.

Karla Escarcega-Loya
Laboratory of Biotechnology, CTAOV, Research Center for Food and Development, CIAD, A.C., Hermosillo, Sonora 83000, Mexico.

Leifa Fan
Institute of Horticulture, Zhejiang Academy of Agricultural Sciences, 139 Shiqiao Road, Hangzhou-ZJ, P.R. China, 310021.

Sascha Habu
Federal University of Paraná, Dept. Bioprocess and Biotechnology. Rua Francisco H. dos Santos - Centro Politécnico, Jardim das Américas - Curitiba/Pr – Brazil.

E. A. Hassan
Department of Chemistry, Faculty of Science, AL-Azhar University, Cairo, Egypt.

A. Jiménez
Analytical Chemistry, Nutrition & Food Sciences Department, University of Alicante, P.O. Box 99, E-03080, Alicante, Spain.
Telephone: +34-965909660
E-mail: alfjimenez@ua.es

Porteen Kannan
Assistant Professor Department of Veterinary Public Health and Epidemiology, Madras Veterinary College, Chennai, India.

E. Chris Kazala
Department of Agricultural, Food & Nutritional Science, University of Alberta, 4-10 Agriculture/Forestry Centre, Edmonton, Alberta, Canada, T6G 2P5.

L. G. Kharuta
Sochi Institute of Russian people's friendship university, 354340, Sochi, Kuibyshev str., 32.
Fax +7 (8622) 411043
E-mail: sfrudn@rambler.ru

Paul P. Kolodziejczyk
Department of Agricultural, Food & Nutritional Science, University of Alberta, 4-10 Agriculture/Forestry Centre, Edmonton, Alberta, Canada, T6G 2P5.
Biolink Consultancy Inc., P.O. Box 430, New Denver, B.C., Canada, V0G 1S0.

E. I. Korotkova
Tomsk Polytechnic University, 30 Lenin Street, 634050, Tomsk, Russia
E-mail: eikor@mail.ru

A. L. Kovarski
Federal state budgetary institution of science Emanuel Institute of Biochemical Physics of Russian Academy of Sciences, Kosygina str., 4, Moscow, 119334, Russia.

V. B. Leonova
Federal state budgetary institution of science Emanuel Institute of Biochemical Physics of Russian Academy of Sciences, Kosygina str., 4, Moscow, 119334, Russia.

Y. López-Franco
Laboratory of Biopolymers, CTAOA, Research Center for Food and Development, CIAD, AC., Hermosillo, Sonora 83000, Mexico.

J. Lizardi-Mendoza
Laboratory of Biopolymers, CTAOA, Research Center for Food and Development, CIAD, AC., Hermosillo, Sonora 83000, Mexico.

B.G. Mane
Assistant Professor, Department of Livestock Products Technology.

A. L. Martínez-López
Laboratory of Biopolymers, CTAOA. Research Center for Food and Development, CIAD, AC., Hermosillo, Sonora 83000, Mexico.

Adriane B. P. Medeiros
Bioprocess Engineering and Biotechnology Department, Federal University of Paraná (UFPR), CEP 81531-990, Curitiba–PR, Brazil.

S. K. Mendiratta
Principal Scientist, Division of Livestock Products Technology, Indian Veterinary Research Institute, Izatnagar-243122, Uttar Pradesh, India.

V. M. Misin
Emanuel Institute of Biochemical Physics, Russian Academy of Sciences, 4 Kosygin Street, 119334 Moscow, Russia
E-mail: misin@sky.chph.ras.ru
E-mail: Natnik48s@yandex.ru

C.O. Mohan
Veraval Research Centre, Central Institute of Fisheries Technology, Gujarat 362 029.
E-mail: comohan@gmail.com

M. S. Mohy Eldin
Polymer Materials Research Department, Advanced Technology and New Materials Research Institute, City for Scientific Research and Technological Applications, New Boarg Elarab City, 21934, Alexandria, Egypt.

A Morales-Ortega
Research Center for Food and Development, CIAD, AC., Hermosillo, Sonora 83000, Mexico,

A. M. Omer
Polymer Materials Research Department, Advanced Technology and New Materials Research Institute, City for Scientific Research and Technological Applications, New Boarg Elarab City, 21934, Alexandria, Egypt.

A. E. Ordyan
Emanuel Institute of Biochemical Physics Russian Academy of Sciences, 4 Kosygin Street, 119334 Moscow, Russia.
E-mail: Natnik48s@yandex.ru

Xiao Qiu
Department of Food and Bioproduct Sciences, University of Saskatchewan, 51 Campus Drive, Saskatoon, Saskatchewan, Canada, S7N 5A8.

Nithya Quintoil
Teaching assistant Department of Veterinary Public Health and Epidemiology, Rajiv Gandhi Institute of Veterinary Education and Research, Puducherry, India.

S. Rakovsky
Institute of Catalysis, Bulgarian Academy of Sciences, Sofia 1113, Bulgaria.

Agustín Rascon-Chu
Laboratory of Biotechnology, CTAOV. Research Center for Food and Development, CIAD, AC., Hermosillo, Sonora 83000, Mexico.

C. N. Ravishankar
Fish Processing Division, Central Institute of Fisheries Technology, Cochin, 628 029.

S. Remya
Veraval Research Centre, Central Institute of Fisheries Technology, Gujarat 362 029.

M. A. Rosenfeld
Federal state budgetary institution of science Emanuel Institute of Biochemical Physics of Russian Academy of Sciences, Kosygina str., 4, Moscow 119334, Russia.

Suzan C. Rossi
Bioprocess Engineering and Biotechnology Department, Federal University of Paraná (UFPR), CEP 81531-990, Curitiba–PR, Brazil.

A. E. Rybalko
Sochi Institute of Russian people's friendship university, 354340, Sochi, Kuibyshev str., 32.
E-mail: sfrudn@rambler.ru
Fax +7 (8622) 411043

Alfonso Sánchez
Laboratory of Biotechnology, CTAOV. Research Center for Food and Development, CIAD, A.C., Hermosillo, Sonora 83000, Mexico

N. N. Sazhina
Emanuel Institute of Biochemical Physics, Russian Academy of Sciences
4 Kosygin Street., 119334 Moscow, Russia.
E-mail: misin@sky.chph.ras.ru
E-mail: Natnik48s@yandex.ru

Saleh Shah
Alberta Innovates-Technology Futures, P.O. Bag 4000, Vegreville, Alberta, Canada, T9C 1T4.

A. B. Sharangi
Department of Spices and Plantation Crops, Faculty of Horticulture, Bidhan Chandra Krishi Viswavidyalaya (Agricultural University), Mohanpur, Nadia, WB (INDIA) ,Pin-741252
*Corresponding author: dr_absharangi@yahoo.co.in

Md. Wasim Siddiqui
Department of Food Science and Technology, Bihar Agricultural University, BAC, Sabour, Bhagalpur, Bihar (813210) India.

K. P. Skipina
Sochi Institute of Russian people's friendship university, 354340, Sochi, Kuibyshev str., 32.
E-mail: sfrudn@rambler.ru
Fax +7 (8622) 411043

Crystal L. Snyder
Department of Agricultural, Food & Nutritional Science, University of Alberta, 4-10 Agriculture/Forestry Centre, Edmonton, Alberta, Canada, T6G 2P5.

Carlos Ricardo Soccol
Bioprocess Engineering and Biotechnology Department, Federal University of Paraná (UFPR), CEP 81531-990, Curitiba–PR, Brazil.
Federal University of Paraná. Dept. Bioprocess and Biotechnology. Rua Francisco H. dos Santos - Centro Politécnico, Jardim das Américas - Curitiba/Pr – Brazil.

E. A. Soliman
Polymer Materials Research Department, Advanced Technology and New Materials Research Institute, City for Scientific Research and Technological Applications, New Boarg Elarab City, 21934, Alexandria, Egypt.

O. N. Sorokina
Federal state budgetary institution of science Emanuel Institute of Biochemical Physics of Russian Academy of Sciences, Kosygina str., 4, Moscow 119334, Russia.

T. K. Srinivasa Gopal
Central Institute of Fisheries Technology, Cochin, Kerala 628 029.

P Torres-Chavez
Department of Food Research & Graduate Program (DIPA), University of Sonora, Hermosillo, Sonora C.P. 83000, Mexico.

Luciana P. S. Vandenberghe
Bioprocess Engineering and Biotechnology Department, Federal University of Paraná (UFPR), CEP 81531-990, Curitiba–PR, Brazil.

Randall J. Weselake
Department of Agricultural, Food & Nutritional Science, University of Alberta, 4-10 Agriculture/Forestry Centre, Edmonton, Alberta, Canada, T6G 2P5.

G. E. Zaikov
N. M. Emanuel Institute of Biochemical Physics Russian Academy of Sciences, Moscow 119334, Russia.

LIST OF ABBREVIATIONS

ACL	Admissible concentration limit
AFLP	Amplified fragment length polymorphism
AO	Antioxidant
AOA	Antioxidant activity
AOPs	Advanced oxidation processes
AP	Active packaging
BAS	Biological active substances
CAT	Capillary agglutination test
CEPM	Continuous electrophoresis with porous membranes
CFUs	Colony forming units
CSNP	Chitosan nanoparticles
DDOS	Deodorized distillate of soybean oil
DM	Dry matter
DSC	Differential scanning calorimetry
EITB	Enzyme linked immunoelectro transfer blot
ELFA	Enzyme-linked fluorescent immunoassay
ELISA	Enzyme-linked immunosorbent assay
ES	Electrical stimulations
FDA	Food and drug administration
FTIR	Fourier-transform infrared spectroscopy
GAC	Granular activated carbon
GC/FID	Gas chromatography with flame-ionization detection
GC/MS	Gas chromatography-mass spectrometry
GFSE	Grapefruit seed extract
GHP	Good hygienic practices
GRAS	Generally recognized-as-safe
HACCP	Hazard analysis of critical control points
HOC	Halogenated organic compounds
HPLC	High-performance liquid chromatography
HS	Hard segments
HTC	Hard-to-cook
ICGFI	International consultative group on food irradiation
IFA	Immunofluorescence assay
INIFAP	Investigation in forestry, agriculture, and animal production
LA	Latex agglutination
LAPS	Light-addressable potentiometric sensors
MAP	Modified atmosphere packaging
MBE	Molecular beam epitaxy

MDSC	Modulated differential scanning calorimetry
MFE	Mercury film electrode
MFI	Myofibrillar Fragmentation Index
MNPs	Magnetic nanoparticles
MWCO	Molecular weight cut off
PACs	Polycyclic aromatic compounds
PCR	Polymerase chain reaction
PTM	Transmembrane pressure
RBPT	Rose Bengal plate test
ROS	Reactive oxygen species
RPLA	Reverse passive latex agglutination
SEM	Scanning electron microscope
SEM	Scanning electron microscopy
SET	Staphylococcal enterotoxin
TAA	Total antioxidant activity
TEAC	Trolox equivalent antioxidant capacity
TGA	Thermogravimetric Analysis
UV	Ultraviolet
VLSI	Very large scale integration
WEAX	Water extractable arabinoxylans
WHO	World Health Organization
WPI	Whey protein isolate

PREFACE

Many foods depend on additives for safety and stability or preservation. Foods are packaged to protect them and keep them in good condition while they are delivered to shops, stacked on shelves, or stored at home. This is a comprehensive advanced level book that provides thorough up-to-date coverage of a broad range of topics in food science and technology and describes avenues of advanced study in the field. The book explores key food commodities and food composition with an emphasis on the functional properties of each commodity.

The so-called HACCP (Hazard Analysis and Critical Control Points) acronym is well known in the food industry in relation to the management of microbiological, chemical, and physical risks. This book is designed to help current and prospective researchers in this field.

This volume introduces and surveys the broad and complex interrelationships among food ingredients and processing, and explores how these factors influence food quality and safety. The book in food science is also a valuable reference for professionals in food processing, as well as for those working in fields that service, regulate, or otherwise interface with the food industry.

This book is divided into 21 chapters:

Thermoplastic polyurethanes bio-based TPUs were synthesized in chapter 1 from a di-functional dimmer fatty acid-based polyol obtained from rapeseed oil, MDI, and BDO at four HS a content that is 10–40 wt%. The polyol characteristics determined the structure and properties of TPUs. The FTIR-ATR spectra confirmed that all the isocyanate groups reacted with hydroxyl groups (from polyol or BDO) during the TPUs synthesis. Thermal studies carried out by TGA, DSC, and MDSC revealed some interactions between hard and soft domains for all TPUs and a degradation behavior closely linked to their HS concentration. Stress-strain uniaxial tests showed that the increase in HS content in TPUs lead to higher tensile modulus and lower elongation at break. The TPU10 and TPU20 showed a strong elastomeric behavior with very high elongation at break (>600%) and very low elastic modulus.

In summary, TPUs partially synthesized from vegetable oils are very promising materials in good agreement with the current tendency for sustainable development, making them very attractive since they are expected to show specific properties which can be easily tailored by selecting the appropriate HS concentration. These materials could also fulfill many industrial requirements for different fields, such as construction, automotive, textile, adhesive, and coatings.

In chapter 2 antioxidant activity of maize bran arabinoxylan micro-spheres were introduced. The comparative analysis of measurements of the total antioxidants content and their activity for juice of 34 different kinds of Kalanchoe (*Kalanchoe* L.) is carried out by two methods in chapter 3: ammetric and chemiluminescence. Results of measurement show good (89%) correlation. Among the studied samples, the two most active kinds of Kalanchoe are exposed: *K. scapigera* and *K. rhombopilosa*. They can appear to be more prospective sources of biologically active components in comparison with kinds which are used now. In chapter 4 it is shown that new applications of enzymes within the food industry will depend of the functional understanding of different enzyme classes. Furthermore, the scientific advances in genome research and their exploitation via biotechnology is leading to a technology driven revolution that will have advantages for the consumer and food industry alike. In chapter 5, it is shown that the membranes are among the most important industrial applications today, and every year, more indications are found for this technology, such as water purification, industrial wastewater treatment, dehydration solvent recovery of volatile organic compounds, protein concentration, and many others.

In chapter 6, the various aspects of meat tenderness—such as process of tenderness of meat, practices of meat tenderness, influences of various conditions on meat tenderness, methods of tenderization of meat and meat products and physicochemical determinants of meat tenderness—are discussed. Biological properties of mushrooms are investigated in chapter 7. Molecular and immunological approaches for the detection of important pathogens in foods of animal origin are investigated in chapter 8.

In chapter 9, Cross-Linking of Ferulated Arabinoxylans Extracted from Mexican Wheat Flour: Rheology and Microstructure of the Gel is presented. Free and ester-linked ferulic acid content in a hard-to-cook pinto bean (*Phaseolus vulgaris* L.) variety is discussed in chapter 10. Chapter 11 discusses polyacrylamide-grafted gelatin: swellable hydrogel delivery system for agricultural applications in detail. The dynamics of bacteria and pathogenic fungi in soil microbiocenosis under the influence of biopreparations used during potato cultivation is introduced in chapter 12.

In chapter 13, the safety of irradiation has been clearly accepted as effective technology, and regulatory authority has been established on a global basis. Consumers' choice is the final preference for taking the technology to the market. In coming years, irradiation will empower the existing processing technologies. Irradiation is therefore providing safety and health as well as minimizing the losses on a large front and emerging as economical processing as well. Antioxidant properties of various alcohol drinks are studied in chapter 14. A study on the potential of oilseeds as a sustainable source of oil and protein for aquaculture feed is presented in chapter 15. Electrochemical methods for estimation of antioxidant activity of various biological objects are investigated in chapter 16. Ozonolysis of chemical and biochemical compounds are reviewed in chapter 17. Antioxidant activity of mint is explained in chapter 18. Wild orchids of Colchis forests to save them as objects of eco education and as producers of medicinal substances are introduced in chapter 19. Chapter 20 is about the fixation of proteins on MNPs, and chapter 21 studies the antimicrobial packaging for food applications.

— A. K. Haghi, PhD, and Elizabeth Carvajal-Millan, PhD

CHAPTER 1

VEGETABLE OILS AS PLATFORM CHEMICALS FOR SYNTHESIS OF THERMOPLASTIC BIO-BASED POLYURETHANES

C. BUENO-FERRER, N. BURGOS, and A. JIMÉNEZ

CONTENTS

1.1   INTRODUCTION

The use of renewable raw materials constitutes a significant contribution to a sustainable development in the plastics production. This strategy is based on the advantages given by nature synthesis potential and green chemistry principles. In this sense, polymers obtained from renewable raw materials have raised some interest in the last years. The development of polymers synthesized from agricultural products such as starch, cellulose, sugars, or lignin has been considerably increased in the last two decades [1]. Among all the possible natural sources for polymers, vegetable oils are considered one of the cheapest and abundant in Nature [2]. They can be used as an advantageous chemical platform to polymer synthesis by their inherent biodegradable condition and low toxicity to humans and the environment. In this context, many efforts are currently going on to propose a great variety of chemical methods to prepare thermoplastics and thermosets based on vegetable oils. This wide range of chemical methods applicable to these natural materials gives rise to many different monomers and polymers with many applications.

Fatty acids are the major chemical entities present in vegetable oils. They are valuable compounds to design specific monomers in the search of polymers with particular properties without any need of important modifications in their native structure. This is an advantageous issue not only in sustainability terms but also in industrial applicability and competitiveness in terms of cost and properties [2-5].

The vegetable oils are mainly formed by triglycerols or triglycerides, mainly composed by three fatty acids bonded to a glycerol molecule. Fatty acids constitute 94–96% of the total triglycerides weight in a vegetable oil and the number of carbon units in their structure is normally between 14 and 22 with zero to three double bonds by fatty acid molecule. The contents in fatty acids in some of the most common vegetable oils are indicated in Table 1.

TABLE 1 Fatty acid distribution in vegetable oils (g fatty acid/100 g oil)

Fatty acid	C:DB	Cotton	Rapeseed	Sunflower	Linseed	Corn	Olive	Palm	Castor	Soybean
Myristic	14:0	0.7	0.1	0.0	0.0	0.1	0.0	1.0	0.0	0.1
Myristoleic	14:1	0.0	0.0	0.0	0.0	0.0	0.0	0.0	0.0	0.0
Palmitic	16:0	21.6	4.1	6.1	5.5	10.9	13.7	44.4	1.5	11.0
Palmitoleic	16:1	0.6	0.3	0.0	0.0	0.2	1.2	0.2	0.0	0.1
Stearic	18:0	2.6	1.8	3.9	3.5	2.0	2.5	4.1	0.5	4.0
Oleic	18:1	18.6	60.9	42.6	19.1	25.4	71.1	39.3	5.0	23.4
Linoleic	18:2	54.4	21.0	46.4	15.3	59.6	10.0	10.0	4.0	53.2
Linolenic	18:3	0.7	8.8	1.0	56.6	1.2	0.6	0.4	0.5	7.8
Ricinoleic	18:1	0.0	0.0	0.0	0.0	0.0	0.0	0.0	87.5	0.0
Arachidic	20:0	0.3	0.7	0.0	0.0	0.4	0.9	0.3	0.0	0.3
Gadoleic	20:1	0.0	1.0	0.0	0.0	0.0	0.0	0.0	0.0	0.0
Eicosadienoic	20:2	0.0	0.0	0.0	0.0	0.0	0.0	0.0	0.0	0.0
Behenic	22:1	0.2	0.3	0.0	0.0	0.1	0.0	0.1	0.0	0.1
Erucic	22:1	0.0	0.7	0.0	0.0	0.0	0.0	0.0	0.0	0.0
Lignoceric	24:0	0.0	0.2	0.0	0.0	0.0	0.0	0.0	0.0	0.0
DB/triglyceride		3.9	3.9	4.7	6.6	4.5	2.8	1.8	2.7	4.6
Iodine index (I)		104–117	91–108	110–143	168–204	107–120	84–86	44–58	82–88	117–143

C, number of carbon atoms; DB, number of C = C double bonds.

The use of fatty acids and vegetable oils either in polymer synthesis or as additives comes from some decades, not only by the raising interest in the search for alternatives to fossil fuels but also by the particular chemical characteristics, that make them adequate for polymerization processes. Triglycerides are molecules with low reactivity and this fact is a disadvantage in their potential application in polymer synthesis. Nevertheless, the introduction of different functionalities in their reactive sites increases largely the synthetic possibilities of triglycerides [3].

At least three different uses of vegetable oils in polymer formulations can be proposed:

(i) As polymer additives (plasticizers, stabilizers, and so on),

(ii) As building blocks to get polymers from them, and

(iii) As units for the thermosets synthesis.

Much work on the use of vegetable oils as additives [6-11] and as thermosets precursors [12-18] has been reported but the development of thermoplastic polymer matrices is still in an early stage of research. Thermoplastics can be easily processed and recycled giving them possibilities in many different applications.

The synthesis of thermoplastics from vegetable oils is still in an early stage of the study and development because of the experimental difficulties to be afforded to get reasonable yields in this process. It is known that thermosets obtained from vegetable oils have been largely studied with many reported work [12-22]. This fact is partially due to the own composition of vegetable oils formed by triglycerides containing different fatty acids with variable number of chain instaurations. Thus, seeds oils are rich in polyunsaturated fatty acids giving highly reticulated rigid and temperature resistant materials [21]. The carbon chains forming the oils can be easily cross-linked by their double bonds. As indicated in Table 1, it can be concluded that most seed oils have fatty acids with 2 or 3 unsaturated bonds as the main component in their lipid profile, except castor, olive and rapeseed oils, which show a monounsaturated fatty acid as their main component. Therefore, thermosets can be easily synthesized from oils rich in polyunsaturated fatty acids, such as those from soya, sunflower or linseed getting polymers with high mechanical and thermal resistance. Castor oil shows high content in ricinoleic acid (87.5% in total oil weight) and the active site is occupied by an alcohol while olive and rapeseed oils show a main content in monounsaturated oleic acid (71.1% and 60.9% in total oil weigh, respectively).

It is known that the potential monomers or polymer building blocks should have at least one (in the case of addition polymerization) or two double bonds in their structure (in the case of condensation polymerization) to get thermoplastic materials. Therefore, triglycerides should be modified of functionalized before polymerization. Nevertheless, there are some examples of thermoplastic biomaterials obtained from naturally functionalized castor oil with homogeneous composition and acceptable polymerization yields. The main thermoplastic materials already synthesized from vegetable oils are thermoplastic polyurethanes (TPUs), polyamides (PA), thermoplastic polyesters, polyesteramides, and polyanhidrides.

1.1.1   TPUS: CHEMISTRY, STRUCTURE, AND PROPERTIES

Polyurethanes are generally synthesized by addition polymerization between a poly-alcohol and a poly-isocyanate. This is an exothermic reaction caused by the release of a proton from the alcohol group followed by a general molecular rearrangement by the formation of the urethane bond. [23]. If both reagents are bi-functional linear polyurethanes are obtained, while if functionalities are increased some cross-linked chains are formed, with the formation of reticulated structures. In summary, one of the most common synthesis routes for TPUs consists basically of the reaction of three main components.

1. Polyols with polyester or polyether functionalities with hydroxyl end groups
2. Di-isocyanate
3. Short-chain diol or diamine used as chain extender

The clear differences in the structure of all these components are essential to get the final properties of the synthesized thermoplastic polyurethanes. The TPUs are block copolymers with $(AB)_n$ segmented structure formed by hard and soft blocks. Soft segments correspond to the elastomeric part of the polymer (polyester chains from the polyol) and they are characterized by a glass transition temperature much lower than ambient which gives them high flexibility. This is the reason why these parts of the polyurethane are known as the soft block. On the other hand, hard segments (HS) are formed by the di-isocyanate and the chain extender forming a rigid structure, mainly formed by the urethane group bonded to aromatic rings. Therefore, some heterogeneity between both blocks in polyurethanes, the HS (with high polarity and melting point) and the soft segment (nonpolar and low melting point), should be expected, leading to phase separation in the copolymer structure, as presented in Figure 1 [24,25].

However, the phase separation is not complete in TPUs at the molecular scale, and it is possible to find soft segments inside the hard region and *vice-versa* (Figure 2), as was reported by Tawa et al. [26]. They indicated that urethane groups from neighbor chains could form hydrogen bonds very easily. These intermolecular bonding leads to the formation of aggregates acting as physical reticulation nodes with crystalline regions dispersed into the soft area in the polymer structure. This would lead to cross-linking with the final result of the increase in the overall rigidity in the TPU. The phase separation between hard and soft segments depends on, among other factors, their affinity, their relative mobility, the chain extender and the isocyanate structural symmetry [25].

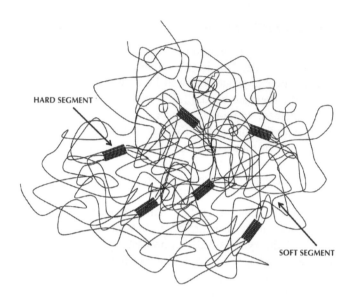

FIGURE 1   Segmented structure of a thermoplastic polyurethane.

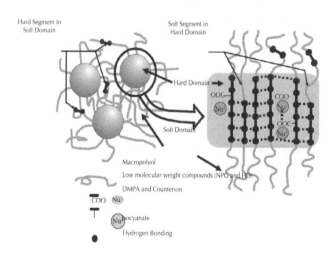

FIGURE 2   Polyurethane biphasic structure at molecular scale [26], with permission.

The TPU structures are generally linear, since, the relative amount of HS is small. Therefore, the most relevant properties of these polymers are conditioned by the secondary and intermolecular interactions (mainly Van der Waals) between the soft segments [24,27]. The elastic properties of polyurethanes mainly depend on the polyol chains mobility which is dependent on their chemical nature and length of the soft segments. The higher the molar mass of the soft segments the higher tensile strength and elongation to break and consequently the more flexible TPU [24,25].

On the other hand, when the HS content is high, the plastic deformation and the polymer general softening are observed after the application of mechanical stresses at high temperature. The thermal stability of polyurethanes is determined by the temperature range where the rigid segments start to melt and consequently their phase separation and segmented structure. Polyurethanes will show thermoplastic behavior at higher temperatures than the melting point of the hard regions [28]. Another important feature of the hard/soft ratio in TPUs is the decrease of molar masses and low phase separation in polymers with high values of that ratio, conditioning their rheological and mechanical properties, giving rise to more rigid materials. In addition, the isocyanate structure also influences the final TPU properties. High volume di-isocyanates lead to polyurethanes with high elastic modulus and tensile strength [25].

The development of TPUs from vegetable oils is the main goal of this study. In the chapter, some work performed by research group for TPUs synthesized from rapeseed oil dimer fatty acids is presented.

1.2   EXPERIMENTAL DETAILS

1.2.1   MATERIALS

The bio-based polyester polyol used in this study was kindly supplied by Croda (Yorkshire, UK) and it is based on dimmer fatty acids from rapeseed oil with purity higher than 98% and weight average molar mass (M_w) around 3000 g mol^{-1}. Hydroxyl and acid values are 40 mg KOH g^{-1} and 0.253 mg KOH g^{-1} respectively, as given by the supplier. 4,4'-diphenylmethane di-isocyanate (MDI) was supplied by Brenntag (Rosheim, France). The 1,4-butanediol (BDO), dibutylamine, toluene, and hydrochloric acid were purchased from Sigma Aldrich (Lyon, France). All reagents were used without any further purification step.

1.2.2   THE TPU SYNTHESIS

Four different TPUs were prepared with a NCO/OH ratio equal to 1 and increasing HS content (10–40 wt%, named TPU10–TPU40, respectively) by the two-step prepolymer process for TPU polymerization (Table 2). In a first step, the polyol reacted with an excess of MDI (ratio 2:1) for 2 hr in a five necked round bottom flask having the provision for nitrogen flushing, mechanical stirring, and temperature control at 80°C. During the synthesis, samples were extracted each 30 min in triplicate and diluted in 20 mL of a standard solution of dibutylamine 0.05 M in toluene for the reaction between the residual di-isocyanate and the amine, to control the NCO consumption during the reaction. Each solution was then stirred at room temperature for 10 hr to ensure the complete reaction of NCO groups with dibutylamine. The excess amine was titrated back with standard aqueous HCl 0.05 M solution using bromophenol green as indicator. For each titration, 25 mL of isopropyl alcohol were added to the solution to ensure compatibility between dibutylamine and the HCl solution. It was calculated from these experiments that 56.1% of NCO groups were consumed at the end of the prepolymer synthesis. This result was used for the addition of the precise amount of diol groups in the processing step. The prepolymer was further melt blended in a second step by reactive processing with the adequate amount of polyol and chain extender, depending on the HS content targeted in the final TPUs. The prepolymer synthesized in the first step and the calculated amount of polyol for a NCO/OH = 1 ratio were directly intro-duced in the feeding zone of an internal mixer (counter-rotating mixer Rheocord 9000, Haake, USA) equipped with a pair of high shear roller type rotors, at 80°C, with a rota-tion speed of 50 rpm and 15 min processing time. Then, the adequate amount of BDO chain extender was added and the temperature was immediately increased to 180°C for 8 min without any catalyst. After polymerization, all systems were cured overnight in an oven at 70°C to ensure the complete reaction of NCO groups which was further checked by attenuated total reflection Fourier-transform infrared spectroscopy (ATR-FTIR). The TPUs were subsequently compression molded in a hot press at 200°C by applying 200 MPa pressure for 5 min and further quenched between two steel plates for 10 min to obtain sheets with 1.5 mm thickness for each system. The expected HS length for each TPU sample was calculated from data in Table 2 by following a proto-col already described by Petrovic et al. [29] to determine the average polymerization degree in segmented polyurethanes. This method was applied to the TPUs synthesized in this study resulting in an HS average polymerization degree of 5–7, 7–9, 11–13, and 19–21 units for TPU10, TPU20, TPU30, and TPU40, respectively. These results could give an estimation of the length of HS for each TPU.

TABLE 2   Calculated HS percentage and reactants amounts for each TPU blend [30], with permission

Sample	HS (wt%)	Polyol (g)	MDI (g)	BDO (g)
TPU10	10	49.5	5.1	0.36
TPU20	20	44.0	9.1	1.94
TPU30	30	38.5	13.0	3.52
TPU40	40	33.0	16.9	5.11

1.3   MATERIALS CHARACTERIZATION

1.3.1   THE ATR-FTIR SPECTROSCOPY

The ATR-FTIR was used to screen the complete reaction between NCO and OH groups and to evaluate the adequate curing of the TPUs. Infrared spectra were collected on a TA Instruments SDT Q600 (Thermo-Nicolet, New Castle, DE, USA) at a resolution of 4 cm^{-1} and 64 scans per run. The ATR accessory was equipped with a germanium (n = 4) crystal and it was used at a nominal incidence angle of 45° yielding 12 interval reflections at the polymer surface.

1.3.2   THERMOGRAVIMETRIC ANALYSIS (TGA)

The four TPU systems as well as the bio-based polyol were analyzed in dynamic mode by using TGA/SDTA 851e Mettler Toledo (Schwarzenbach, Switzerland) equipment. Approximately, 7 mg samples were weighed in alumina pans (70 µL) and they were heated from 30°C to 700°C at 10°C min^{-1} under nitrogen atmosphere (flow rate 30 mL min^{-1}). In the case of TPUs, the initial degradation temperature was calculated as the temperature where 5 wt% of the initial mass was lost ($T_{5\%}$).

1.3.3   DIFFERENTIAL SCANNING CALORIMETRY (DSC)

The thermal and structural characterization of TPUs and the polyol was carried out by using a TA Instruments Q2000 (New Castle, DE, USA) equipment. Approximately, 5 mg of each sample were weighed in aluminum pans (40 µL) and they were subjected to a first heating stage from 30°C to 240°C with a further cooling from 240°C to −90°C and a subsequent heating from −90°C to 240°C. All steps were carried out at 10°C min^{-1} under nitrogen (flow rate 50 mL min^{-1}). All tests were performed in duplicate. Glass transition temperatures (T_g) were determined on the second heating scan. The T_g of the polyol was determined by using modulated differential scanning calorimetry (MDSC) during a cooling scan from 30°C to −90°C at 2°C min^{-1}, with 60 sec period and heat only mode.

1.3.4   UNIAXIAL MECHANICAL TESTS

Tensile properties of TPUs were determined with an Instron tensile testing machine (model 4204, USA), at 25°C and 50% relative humidity at a rate of 20 mm min^{-1}, using dumbbell specimens (dimensions: 30 × 10 × 1.5 mm^3). For each formulation at least five samples were tested.

1.4   DISCUSSION AND RESULTS

1.4.1   THE ATR-FTIR SPECTROSCOPY

The adequate curing of all TPUs is a key point prior to the materials characterization, since the presence of residual NCO groups in the final polymer gives an indication of an incomplete synthesis. In this work, ATR-FTIR was used to confirm the complete reaction between NCO and OH groups. The TPUs spectra are shown in Figure 3 and they could be used to highlight the main structural differences between them. No peak was found at 2270 cm^{-1} (NCO stretching band) suggesting that the reaction was complete in all cases. As expected, the main variations are related to the increasing content in HS and consequently, the higher concentration in urethane groups (-NH-CO-O). In Figure 3, vibrations at 3335 and 1550 cm^{-1} corresponded to -NH stretching and bending, respectively. Besides, the peak for the C=O stretching from the urethane group could be observed at ≈1700 cm^{-1} [31-33]. All these bands, assigned to the urethane groups, increased in their intensity from TPU10 to TPU40, with confirmation of the higher concentration in carbonate groups at higher HS contents. Nevertheless, the absorption band at 1735 cm^{-1} was assigned to the C=O group stretching in the polyol, and it was similar for all TPUs, except for TPU10 where this band was broader and almost no discernible from the band at 1700 cm^{-1}, certainly due to the lower content in urethane groups in this material.

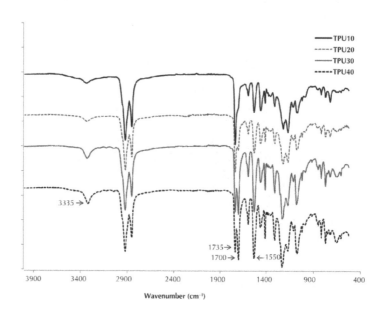

FIGURE 3   The ATR-FTIR spectra of TPU 10–40 wt% of HS and main peak assignments (cm^{-1}) [30] with permission.

1.4.2   THERMOGRAVIMETRIC ANALYSIS (TGA)

The thermal stability of the polyol and all bio-based TPUs were studied by dynamic TGA. Figure 4 shows their mass losses and derivative curves. The bio-based polyol showed a narrow derivative peak (Figure 4 (b), left) due to its purity, while Figure 4 (b) (right) clearly shows the derivative curves of TPUs with lower intensity peaks. It is known that degradation of polyurethanes is a complex and multistep process, as observed in Figure 4. An important parameter, the degradation onset, is dependent on the thermal stability of the less thermally stable part on the polyurethane chains [34,35]. Initial degradation temperatures ($T_{5\%}$) of each step were also studied in TPUs and, together with mass loss percentages, allowed to study the differences between samples depending on their HS content. It was noted that the $T_{5\%}$ value of the pure rapeseed oil-based polyol was higher than in the case of TPU systems, as it was expected, but also higher than the $T_{5\%}$ value reported for castor oil [36] and cashew nut shell liquid-based polyols [37]. The TPUs also showed higher thermal stability than polymers with similar structures [38]. It has been reported that their first degradation stage is related to urethane bond decomposition into isocyanate and alcohol with possible formation of primary and secondary amines [39,40]. Nevertheless, the complexity of this stage is also related with the HS content. In this way, when this content increases $T_{5\%}$ decreases, making those materials more susceptible to degradation and suggesting that the starting point of degradation takes place predominantly within HS.

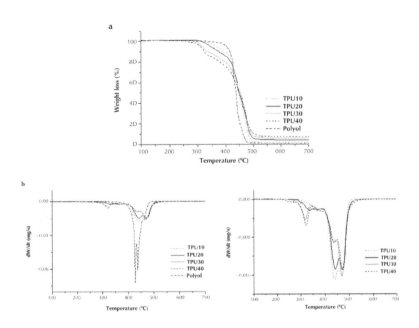

FIGURE 4   The TGA curves for mass loss (a) and derivative (b) versus temperature for bio-based polyol and TPUs [41] with permission.

1.4.3   DIFFERENTIAL SCANNING CALORIMETRY (DSC)

The structure of TPU samples was investigated by DSC while the bio-based polyol was studied both by DSC and MDSC. The main results of this study are shown in Table 3 and Figure 5. This technique is valuable for a precise determination of T_g and from these values it is possible to estimate the real amounts of HS in the unorganized and organized microphases. It has been indicated that HS in TPUs do not fully belong to the hard domains, since, some of them can be found in the soft regions and *vice versa*. Moreover, when MDI is not bonded to the chain extender but to the polyol, presumably in systems with high polyol amount [42], such as in the case of TPU10, this trend is more clearly observed. This phenomenon was also clearly evidenced in the thermal stability of the TPUs, since, the initial degradation temperatures ($T_{5\%}$) of TPU30 and TPU40 fell significantly with respect to those obtained for TPU10 and TPU20. This behavior could be associated to the higher HS content. Moreover, it should be mentioned that the mass loss associated with this first degradation stage could be also correlated with the HS content. In samples with higher HS content, such as TPU30 and TPU40, a peak and a shoulder in their derivative curves were observed, both associated with the first stage of the thermal decomposition of urethane bonds (Figure 4 (b)).

TABLE 3   Glass transition temperatures (T_g) and HS content (%) for TPUs

Sample	T_{gs} SS (°C)	T_{gh} HS (°C)
TPU10	–47.0	---
TPU20	–47.8	122.8
TPU30	–50.0	120.7
TPU40	–51.3	118.1

In a preliminary step the bio-based polyol was studied by conventional DSC but the glass transition was not clearly observed since some melting transitions were superposed in this temperature range, probably associated to the crystalline polymorphism of some oils and fats, such as the case of rapeseed oil. Modulation of DSC data (MDSC) is a powerful tool to separate transitions and to get higher resolutions in particular thermal events. In the case of the polyol, MDSC was used during the cooling cycle and the T_g transition was determined in the reversing phase curve at –61.8°C. It was also observed that glass transition temperatures of the SS for all TPUs were higher than that of the polyol (Table 3). The T_g values of SS were slightly lower with increasing HS concentrations as it was reported by Xu et al. [43]. Soft domains have higher mobility when larger HS are present and this fact could be related with a better microphase separation at high HS content. Besides, the TPUs with lower molar masses and the highest concentrations in HS could contribute to the decrease in T_g values. Higher amount of end chains could result in higher free volume increasing the mobility of the amorphous phase. Moreover, as it is shown in Figure 5, T_g transition of the MDI-BDO HSs was very difficult to detect due to their stiffness and low mobility [42]. The T_g of HS of TPUs slightly increased when their concentration decreased. This result could be attributed to the higher concentration in HS inside the soft domains as explained. Moreover, the intensity of the glass transition of HS is larger at higher HS content but the T_g value is lower, suggesting some interactions between hard and soft phases, that is the low T_g of SS observed for TPU40 which was previously attributed to the microphase separation, could also help to the decrease in T_g of HS in this material due to the higher interaction of soft domains in the hard phase.

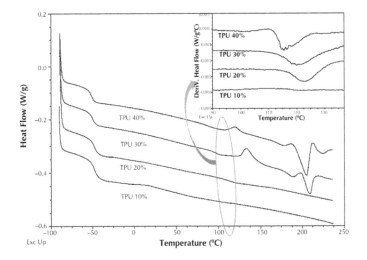

FIGURE 5 The DSC curves from TPU samples and zoom in the zone of T_g of HS [30] with permission.

1.4.4 UNIAXIAL MECHANICAL TESTS

Tensile properties were determined by uniaxial tensile tests and data are summarized in Table 4 and Figure 6. Results showed that the increase in HS content lead to a brittle material with higher tensile modulus and lower elongation at break (around 25% for TPU40), as expected, while this material showed lower tensile strength than TPU30 and TPU20. This behavior could be attributed to the fragility provided by the crystalline macro-structures with sizes between 20 and 30 μm that were observed in a morphological study [30]. The presence of these macro-structures in TPU40 increased the segregation phase size and may promote points of stress concentration, due to their boundary impingement, which could influence properties such as tensile strength and elongation at break [43]. As expected, TPU40 also exhibited the higher modulus (11.1 MPa), leading to the conclusion that the HS higher crystallinity has a significant effect on the mechanical properties of these TPUs.

TABLE 4 Uniaxial tensile properties of TPUs [41] with permission

Sample	Tensile Strength (MPa)	Elongation at Break (%)	Young Modulus (MPa)
TPU10	1.3 ± 0.1	>600	0.7 ± 0.0
TPU20	3.7 ± 0.1	>600	2.5 ± 0.1
TPU30	5.6 ± 0.3	430 ± 40	8.6 ± 0.5
TPU40	1.5 ± 0.4	25 ± 7	11.1 ± 1.2

Figure 6 shows the elastomeric behavior of those TPUs with lower HS content as demonstrated by their lower moduli. It was also observed in Figure 6 that TPU20 was the only material with a yield point in the range of deformation 250–300%. This particular behavior, presented as a rubbery region in Figure 6, could be related to a good resistance to the permanent deformation.

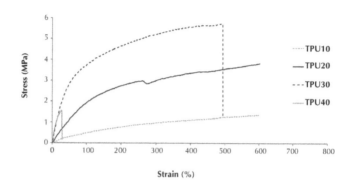

FIGURE 6   Stress-strain curves of TPUs [41] with permission.

1.5   CONCLUSION

Bio-based TPUs were synthesized from a di-functional dimmer fatty acid-based polyol obtained from rapeseed oil, MDI, and BDO at four HS contents that is 10–40 wt%. The polyol characteristics determined the structure and properties of TPUs. The FTIR-ATR spectra confirmed that all the isocyanate groups reacted with hydroxyl groups (from polyol or BDO) during the TPUs synthesis. Thermal studies carried out by TGA, DSC, and MDSC revealed some interactions between hard and soft domains for all TPUs and a degradation behavior closely linked to their HS concentration. Stress-strain uniaxial tests showed that the increase in HS content in TPUs lead to higher tensile modulus and lower elongation at break. The TPU10 and TPU20 showed a strong elastomeric behavior with very high elongation at break (>600%) and very low elastic modulus.

In summary, TPUs partially synthesized from vegetable oils are very promising materials in good agreement with the current tendency for sustainable development and making them very attractive since they are expected to show specific properties which can be easily tailored by selecting the appropriate HS concentration. These materials could also fulfill many industrial requirements for different fields such as construction, automotive, textile, adhesive, and coatings.

KEYWORDS

- Bio-based polyols
- Dimmer fatty acids
- Hard segment distribution
- Polyurethanes
- Structure-property relationships

ACKOWLEDGMENT

Authors thank the Spanish Ministry of Economy and Competitiveness (MAT2011-28648-C02-01) for financial support.

REFERENCES

1. Lligadas, G., Ronda, J. C., Galià, M., and Cádiz, V. Plant Oils as Platform Chemicals for Polyurethane Synthesis: Current-State-of-the-Art. *Biomacromolecules*, **11**, 2825–2835 (2010).
2. Biermann, U., Friedt, W., Lang, S., Luhs, W., Machmuller, G., Metzger, J. O., Klaas, M. R. G., Schafer, H. J., and Scheider, M. P. New Syntheses with Oils and Fats as Renewable Raw Materials for the Chemical Industry. *Angewandte Chemie International Edition*, **39**, 2206–2224 (2000).
3. Montero de Espinosa, L. and Meier, M. A. R. Plant oils: The perfect renewable resource for polymer science?, *European Polymer Journal*, **47**, 837–852 (2011).
4. Behr, A. and Gomes, J. P. The refinement of renewable resources: New important derivatives of fatty acids and glycerol. *European Journal of Lipid Science and Technology*, **112**(1), 31–50 (2010).
5. Kockritz, A., Khot, S. N. M., Lascala, J. J., Can, E., Morye, S. S., Williams, G. I., Palmese, G. R., Kusefoglu, S. H., and Wool, R. P. Development and application of triglyceride-based polymers and composites. *Journal of Applied Polymer Science*, **82**(3), 703–723 (2001).
6. Benaniba, M. T., Belhaneche-Bensemra, N., and Gelbard, G. Stabilizing effect of epoxidized sunflower oil on the thermal degradation of poly(vinyl chloride). *Polymer Degradation and Stability*, **74**, 501–505 (2001).
7. Boussoum, M. O., Atek, D., and Belhaneche-Bensemra, N. Interactions between poly(vinyl chloride) stabilised with epoxidised sunflower oil and food simulants. *Polymer Degradation and Stability*, **91**, 579–584 (2006).
8. Choi, J. S. and Park, W. H. Thermal and mechanical properties of Poly(3-hydroxybutyrate-co-3-hydroxyvalerate) plasticized by biodegradable soybean oils. *Macromolecular Symposium*, **197**, 65–76 (2003).
9. Semsarzadeh, M. A., Mehrabzadeh, M., and Arabshahi, S. S. Mechanical and Thermal Properties of the Plasticized PVC-ESBO. *Iranian Polymer Journal*, **14**(9), 769–773 (2005).
10. Ali, F., Chang, Y. W., Kang, S. C., and Yoon, J. Y. Thermal, mechanical, and rheological properties of poly(lactic acid)/epoxidized soybean oil blends. *Polymer Bulletin*, **62**(1), 91–98 (2009).

11. Karmalm, P., Hjertberg, T., Jansson, A., and Dahl, R. Thermal stability of poly(vinyl chloride) with epoxidised soybean oil as primary plasticizer. *Polymer Degradation and Stability*, **94**, 2275–2281 (2009).

12. Meier, M. A. R., Metzger, J. O., and Schubert, U. S. Plant oil renewable resources as green alternatives in polymer science. *Chemical Society Re*views, **36**, 1788–1802 (2007).

13. Güner, F. S., Yagci, Y., and Erciyes, A. T. Polymers from trygliceride oils. *Progress in Polymer Science*, **31**, 633–670 (2006).

14. Lu, Y. and Larock, R. C. Novel polymeric materials from vegetable oils and vinyl monomers: Preparation, properties, and applications. *ChemSusChem*, **2**(2), 136–147 (2009).

15. Khot, S. N., Lascala, J. J., Can, E., Morye, S. S., Williams, G. I., Palmese, G. R., Kusefoglu, S. H., and Wool Development and application of triglyceride-based polymers and composites. *Journal of Applied Polymer Science*, **82**(3), 703–723 (2001).

16. Hablot, E., Zheng, D., Bouquey, M., and Averous, L. Polyurethanes Based on Castor Oil: Kinetics, Chemical, Mechanical, and Thermal Properties. *Macromolecular Materials and Engineering*, **293**(11), 922–929 (2008).

17. Liu, Z. and Erhan, S. Z. "Green" composites and nanocomposites from soybean oil. *Materials Science and Engineering A*, **483–484**, 708–711 (2008).

18. Doll, K. M. and Erhan, S. Z. The improved synthesis of carbonated soybean oil using supercritical carbon dioxide at a reduced reaction time. *Green Chemistry*, **7**, 849–854 (2005).

19. Meiorin, C., Aranguren, M. I., and Mosiewicki, M. A. Smart and structural thermosets from the cationic copolymerization of a vegetable oil. *Journal of Applied Polymer Science*, **124**(6), 5071–5078 (2011).

20. Kong, X., Omonov, T. S., and Curtis, J. M. The development of canola oil based bioresins. *Lipid Technology*, **24**(1), 7–10 (2012).

21. Kim, J. R. and Sharma, S. The development and comparison of bio-thermoset plastics from epoxidized plant oils. *Industrial Crops and Products*, **36**(1), 485–499 (2012).

22. Quirino, R. L., Woodford, J., and Larock, R. C. Soybean and linseed oil-based composites reinforced with wood flour and wood fibers. *Journal of Applied Polymer Science*, **124**(2), 1520–1528 (2012).

23. Seymour, R. B. and Carraher, C. E. *Polymer Chemistry - An introduction. 2nd Edition.* Marcel Dekker, Basel, Suiza (1988).

24. Ionescu, M. *Chemistry and Technology of Polyols for Polyurethanes*. Rapra Technology Limited, Shropshire, United Kingdom (2005).

25. Meier-Westhues, U. *Polyurethanes. Coatings, adhesives, and sealants*. Vincentz Network, Hannover, Alemania (2007).

26. Tawa, T. and Ito, S. The role of hard segments of aqueous polyurethane-urea dispersion in determining the colloidal characteristics and physical properties. *Polymer Journal*, **38**(7), 686–693 (2006).

27. Wool, R. P. and Sun, X. S. *Bio-Based Polymers and Composites*. Elsevier Academic Press, Burlington, MA, United States (2005).

28. Yamasaki, S., Nishiguchi, D., Kojio, K., and Furukawa, M. Effects of Polymerization Method on Structure and Properties of Thermoplastic Polyurethanes. *Journal of Polymer Science B: Polymer Physics*, **45**, 800–814 (2007).

29. Petrovic, Z. S., Cevallos, M. J., Javni, I., Schaefer, D. W., and Justice, R. Soy-oil-based segmented polyurethanes. *Journal of Polymer Science B*, **43**, 3178–3190 (2005).

30. Bueno-Ferrer, C., Hablot, E., Perrin-Sarazin, F., Garrigós, M. C., Jiménez, A., and Avérous, L. Structure and morphology of new bio-based thermoplastic polyurethanes obtained from dimer fatty acids. *Macromolecular Materials and Engineering*, **297**(8), 777–784 (2012).

31. Irusta, L. and Fernandez-Berridi, M. J. Aromatic poly(ester–urethanes): effect of the polyol molecular weight on the photochemical behaviour. *Polymer*, **41**, 3297–3302 (2000).

32. Irusta, L., Iruin, J. J., Mendikute, G., and Fernández-Berridi, M. J. Infrared spectroscopy studies of the self-association of aromatic urethanes. *Vibrational Spectroscopy*, **39**, 144–150 (2005).

33. Silva, B. B. R., Santana, R. M. C., and Forte, M. M. C. **A solvent less castor oil-based PU adhesive for wood and foam substrates**. *International Journal of Adhesion and Adhesives*, **30**, 559–565 (2010).

34. Javni, I., Petrovic, Z., Guo, A., and Fuller, R. Thermal stability of polyurethanes based on vegetable oils. *Journal of Applied Polymer Science*, **77**, 1723–1734 (2000).

35. Król, P. Synthesis methods, chemical structures and phase structures of linear polyurethanes. Properties and applications of linear polyurethanes in polyurethane elastomers, copolymers and ionomers. *Progress in Materials Science*, **52**, 915–1015 (2007).

36. Corcuera, M. A., Rueda, L., Fernandez d'Arlas, B., Arbelaiz, A., Marieta, C., Mondragon, I., and Eceiza, A. Microstructure and properties of polyurethanes derived from castor oil. *Polymer Degradation and Stability*, **95**, 2175–2184 (2010).

37. Bhunia, H. P., Nando, G. B., Chaki, T. K., Basak, A., Lenka, S., and Nayak, P. L. Synthesis and characterization of polymers from cashewnut shell liquid (CNSL), a renewable resource II. Synthesis of polyurethanes. *European Polymer Journal*, **35**, 1381–131 (1999).

38. Yeganeh, H. and Mehdizadeh, M. R. Synthesis and properties of isocyanate curable millable polyurethane elastomers based on castor oil as a renewable resource polyol. *European Polymer Journal*, **40**, 1233–1238 (2004).

39. Hablot, E., Zheng, D., Bouquey, M., and Avérous, L. Polyurethanes based on castor oil: kinetics, chemical, mechanical and thermal properties. *Macromolecular Materials and Engineering*, **293**, 922–929 (2008).

40. Hojabri, L., Kong, X., and Narine, S. S. Fatty acid-derived diisocyanate and biobased polyurethane produced from vegetable oil: synthesis, polymerization and characterization. *Biomacromolecules*, **10**, 884–891 (2009).

41. Bueno-Ferrer, C., Hablot, E., Garrigós, M. C., Bocchini, S., Avérous, L., and Jiménez, A. Relationship between morphology, properties and degradation parameters of novative biobased thermoplastic polyurethanes obtained from dimer fatty acids. *Polymer Degradation and Stability*, **97**(10), 1964–1969 (2012).

42. Bagdi, K., Molnar, K., Pukanszky, Jr. B., and Pukanszky, B. Thermal analysis of the structure of segmented polyurethane elastomers. *Journal of Thermal Analysis and Calorimetry*, **98**, 825–832 (2009).

43. Xu, Y., Petrovic, Z., Das, S., and Wilkes, G. L. Morphology and properties of thermoplastic polyurethanes with dangling chains in ricinoleate-based soft segments. *Polymer*, **49**, 4248–4258 (2008).

CHAPTER 2

ANTIOXIDANT ACTIVITY OF MAIZE BRAN ARABINOXYLAN MICROSPHERES

A. L. MARTHNEZ-LÓPEZ, E. CARVAJAL-MILLAN,
Y. L. LÓPEZ-FRANCO, J. LIZARDI-MENDOZA, and A. RASCÓN-CHU

CONTENTS

2.1 INTRODUCTION

Ferulatedarabinoxylans (FAX) present antioxidant properties which appear to be correlated with their ferulic acid (FA) content. The aim of this research was to investigate the effect of oxidative coupling of FA on the antioxidant properties of FAX microspheres.

After gelation, the FA content in FAX decreased from 0.255 to 0.045 µg/mg of FAX. The Fourier transform infrared (FTIR) spectrum of FAX before and after crosslinking presented changes in the band position of carbonyl vibrations suggesting the formation of ferulated structures unreleased by mild alkaline hydrolysis. The FAX microspheres presented an average pore size of 531 µm and a swelling ratio value (q) of 18 g water/g FAX. Microstructure and textural properties of dried FAX microspheres were studied by scanning electron microscopy (SEM) and nitrogen adsorption/desorption isotherms, respectively, showing a heterogeneous mesoporous and macroporous structure throughout the network. The antioxidant capacity of FAX after cross-linking process decreases by 47% in relation to non-cross-linked FAX. These results suggest that FAX microspheres exhibited antioxidant activity indicating their potential application as microencapsulation systems based in antioxidants for food, pharmacy, and cosmetics applications.

The development of microencapsulation technologies and use of natural antioxidant have been the focus of extensive research in the last decade as several functional food ingredients are highly sensitive to elevated temperatures, oxygen, and light and thus are easily oxidized when exposed to air. Other food ingredients are healthy but taste badly in the desired foodstuff [1]. The microencapsulation of food ingredients in polysaccharide-based hydrogels offers a number of advantages such as protection against oxidation and targeted release inside the human body due to their chemical and three-dimensional structure, good mechanical properties, and biocompatibility [2,3].

The FAX are non-digestible polysaccharides which resist digestion and absorption in the human small intestine and are fermented in the large intestine [4]. The FAX consist of a linear backbone chain of xylose units containing arabinose substituents attached through O-2 and/or O-3 [5]. Some of the arabinose residues are ester-linked on (O)-5 to FA [6]. The FAX can form covalent hydrogels by oxidative coupling of FA resulting in the formation of dimers and trimers of FA (diFA and triFA) as covalent cross-linking structures. The FAX covalent hydrogels generally present high water absorption capacity (up to 100 g of water per gram of dry polymer) and absence of pH or electrolyte susceptibility and porous structure [7,8]. Previous studies have demonstrated that cross-linked FAX could be employed for controlled release of proteins [9,10], methyl xanthine [11], and lycopene [12], making FAX networks good candidates for the design of novel controlled delivery systems.

On the other hand, polymers presenting antioxidant activity have received increasing attention for the development of microencapsulation systems [13,14]. The FAX present antioxidant properties which appear to be correlated with their FA content [15]. However, to knowledge, there are no reports about the antioxidant activity of FAX microspheres. The aim of this chapter was to investigate the effect of oxidative coupling of FAX in antioxidant properties of the polysaccharide.

2.2 MATERIALS AND METHODS

2.2.1 MATERIALS

The FAX from maize bran were obtained and characterized as previously reported [17]. The FAX contained 85% dry basis (d.b.) of pure AX. The FAX presented an A/X ratio of 0.72, a Mw of 273 KDa, and a [η] of 2.98 dL/g. The FA, di-FA, and tri-FA content in FAX were 0.25, 0.14, and 0.07 μg/mg of FAX, respectively. The Laccase (benzenediol: oxygen oxidoreductase, E.C.1.10.3.2) from *Trametesversicolor* and all other chemical products were purchased from Sigma Chemical Co. (St. Louis, MO, USA).

2.2.2 METHODS

FAX MICROSPHERES

The FAX microspheres were prepared as described [16].

RHEOLOGICAL TEST

The rheological tests were performed by small amplitude oscillatory shear by using a strain-controlled rheometer (Discovery HR-3 rheometer, TA Instruments, New Castle, DE, USA) as reported before [18]. The FAX gelation was studied for 6 hr at 25°C. All measurements were carried out at a frequency of 0.25 Hz and 5% strain (linearity range of visco-elastic behavior).

PHENOLIC ACIDS CONTENT

The FA, di-FA, and tri-FA contents in FAX microspheres were quantified by RP-HPLC after a de-esterification step as described elsewhere[18]. An Alltima (Alltech, Deerfield, IL, USA) C18 column (250 × 4.6 mm) and a photodiode array detector Waters 996 (Millipore Co., Milford, MA, USA) were used. Detection was followed by Ultraviolet (UV) absorbance at 320 nm.

FTIR SPECTROSCOPY

The FTIR spectra of dry FAX powder and FAX microspheres were recorded on a Nicolet FTIR spectrophotometer (Nicolet Instruments Corp. Madison, WI, USA) in the form of KBr pellets. The pellets were investigated in transmission mode from 400–4000 cm^{-1} resolution.

THE SEM

The surface morphology and shape of the freeze-dried FAX microspheres was studied by field emission SEM (JEOL JSM-7401F, Peabody, MA, USA) without coating at low voltage (20 kV). The SEM images were obtained in secondary and backscattered electrons image mode.

SWELLING TESTS

After gelation, the FAX microspheres were recovered by filtration placed in glass vials and weighted. The FAX microspheres were allowed to swell as described elsewhere [19]. The equilibrium swelling was reached when the weight of the samples changed by no more than 3% (0.06 g). The swelling ratio (q) was calculated by Equation 1.

$$q = (Ws - W_{FAX})/W_{FAX} \tag{1}$$

TEXTURAL ANALYSIS

The textural analysis was conducted by adsorption/desorption of nitrogen. Surface area was determined by using nitrogen adsorption at their condensation temperature (77.35K) and at a relative pressure (p/p$_0$) of 0.5–0.22 by the Brunauer Emmett–Telle (BET) method [20]. A surface characterization Autosorb-1 (Quantachrome Instruments, Boynton Beach, FL, USA) was used. The surface of the samples was cleaned at 100°C for 2 hr under vacuum.

ANTIOXIDANT CAPACITY

The antioxidant capacity of non-cross-linked FAX and FAX microspheres was measured using the 2,2'-azino-*bis*(3-ethylbenzothiazoline-6-sulphonic acid) (ABTS$^+$) method as reported elsewhere [21,22]. The absorbance of the optically clear supernatant was measured at 734 nm. All measurements were performed exactly 15 and 30 min after mixing the samples with ABTS reagent. The antioxidant activity was expressed as μmol of Trolox equivalent antioxidant capacity (TEAC) per gram sample by means of a dose-response curve for Trolox. The analyses were carried out in triplicate for each sample.

2.3 DISCUSSION AND RESULTS

2.3.1 OXIDATIVE GELATION OF FAX

Figure 1 shows the rheological response of a FAX solution during gelation with a rapid initial rise in store modulus (G') followed by a plateau region reaching a value of 215 Pa, a los modulus (G'') value prevailing over G' and a mechanical spectrum typical of solid-like material. This behavior reflects an initial formation of covalent linkages between FA of adjacent FAX molecules producing a three-dimensional network. The high G' value of FAX gel has been attributed to the covalent cross-linking content and to the physical entanglement of FAX chains [23].

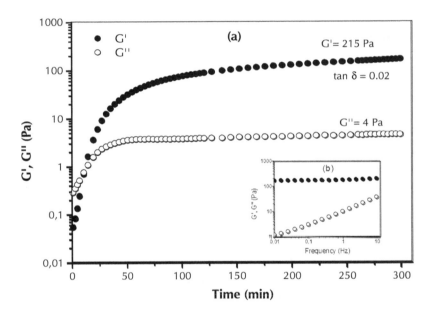

FIGURE 1 Store (G') and loss (G'') modulus during gelation (a) and mechanical spectrum after gelation (b) of 4 % (w/v) FAX solution treated with 1.67 nKat/mg FAX of laccase at 25°C. Data obtained at 0.25 Hz and 5% strain.

The extent of covalent crosslinking in FAX gels was determined by the content of ferulate monomer and total di-FA and tri-FA before and after 6 hr of gelation (Table 1). The FA was oxidized (82% of initial FA content) during the gelation process. After gelation, the di-FA content in FAX gels did not increase but rather decreased from 0.135 to 0.030 µg/mg FAX. The tri-FA was present only in trace quantities (0.003 µg/mg FAX). Nevertheless, the tan δ (G''/G') values confirm the formation of a true gel after laccase treatment (Figure 1 (b)). This behavior in FAX gels has been previously reported by several authors [8,17,24,25]. These authors attributed this result to the formation of ferulated cross-linking structures which cannot be released by mild alkaline hydrolysis and/or to the participation of lignin residues in the formation of FAX gel.

In this chapter, IR spectroscopy was used to compare structural features between non-cross-linked FAX and FAX microspheres. The FTIR spectrum of FAX powder and lyophilized FAX microspheres are presented in Figure 2. Both spectra showed a typical absorbance region at 1200 and 900 cm^{-1} characteristic of arabinoxylans-type-polysaccharide. A broad band with a prominent shoulder at 1537 cm^{-1} as well as a band at 1414 cm^{-1} are observed in the FTIR spectrum of FAX microspheres. The absorption band found at 1573 cm^{-1} has been associated with stretching of carboxylic anions indicating the formation of monoesters. This band possibly originated from aromatic skeletal vibrations in association with lignin [15]. In addition, the intensity of the absorption band at 1414 cm^{-1} for CO asymmetric stretching increased after cross-linking, suggesting the presence of ester groups. These significant changes of the absorption bands suggest that covalent cross-linking of FAX chains occurs during FAX microspheres formation under the conditions given.

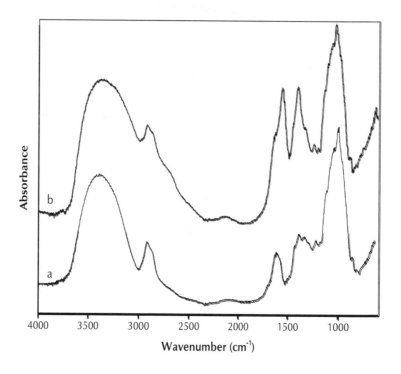

FIGURE 2   A comparisons of FTIR spectra of FAX—(a) non-cross-linked FAX and (b) FAX microspheres obtained by oxidative cross-linked of FA.

2.3.2   STRUCTURAL FEATURES AND PROPERTIES OF FAX MICROSPHERES

The secondary electron SEM images of FAX microspheres are shown in Figure 3. The FAX microspheres show a spherical shape, rough surface, and an average diameter of particles at 531 μm. The microsphere morphology presented a three-dimensional and heterogeneous network structurewith irregular pores size and geometries. It has been reported that for gels cross-linked via phenolic groups and the pore size is determined by the presence of nodular clusters [26]. Thus, the heterogeneous microstructure of FAX microspheres could be attributed to the content and distribution of FA covalent crosslinking structures forming the FAX network. The calculation of pore size distribution in FAX microspheres was based on the Barrett-Joyner-Halenda (BJH) method using the desorption isotherms (data not shown). The average pore size in microspheres was 30.8 nm and denoting the presence of mesopores. However, it should be emphasized that the absence of a plateau at high pressures in the isotherms and a non-Gaussian pore diameter distribution indicates a macroporous material. Nevertheless, it could also be attributed to apparent porosity which could produce macrocavities. These results confirm the heterogeneous microstructure observed in the SEM micrographs (Figure 3 (b)).

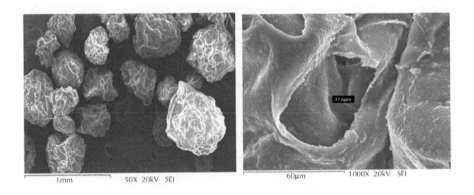

FIGURE 3   The SEM photomicrograph showing shape (a) and pore size (b) of FAX microspheres.

The swelling behavior of FAX microspheres was followed for 30 min at 25°C, with the equilibrium being reached 8–12 min. This water dissolution mechanism of the FAX microspheres was fitted to the Fick model. The n value was 0.19, indicating that the dissolution is due to a Fickian diffusion, where water penetration rate in the microsphere is lower than the polymer chain relaxation rate. The swelling ratio value (q) in FAX microspheres was 18 g water/g FAX which is similar to the value reported for FAX hydrogels (20 g water/g AX) [27]. Nevertheless, FAX microspheres showed a very short swelling time to reach equilibrium (12 min) compared to that reported for cylindrical FAX hydrogels (15 hr) [27]. This behavior could be explained in terms of an increase in the surface areatovolume ratio in FAX microspheres.

Table 1 shows the TEAC values of the non-cross-linked FAXand FAX microspheres as determined by the procedure using ABTS radical cation solution. The results show that the cross-liking process reduced their antioxidant activity by 46%. This decrease in antioxidant activity in the FAX microspheres could be due to coupling of FA and lower movement of chains by the rigidity of gel or formation of structural without antioxidant properties. However, the values obtained for the microspheres are similar than those reported in water-soluble polysaccharide fractions (6 μmol TEAC/g) with a higher total phenolic content [15]. This antioxidant capacity with a low content of FA could be explained in terms of the specific surface area (S_a) of FAX microspheres (Table 1) as this textural feature could allow higher exposure of phenolic groups and therefore, retain their antioxidant capacity. Previous studies have demonstrated that the increase in the antioxidant capacity is correlated with increase in the specific surface of wheat bran [22]. The antioxidant activity results obtained in study clearly established the antioxidant potency of the FAX and FAX microspheres. However, the relationship between the structure of FAX and antioxidant mechanisms require further studies.

TABLE 1   Antioxidant capacity, textural characteristic, and phenolic acid content of FAX and FAX microspheres

Sample	FA	di-FA	tri-FA	S_a	$TEAC_{/ABTS}^{+}$
	(μg/mg FAX)	(μg/mg FAX)	(μg/mg FAX)	(m²/g)	(μmol/g)[a]
Non-cross-linked FAX	0.255 ± 0.017	0.135 ± 0.011	0.064 ± 0.010	n.p	24.61± 0.210
FAX micro-spheres	0.045 ± 0.002	0.03 ± 0.001	Traces	11.19	13.24± 1.9

[a]TEAC, in μmol/g FAX or FAX microspheres

np, Not presented. All values are means ± standard deviation of three repetitions.

2.4   CONCLUSION

This chapter demonstrated that spherical and porous FAX microspheres with antioxidant activity can be prepared by enzymatic cross-linking. The FAX cross-linking process decreases only 47% of the polysaccharide antioxidant capacity. The results suggest that FAX microspheres exhibited antioxidant activities in the ABTS radical scavenging test indicating their potential application as microencapsulation systems based in antioxidants for food, pharmacy, or cosmetics among other possible applications.

KEYWORDS

- Covalent cross-linking
- Ferulated arabinoxylans
- Rheological test
- Scanning electron microscopy
- Swelling test

ACKNOWLEDGMENTS

This research was supported by FondoInstitucional SEP-CONACYT, Mexico (Grant 134301 to E. Carvajal-Millan).

REFERENCES

1. Dai, C., Wang, B.,and Zhao, H. Microencapsulation peptide and protein drugs delivery system. *Colloids and Surfaces B: Biointerfaces*, **41**, 117–120 (2005).
2. Peppas, N. A. and Merrill, E. W. Poly(vinyl alcohol) hydrogels: Reinforcement of radiation-crosslinked networks by crystallization. *Journal of Polymer Science Polymer Chemistry Edition*, **14**, 441–457 (1976).
3. Peng, X., Ren, J., Zhong, L.,and Sun, R. Nanocomposite films based on xylan-rich hemicelluloses and cellulose nanofibers with enhanced mechanical properties. *Biomacromolecules*, **12**, 3321–9 (2011).
4. Hughes, S. A, Shewry, P. R., Li, L., Gibson, G. R., Sanz, M. L.,and Rastall, R. a In vitro fermentation by human fecal microflora of wheat arabinoxylans. *Journal of agricultural and food chemistry*, **55**, 4589–95 (2007).
5. Izydorczyk, M. S. and Biliaderis, C. G. Cereal arabinoxylans: advances in structure and physicochemical properties. *Carbohydrate Polymers*, **28**, 33–48 (1995).
6. Smith, M. M. and Hartley, R. D. Occurrence and nature of ferulic acid substitution of cell-wall polysaccharides in graminaceous plants. *Carbohydrate Research*, **118**, 65–80 (1983).
7. Carvajal-Millan, E., Guilbert, S., Morel, M.,and Micard, V. Impact of the structure of arabinoxylan gels on their rheological and protein transport properties. *Carbohydrate Polymers*, **60**, 431–438 (2005).
8. Niño-Medina, G., Carvajal-Millán, E., Rascon-Chu, A., Marquez-Escalante, J. A., Guerrero, V.,and Salas-Muñoz, E. Feruloylated arabinoxylans and arabinoxylan gels: structure, sources and applications. *Phytochemistry Reviews*, **9**, 111–120 (2009).
9. Berlanga-Reyes, C. M., Carvajal-Millán, E., Lizardi-Mendoza, J., Rascón-Chu, A., Marquez-Escalante, J. A,and Martínez-López, A. L. Maize arabinoxylan gels as protein delivery matrices. *Molecules (Basel, Switzerland)*, **14**, 1475–82 (2009).
10. Carvajal-Millan, E., Guilbert, S., Doublier, J. L.,and Micard, V. Arabinoxylan/protein gels: Structural, rheological and controlled release properties. *Food Hydrocolloids*, **20**, 53–61 (2006).
11. Iravani, S., Fitchett, C. S.,and Georget, D. M. R. Physical characterization of arabinoxylan powder and its hydrogel containing a methyl xanthine. *Carbohydrate Polymers*, **85**, 201–207 (2011).

12. Hernández-Espinoza, A. B., Piñón-Muñiz, M. I., Rascón-Chu, A., Santana-Rodríguez, V. M.,and Carvajal-Millan, E. Lycopene/arabinoxylan gels: rheological and controlled release characteristics. *Molecules (Basel, Switzerland)*, **17**, 2428–36 (2012).

13. Trombino, S., Cassano, R., Ferrarelli, T., Barone, E., Picci, N.,and Mancuso, C. Transferulic acid-based solid lipid nanoparticles and their antioxidant effect in rat brain microsomes. *Colloids and Surfaces B: Biointerfaces*, **109**, 273–279 (2013).

14. Trombino, S., Cassano, R., Muzzalupo, R., Pingitore, A., Cione, E.,and Picci, N. Stearyl ferulate-based solid lipid nanoparticles for the encapsulation and stabilization of β-carotene and α-tocopherol. *Colloids and Surfaces B: Biointerfaces*, **72**, 181–187 (2009).

15. Hromádková, Z., Paulsen, B. S., Polovka, M., Košťálová, Z.,and Ebringerová, A. Structural features of two heteroxylan polysaccharide fractions from wheat bran with anti-complementary and antioxidant activities. *Carbohydrate Polymers*, **93**, 22–30 (2013).

16. Martinez-Lopez, A.L.,Carvajal-Millan, E.,Lizardi-Mendoza, J.,Rascón-Chu, A.,López-Franco, Y.L.,and Salas-Muñoz, E. FerulatedArabinoxylans as by-Product from Maize Wet-Milling Process: Chacterization and Gelling Capability. J.C. Jimenez-Lopez(Ed.), Nova Science Publisher:Granade, Spain, In *Maize: Cultivation, Uses and Health Benefits*,pp. 65–74 (2012).

17. Martínez-López, A. L., Carvajal-Millan, E., Miki-Yoshida, M., Alvarez-Contreras, L., Rascón-Chu, A., Lizardi-Mendoza, J.,and López-Franco, Y. Arabinoxylan Microspheres: Structural and Textural Characteristics. *Molecules*, **18**, 4640–4650 (2013).

18. Vansteenkiste, E., Babot, C., Rouau, X.,andMicard, V. Oxidative gelation of feruloylated arabinoxylan as affected by protein. Influence on protein enzymatic hydrolysis. *Food Hydrocolloids*, **18**, 557–564 (2004).

19. Carvajal-Millan, E., Guilbert, S., Morel, M. H.,and Micard, V. Impact of the structure of arabinoxylan gels on their rheological and protein transport properties. *Carbohydrate Polymers*, **60**, 431–438 (2005).

20. Brunauer, S., Emmett, P.H.,and Teller, E. Asorption of gases in multimolecular layers. *J. Am. Chem. Soc.*, **60**, 309–319 (1938).

21. Re, R., Pellegrini, N., Proteggente, A., Pannala, A., Yang, M.,and Rice-Evans, C. Antioxidant activity applying an improved (ABTS) radical cation decolorization assay. *Free Radical Biology and Medicine*, **26**, 1231–1237 (1999).

22. Rosa, N. N., Barron, C., Gaiani, C., Dufour, C.,and Micard, V. Ultra-fine grinding increases the antioxidant capacity of wheat bran. *Journal of Cereal Science*, **57**, 84–90 (2013).

23. Carvajal-Millan, E., Landillon, V., Morel, M. H., Rouau, X., Doublier, J. L.,and Micard, V. Arabinoxylan gels: impact of the feruloylation degree on their structure and properties. *Biomacromolecules*, **6**, 309–17 (2005).

24 Carvajal-Millan, E., Rascón-Chu, A., Márquez-Escalante, J. A., Micard, V., León, N. P. De,and Gardea, A. Maize bran gum: Extraction, characterization and functional properties. *Carbohydrate Polymers*, **69**, 280–285 (2007).

25. Berlanga-Reyes, C. M., Carvajal-Millan, E., Lizardi-Mendoza, J., Islas-Rubio, A. R.,and Rascón-Chu, A. Enzymatic Cross-Linking of Alkali Extracted Arabinoxylans: Gel Rheological and Structural Characteristics. *International Journal of Molecular Sciences*, **12**, 5853–5861 (2011).

26. Fundueanu, G., Esposito, E., Mihai, D., Carpov, A., Desbrieres, J., Rinaudo, M.,and Nastruzzi, C. Preparation and characterization of Ca-alginate microspheres by a new emulsification method. *International journal of pharmaceutics*, **170**, 11–21 (1998).

27. Berlanga-Reyes, C. M., Carvajal-Millán, E., Caire Juvera, G., Rascón-Chu, A., Marquez-Escalante, J. A.,and Martinez-Lopez, A. L. Laccase induced maize bran arabinoxylans gels: structural and rheological properties. *Food Science and Biotechnology*, **18**, 1027–1029 (2009).

CHAPTER 3

COMPARATIVE ESTIMATION OF KALANCHOE JUICE ANTIOXIDANT PROPERTIES

N. N. SAZHINA

CONTENTS

3.1   INTRODUCTION

The comparative analysis of measurements of the total antioxidants content and their activity for juice of 34 different kinds of *Kalanchoe* (*Kalanchoe* L.) is carried out by two methods: Ammetric and chemiluminescence (CL). The measurement results show good (89%) correlation. Among the studied samples two most active kinds of *Kalanchoe* are exposed: *K.scapigera* and *K.rhombopilosa*. They can appear more perspective sources of biologically active components in comparison with kinds which are used now.

One of pharmaceutical science problems is studying biological, including antioxidant, activities of various herbs for the purpose of search among them the most active sources of biologically active substances. In medicine preparations from components of some succulent plants widely apply, in particular some kinds of the genus *Kalanchoe*. The genus Kalanchoe (*Kalanchoe Adans.*) haves now more than 130 different kinds. These plants are widely cultivated worldwide. Representatives of the genus *Kalanchoe* are succulent plants with juicy water reserving leaves of a freakish form [1]. Generally, they are used as decorative but some types are applied and in the medical purposes since their leaves contain useful mineral salts, organic acids, and numerous phenolic compounds [2,3]. Education and accumulation of these compounds in the course of a secondary metabolism depends on genetic features of a plant and numerous factors of environment. Phenolic compounds cause, mainly, biological, including antioxidant activity (AOA) of this or that species of a plant, that is ability its component to inhibit oxidizing free radical processes. The biochemical structure and curative properties of juice and extracts of *Kalanchoe pinnata* (*K.pinnata*) and *Kalanchoe Daigremontiana* (*K.daigremontiana*) are studied especially well now [2-4]. However, results of purposeful scientific researches of antioxidant properties of these and other *Kalanchoe* kinds practically are not present [5,6]. Modern methods of these properties research allow to study them at higher level and to comprehend the medicinal value of this or that plants.

In the present chapter the comparative analysis of measurement results of the total antioxidants content and their activity for juice of various kinds of *Kalanchoe Kalanchoe* L.) is carried out by ammetric and CL methods for the purpose of identification among them the most active producers of biologically active compounds.

3.2   EXPERIMENTAL DETAILS

The objects of research were juice samples of 34 kinds of the genus *Kalanchoe* grown up in a succulent collection in Timiryazev Institute of plant physiology of the Russian Academy of Sciences in Moscow. Juice was squeezed from leaves of these plants, was stored in the refrigerator at a temperature $-12°C$, and was de-frozen for measurements to room temperature, and if necessary, diluted with the distilled water. The measurements of the total antioxidant content in samples were carried out by an electrochemical (ammetric) method and of the total AOA by CL method.

The ammetric method allows defining the total phenol type antioxidant content in investigated samples [7]. The essence of this method consists in measurement of the electric current arising at oxidation of investigated substance on a surface of a working electrode at certain electric potential (0–1.3V). At such values of potential, oxidation of only OH-groups of natural phenolic antioxidants (R-OH) takes place.. The electrochemical oxidation proceeding under scheme R-OH $\rightarrow$ R-O$\cdot$ + e$^-$ + H$^+$ can be used as model at measurement of free radical absorption activity. The capture of free radicals is carried out according to reaction R-OH $\rightarrow$ R-O$\cdot$ + H$\cdot$. Both reactions include the rupture of the same bond O-H. In this case, the ability of same phenol type antioxidants to capture free radicals can be measured by value of the oxidizability of these compounds on a working electrode of the ammetric detector [8]. The integral signal (the area under a current curve) is compared to the signal received in same conditions for the comparison sample with known concentration. Gallic acid (GA) was used in work as the comparison sample. The total antioxidant (AO) content is determined by calibration dependence of the oxidizability of GA on its concentration in mg/l of GA. The method does not use of model chemical reaction. The error in determination of the AO content including the error by reproducibility of results was within 10%.

In CL method of AOA definition the scheme of oxidation system "hemoglobin–hydrogen peroxide–luminol" was used. The detailed measurement technique is given [9]. Distinctive feature of this system from other oxidation systems is that the formed in it radicals can initiate free radical oxidation reactions *in vivo*. Interaction of hydrogen peroxide (H_2O_2) with metHb is accompanied by gem destruction and by exit from it of iron ions which participate in education of OH$^{\cdot-}$ radicals. Besides, as a result of this interaction active ferril-radicals (Hb($^{\cdot+}$)–Fe^{4+} = O) are formed. Being formed radicals initiate the luminol oxidation in the process of which a luminol-endoperoxide $(LO_2)^{2-}$ is formed, and further an aminophthalate anion in excited state (AP^{2-})* upon which transition to the main state light quantum with a wave length 425 nm is highlighted. Introduction of antioxidants in «metHb-H_2O_2-luminol» system leads to change of kinetics of its CL and increase in the latent period (t) which is directly proportional to the concentration of added antioxidants [9]. For realization of this method in the present work the device "Lum-5373" (OOO"DISoft", Russia) was used. The latent period (t) was calculated from introduction time of a hydrogen peroxide in CL-cell to a point of intersection of a tangent in a point of a maximum of the first derivative of a CL-curve with a time axis. A tangent of an inclination angle (k) for a straight line describing dependence of the relative latent period (t/t$_0$) on the sample mass (m, mg), entered into a cell, was accepted to AOA criterion of studied samples (k, mg^{-1}): t/t0 = km + 1; t$_0$—the latent period without a sample. The error of the AOA determination by this method including the error by reproducibility of results is not more 15%. In Figure 1, the characteristic dynamics of a luminescence for various values (m) is presented. In Figure 2, dependences of t/t$_0$ on m for some samples of juice are given by which values of AOA (k) are determined.

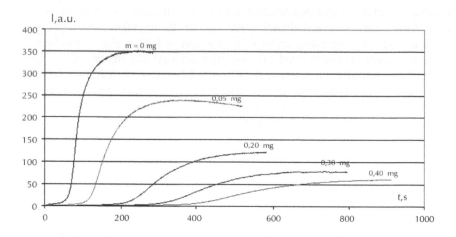

FIGURE 1 Characteristic dynamics of a CL for various values of a sample mass (m), entered into a CL cell.

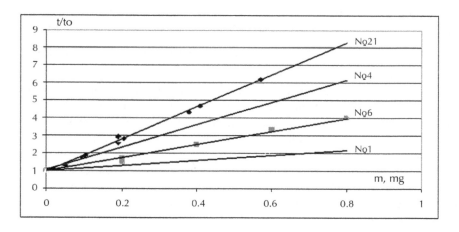

FIGURE 2 Dependence of the relative latent period (t/t_0) of CL on a sample mass (m), entered into a cell, for juice of various *Kalanchoe* kinds.

3.3   DISCUSSION AND RESULTS

Table 1 shows results of measurements of the total phenol type antioxidant content in units of GA concentration (C, mg/l GA) obtained by ammetric for juice of 34 various kinds of *Kalanchoe*. Juice of such *Kalanchoe* kinds as *K.scapigera* (№ 11) and *K.rhombopilosa* (№ 21) have the greatest AO content values, the smallest—*K.bloss-feldiana* juice (№ 30) and *K.laetivirens* (№ 32) juice. The most widespread and studied

Kalanchoe kinds, such as *K.pinnata* (№ 1) and *K.daigremontiana* (№ 6) show rather low levels of phenolic AO accumulation (294 and 550 mg/l GA, respectively).

TABLE 1 The total phenol type AO content in juice samples of various kinds of *Kalanchoe* (C, мg/l GA)

No sample	*Kalanchoe*'s kind name	Ammetry, C, mg/l GA	No sample	*Kalanchoe*'s kind name	Ammetry, C, mg/l GA
1	*K.pinnata*	294	18	K.millotii	1070
2	*K.beauverdii*	730	19	K.fedtschenkoi	604
3	*K.orgyalis*	393	20	K.serrata	383
4	*K.imperialis*	878	21	K.rhombopilosa	1911
5	*K.tomentosa*	593	22	K.kalandiva	505
6	*K.daigremontiana*	550	23	K.laciniata	358
7	*K.thyrsiflora*	220	24	K.tubiflora	423
8	*K.hildebrandtii*	1265	25	K.longiflora	330
9	K.velutina	545	26	K.bracteata	265
10	K.beharensis v.aureoaeneu	368	27	K.crenata	380
11	K.scapigera	1981	28	K.nyikae	514
12	K.citrina	328	29	K.sedoides	342
13	K.X kewensis	301	30	K.blossfeldiana	205
14	K.serratifolia	355	31	K.blossfeldiana mini	278
15	K.germanae	321	32	K.laetivirens	186
16	K.syncepala	247	33	K.figuereidoi	505
17	K.gastonis-bon-nieri	403	34	K.pubescens	388

For comparison of the total antioxidant content measurement results in *Kalanchoe* juice samples presented in the Table 1, with results of AOA measurements by a CL method 14 samples having a different range of values C (in the table they are allocated) were chosen, and for them CL measurements were carried out.

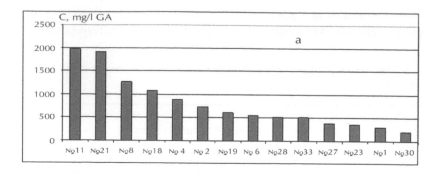

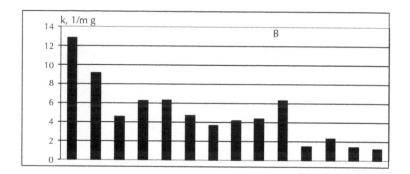

FIGURE 3   (a) The total phenol type AO content in juice samples of various *Kalanchoe* kinds (C, мg/l GA) (ammetric method) and (b) AOA of *Kalanchoe* juice (k, 1/mg) (CLmethod). Numbers of samples in Table 1.

In Figure 3 results of measurements of the total antioxidant content (C) and their activities in 14 samples of juice from Table 1 are presented by two methods. For an ammetry C (C, mg/l GA) is constructed in decreasing order (Figure 3 (a)). Results of measurements of inhibition degree of «metHb-H_2O_2-luminol» oxidation system by specified juice (AOA) are given in Figure 3 (b) (k, 1/mg). It is visible that monotony of AOA decreasing in this cause is a little broken, generally at the expense of samples №18, №4, and №33. However, correlation of results received by two methods is rather high (r = 0. 89).

In works [2,3] the biochemical composition of *K.pinnata* and *K.daigremontiana* tinctures was carefully studied by different methods including the high-performance liquid chromatography (HPLC) method. Leaves and stalks of the specified Kalanchoe kinds contain flavanoids: quercetine, kaempferol and its glycosides, routines, hyperoside, many organic acids, basic of which apple, isolemon and lemon. The 15 amino acids among which coumaric acid, asparagin acid, and glycine are identified. *Kalanchoes* are rich with polysaccharides (35–40%), there are at him condensed tannins, derivatives of cholesterol, methyl ether, and also vitamins: ascorbic acid, thiamine, riboflavin, and microcells: zinc, iron, calcium, sodium, and potassium [2]. The conclusion about considerable similarity of qualitative structure of phenolic and other components in these two *Kalanchoe* kinds is drawn [3],as it belongs to other kinds. The increased values of AOA for juice of some *Kalanchoe* kinds, received by CL method, can testify to existence in them any other components not phenol type, having rather high antioxidant activity. Probably, they are any enzymes or metal complexes.

It should be noted that both methods used in the present study show much higher antioxidant properties for juice of two *Kalanchoe* kinds: *K.scapigera* and *K.rhombopilosa* in comparison with above mentioned popular and widely used in medical purposes of *K.scapigera* (№1) and *K.rhombopilosa* (№6). Expansion of possibility to use the specified *Kalanchoe* kinds as sources of biologically active compounds requires carrying out additional researches of these plant juices for studying of their biochemical structure and also antibacterial, antimicrobic, phytoregulating, and other properties of their compounds. Perhaps, they will appear more perspective for their use in medicine.

3.4   CONCLUSION

The total antioxidant content in juice samples of 34 various kinds of the genus *Kalanchoe* was measured by an electrochemical (ammetric) method. The comparative analysis of measurement results of the total antioxidants content and their activity in juice of 14 kinds of Kalanchoe (Kalanchoe L.) is carried out by ammetric and CL methods. Correlation of measurement results is r = 0.89. Among the studied samples two most active from the point of view of their antioxidant properties Kalanchoe kinds—K. scapigera and K.rhombopi*losa* are revealed. They can appear more perspective in pharmacy and medicine sources of biologically active substances in comparison with the *Kalanchoe* kinds which are used now.

KEYWORDS

- **Ammetric method**
- **Antioxidant activity**
- **Chemiluminescence method**
- **Gallic acid**
- **Kalanchoe species**

ACKNOWLEDGMENT

We thank sciences of Timiryazev Institute of plant physiology of the RAS N. Zagoskina and P. Lapshin for supplying samples.

REFERENCES

1.   Semenova, L. V. and Yampolsky, Yu. V. *Medicinal exotic plants*. Vol.1. Sankt-Petersburg (2003).

2.   Volzhanova, M. I., Baylman, R. A., Suslina, S. N., and Bykov, V. A. *Kalanchoe pinnata* and *daigremontiana*: chemical composition, application in medicine (review). *Questions of biological, medical and pharmaceutical chemistry*, **7**, 14–20 (2010).
3.   Volzhanova. M. I., Suslina, S. N., and Bykov, V. A. Comparative research of target part metabolomic of close kinds of Kalanchoe by chromatography methods. *Questions of biological, medical, and pharmaceutical chemistry*, **8**, 3–6 (2011).
4.   Anisimov, M. M., Gerasimenko, N. I., Tchaikin, E. L., and Serebryakov, YU. M. Biological activity of metabolites of a herb of *Kalanchoe daigremontiana* (Hamet de la Bbthie) of Jacobs et Perr. *Isvestia RAS. Series biological*, **6**, 669–676 (2009).
5.   Misin, V. M., Sazhina, N. N., Vandyshev, V. V., and Homik, A. S. The comparative analysis of the total antioxidants content and level of their activity in juice of succulent plants by ammetric and voltammetric methods. *Questions of biological, medical, and pharmaceutical chemistry*, **12**, 1–5 (2010).
6.   Olennikov, D. N., Zilfikarov, I. N., Toropova, A. A., and Ibragimov, T. A. Chemical composition of juice of a collision fragrant and its antioxidant activity. *Chemistry of vegetable raw materials*, **4**, 95–1001 (2008).
7.   Yashin, A. Ya. Inject-flowing system with ammetric detector for selective definition of antioxidants in foodstuff and drinks. *Russian chemical magazine*, **52**(2), 130–135 (2008).
8.   Peyrat-Maillard, M. N., Bonnely, S., and Berset C. Determination of the antioxidant activity of phenolic compounds by coulometric detection. *Talanta*, **51**, 709–714 (2000).
9.   Teselkin, YU. O., Babenkova, I. V., Lyubitsky, O. B., Klebanov, G. I., and Vladimirov, Yu. A. Inhibition of oxidation of luminol in the presence of hemoglobin and hydrogen peroxide by serum antioxidants. *Questions of medical chemistry*, **43**(2), 87–93 (1997).

CHAPTER 4

ENZYMES FOR FLAVOR, DAIRY, AND BAKING INDUSTRIES

ADRIANE B. P. MEDEIROS, SUZAN C. ROSSI, MÁRIO C. J. BIER, LUCIANA P. S. VANDENBERGHE, and CARLOS R. SOCCOL

CONTENTS

4.1   INTRODUCTION

Enzymes from different sources (plants, animals, and microorganisms) have wide application in different sectors. Microbial enzymes are often more stable and useful than enzymes derived from plants or animals because of the great variety of catalytic activities available, the high yields possible, ease of genetic manipulation, regular supply due to absence of seasonal fluctuations, and rapid growth of microorganisms on inexpensive media [1]. According to a report from Business Communications Company Inc. the global market for industrial enzymes increased from $2.2 billion in 2006 to an estimated $2.3 billion by the end of 2007. It should reach $2.7 billion by 2012, a compound annual growth rate of 4%. The total industrial enzyme market in 2009 is expected to reach nearly $2.4 billion [2]. Food industry is the largest consumer of enzymes and approximately 45% of bulk share goes to it [3].

Enzymes are commonly used in food processing and the production of food ingredients. As indicated in Table 1, applications of enzymes in the food industry are many and diverse, including the production of food quality attributes such as flavors, control of color, texture, and appearance. Several advances have been made in optimization enzymes for existing applications and in the use of recombinant protein production to provide efficient mono-component enzymes that do not result in harmful side effects [4].

TABLE 1   Applications of enzymes in the food industry

Enzyme	Application
Protease	Milk clotting, infant formulas (low allergenic), biscuits, cookies, flavor
Lipase	Cheese flavor, dough stability
Lactase	Lactose removal (milk)
Pectin methyl esterase	Firming fruit-based products
Pectinase	Fruit-based products
Transglutaminase	Modify visco-elastic properties, laminated dough strengths
Amylase	Bread softness and volume, flour adjustment
Xylanase	Dough conditioning
Phospholipase	Dough stability and conditioning (*in situ* emulsifier)
Glucose oxidase	Dough strengthening
Lipoxygenase	Dough strengthening, bread whitening

4.2 ENZYMES INVOLVED IN AROMA PRODUCTION

The aroma formulator consists in constructing a flavor recalling a true and original aroma in a processed food product with a specific texture and composition. To be able to formulate this flavor, a technologist needs a wide spectrum of different aroma components. These compounds can be extracted from fruits or vegetables but as they are required in the product in concentrations comparable to those in the source material, this utilizes high amounts of materials and is generally not economically realistic. Most food flavoring compounds can also be produced via chemical synthesis. However, recent market surveys have demonstrated that consumers prefer foodstuff that can be labeled "natural". The use of biotechnology appears to be attractive in various ester preparations under milder conditions and the product may be given the natural label [5].

Biotechnology proposes to use enzymes or whole cells to produce aroma compounds from natural substrates in a way often inspired by biochemical pathways encountered in nature. The challenge is to put a naturally rich source of substrate in contact with highly active enzymes. In adequate conditions, this can result in the production of flavor compounds in mass fractions of the order of several g/kg, instead of mg/kg encountered in raw materials. The resulting flavor compounds are called natural since they are produced from agro-products through natural biological activities [6]. Besides the advantage of being recognized as natural, the aromas produced by microorganisms have a great economic potential for obtaining a wide range of biomolecules of interest in the food industry [8,9]. The industrial production of aromas corresponds of 25% of the food additives world market [7]. The world market of flavors and fragrances has a current volume of U$18.6 billion [10]. Still 10% of the supply is derived from bioprocesses. Examples, such as the Bartlett pear impact compound, ethyl 2,4-(E,Z)-decadienoate, which is cheaper to produce using enzyme catalysis than chemosynthesis, should encourage further research [26].

The majority method for the production of natural aromas has enzymes involved in the biosynthesis. A number of enzymes (lipases, proteases, and glycosidases) catalyze the production of aroma-related compounds from precursor molecules. Despite the higher costs involved, microbial enzymes – rather than cells can offer high stereo and enatioselectivity towards substrate conversion Techniques such as enzyme immobilization and eventually coenzyme regeneration might result in highly efficient and specific biocatalytic processes for flavor synthesis [9,25].

4.2.1 LIPASES

Among the hydrolytic enzymes of interest are the lipases (triacylglycerohydrolases, EC 3.1.1.3). These enzymes had been used for the hydrolysis of acylglycerides, which are versatile biocatalysts that can catalyze different reactions in both aqueous and organic media, with limited water content. Among the lipases of plant, animal and microbial, these are the most used and, mostly, are not harmful to human health and are recognized as "Generally Regarded as Save – GRAS" [12].

From the industrial point of view, the fungi are especially valued because the enzymes they produce are usually extracellular, which facilitates their recovery from the fermentation medium.

Lipases have been used in a variety of biotech sectors, as in the food (flavor development and cheese ripening), detergent, oleochemical (hydrolysis of fats and oils, synthesis of biosurfactants); and for treatment of oily wastes emanating from the leather and paper [13]. This enzyme mediated synthesis of flavor esters has the potential of satisfying the increasing demand for natural flavor esters. Enzymatic esterification of flavor esters has the advantage of catalyzing reactions more specifically than chemical synthesis under mild condition and higher yield over microbiological production.

In recent years, lipases have been found to catalyze reversible reactions such as esterifying reaction in aqueous, non-aqueous, and solvent-free phases. Lipase-mediated synthesis of aliphatic esters of longer chain substrates has shown the potential of their esterification abilities. The synthesis of low molecular flavor ethyl esters from shorter chain substrate has been attempted. However, this attempt comparatively received less attention with no satisfaction as short fatty-acids easily strip the essential water around enzymes to cause their deactivation or cause dead-end inhibition reacting with the serine residue at the active site of lipase. Almost all researchers did not succeed in synthesizing them with high yields (>80%) at more than 0.5M substrate concentration , which keeps enzymatic pathway for synthesis of ethyl esters of short-chain fatty acids far from industries [14].

The lipases, both free and immobilized, have a high denature capability. Strains capable of producing active and stable lipases to catalyze esterification of ethanol with short-chain fatty acids in a non aqueous phase should be screened for the synthesis of flavor esters. Lipases have been used for addition to food to modify flavor by synthesis of esters of short chain fatty acids and alcohols, which are known flavor and fragrance compounds because psychotropic Gram-negative bacteria, such as *Pseudomonas* species, pose a significant spoilage problem in refrigerated meat, and dairy products due to secretion of hydrolytic enzymes, especially lipases and proteases.

Lipases have earlier been used in production of leaner meat such as in fish. The fat is removed during the processing of the fish meat by adding lipases and this procedure is called biolipolysis. The lipases also play an important role in the fermentative steps of sausage manufacture and to determine changes in long-chain fatty acid liberated during ripening. Earlier, lipases of different microbial origin have been used for refining rice flavor, modifying soybean milk, and for improving the aroma and accelerating the fermentation of apple wine [1]. These are of great importance for the dairy industry for the hydrolysis of milk fat. Present applications include the flavor enhancement of cheeses, the acceleration of cheese ripening, the manufacturing of cheese like products, and, the lipolysis of butterfat and cream. The free fatty acids resultants by the action of lipases on milk fat endow many dairy products with their specific flavor characteristics.

A whole range of microbial lipase preparations has been developed for the cheese manufacturing industry: *Mucor meihei* (Piccnate, Gist-Brocades and Palatase M, Novo Nordisk), *A. niger* and *A. oryzae* (Palatase A, Novo Nordisk; Lipase AP, Amano; and Flavor AGE, Chr. Hansen), and several others [1].

Ester synthesis by means of lipase is an interesting alternative considering that there are many well known flavor esters in the natural aroma of fruits, traditionally obtained by extraction or by chemical synthesis [15].

Ethyl and hexyl esters are the important and versatile components of natural flavors and fragrances. The worldwide market for natural "green notes" is estimated to be 5–10 metric tonnes per year. Ethyl valerate with a typical fragrance compound of green apple and hexyl acetate with a pear flavor property are in high demand and are widely used in food, cosmetic, and pharmaceutical industries. Traditionally, this kind of compounds have been isolated from natural sources or produced by chemical synthesis. Nowadays, there is a growing demand for natural flavors.

Lipase catalyzed biotransformations are gaining importance because of their regio-, stereo-, and substrate specificity's, their milder reaction conditions and the relatively lower energy requirement. Enzymatic reactions can be efficiently accomplished by employing lipases in an adequate organic solvent, such as heptane and hexane, thereby shifting the reaction equilibrium toward esterification rather than hydrolysis. Lipases have been employed for direct esterification and transesterification reactions in organic solvents to produce esters of glycerol, aliphatic alcohols, and terpene alcohol's. However, the esterification of short chain fatty acids and alcohol's has not received much attention. Moreover, short-chain (C5) acids and alcohols are found to exert inhibitory effects on the enzyme. Data generation for esterification with such substrates is important because short-chain esters are vital components of many fruit flavors.

Few attempts have been made to establish the feasibility of synthesizing isoamyl acetate by employing lipases from different sources. However, the uses of high enzyme concentration and low yields have been the significant drawbacks. Also the lipase catalyzed esterification reaction has not been examined thoroughly. Reports on the use of high enzyme concentration leading to 80% yields are available. Considering the high demand and benefits, an optimized process for high yield enzymatic synthesis of isoamyl esters, utilizing low enzyme contents, is important. The relationships between the important esterification variables (substrate and enzyme concentrations, and incubation period) and ester yield to determine optimum conditions for the synthesis of isoamyl acetate catalyzed by Lipozyme.

4.2.2 ESTERASES

Carboxylesterases (E.C. 3.1.1.1, carboxyl ester hydrolases) are enzymes widely distributed among all forms of life; their physiological functions have been implicated in carbon source utilization, pathogenicity, and detoxification [17]. This has a number of unique enzyme characteristics such as substrate specificity, regiospecificity, and chiral selectivity. These enzymes preferably catalyze the hydrolysis of esters composed of short-chain fatty acids are preferably catalyzed by the enzymes, but they also can catalyze other reactions, like ester synthesis, and transesterification reactions. Notably, the interest of a broad range of industrial fields like foods, pharmaceuticals, and cosmetics has been attracted by the potential application of these enzymes for the synthesis of short-chain esters. Compounds with a great application due to their characteristic fragrance and flavor are constituted by flavor acetates from primary alcohols, which are included among the present esters.

Bacillus species has the advantages of extracellular enzyme production and lack of toxicity (most of *Bacillus* species are considered GRAS by FDA, USA). Additionally, species of the genus *Bacillus* have a history of safe use in the elaboration of traditional fermented food products, such as condiments and sauces produced from grains, legumes and seafood. Also, safety industrial use of these bacteria and their enzymes was reported, including the production of food additives and probiotic products. [18] produced and characterized a bacterial esterase, from a wild type organic solvent-tolerant *Bacillus licheniformis* S-86. One of the esterases from *B. licheniformis* S-86, called type II esterase, was previously purified in a five-step procedure. This enzyme was demonstrated to be a carboxylesterase specific for short-chain acyl esters, and stable and active in the presence of hydroxylic organic solvents.

The most important features about this enzyme are its proven utility for the synthesis in non-aqueous media of a valuable flavor compound with a great application in food industries. The use of type II esterase from *B. licheniformis* S-86 can be an alternative to be explored in order to obtain isoamyl acetate by "green chemistry technology" [18].

In Japanese sake, esters are the most important flavor compounds, and the mechanisms of their formation have been investigated in great detail. Vinegar also contains many flavor components, including esters. However, little is known about how esters are produced in a particular fermentation environment in vinegar.

During vinegar production, esters are the major flavor compounds produced by *Acetobacter* sp. [19] on analyzing the relationship between ethyl acetate production and the extracellular ethanol and acetic acid concentrations; found that the highest amount of ethyl acetate was produced when the molar ratio of ethanol and acetic acid was 1:1. These results indicate that the ester production by *Acetobacter* sp. is mostly catalyzed by the intracellular esterase, esterase-1, with ethanol and acetic acid used as the substrates.

4.2.3 ALCOHOL ACETYLTRANSFERASES

Esters generally have a low odor threshold: 20–30 ppm for ethyl acetate and around 1.2 ppm for isoamyl acetate. Because the concentration of most esters formed during natural fermentations hovers around their respective threshold values, slight variations in their concentration may have dramatic effects on the beer flavor. Therefore, understanding the mechanisms of their formation to control their levels in the end product is of major industrial interest. As a consequence, the biochemical background of ester synthesis has been intensively studied.

Acetate ester formation occurs intracellularly through an enzyme-catalyzed reaction between acetyl-CoA and an alcohol. In the brewer's yeast *Saccharomyces cerevisiae*, two ester synthesizing enzymes have been identified: the alcohol acetyltransferases I and II (AATase I and II; EC number: 2.3.1.84). The corresponding genes were cloned and named ATF1 and ATF2, respectively. *Saccharomyces pastorianus, Saccharomyces carlsbergensis, Saccharomyces bayanus* are other sources that can produce

AATase. Furthermore, the activity of the enzymes was not limited to the acetylation of isoamyl alcohol (resulting in the formation of the important banana flavour isoamyl acetate), but also other alcohols like propanol, butanol, and phenyl ethanol are esterified by Atf1 and Atf2. This opens up the exciting possibility to tailor yeast's aroma production to meet consumer preferences. Not only *Saccharomyces* produced AATase, but *Kluyveromyces lactis* is commonly used in mould surface ripened cheese, and this yeast appeared to induce fruity characteristics when cultured in a cheese model medium. The alcohol acetyltransferase orthologue of *K. lactis* is most likely responsible for part of this fruity flavor formation [20].

The characteristic fruity odors of wine, brandy, and other grape-derived alcoholic beverages are primarily due to a mixture of hexyl acetate, ethyl caproate (apple-like aroma), isoamyl acetate (banana-like aroma), ethyl caprylate (apple-like aroma), and 2-phenylethyl acetate (fruity, flowery flavor with a honey note). The synthesis of acetate esters such as isoamyl acetate and ethyl acetate in *S. cerevisiae* is ascribed to at least three acetyltransferase activities, namely alcohol acetyltransferase (AAT), ethanol acetyltransferase, and isoamyl AAT. These acetyltransferases are sulfhydryl enzymes which react with acetyl coenzyme A (acetyl-CoA) and, depending on the degree of affinity, with various higher alcohols to produce esters. It has also been shown that these enzymatic activities are strongly repressed under aerobic conditions and by the addition of unsaturated fatty acids to a culture. The ATF1-encoded AAT activity is the best-studied acetyltransferase activity in *S. cerevisiae*. It has been reported that the 61-kDa ATF1 gene product (Atf1p) is located within the yeast's cellular vacuomes and plays a major role in the production of isoamyl acetate and to a lesser extent ethyl acetate during beer fermentation. The AAT in wine and brandy composition, has cloned, characterized, and mapped the ATF1 gene from a widely using commercial wine yeast strain, VIN13, was to over express the ATF1 gene during fermentation to determine its effect on the yeast metabolism, acetate ester formation, and flavor profiles of Chenin blanc wines and distillates from Colombar base wines. Ultimately lead to the development of a variety of wine yeast strains for the improvement of the flavor profiles of different types and styles of wines and distillates, especially of those products deficient in aroma and lacking a long, fruity shelf life [21].

4.2.4 ACYL-COA

Acyl CoA is involved to generation of aroma compounds in this case, γ-decalactone through ß-oxidation is the classical biochemical pathway involved in fatty acid degradation, it acts on an acyl-CoA and consists of a four-steps reaction sequence, yielding an acyl-CoA which has two carbons less and an acetyl-CoA. This can lead to a variety of volatile compounds (Figure 1) [6].

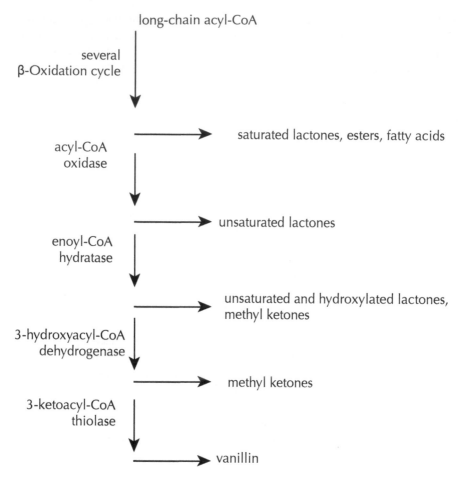

FIGURE 1   β-Oxidation cycle and accumulation of aroma compounds.

4.3   ENZYMES FOR DAIRY INDUSTRY

Since, the origin of cheese production, at least 4000 years, the manufacturing process has been adapted to current technologies. Rennet, also called rennin, is a mixture of chymosin and pepsin obtained from the 4th stomach of suckling calves. Over the years there was an imbalance between the production of cheese and availability of animal rennet. Microbial enzymes have therefore been introduced as alternative. Chymosin is an acid protease that cleaves specifically the peptide bond between the phenylalanine and methionine, from N-terminus of the protein content. Submerged fermentation with fungi such as *Rhizomucor miehei* produces microbial rennets with similar properties to those of chymosin. In practice, microbial enzymes have lower milk coagulating activity and proteolytic activity than pure calf rennet [11].

Proteases are used in order to influence the organoleptic properties (aroma and flavor) of cheese, which in turn depend on the relative amounts of amino acids and fatty acids throughout the product. Proteases are also used for accelerating cheese ripening, for modifying the functional properties of cheese, and for modifying milk proteins to reduce the allergic properties of dairy products.

Other components of milk are also allergens such as lactose. The enzyme lactase (beta-galactosidase) is used to hydrolyze lactose in order to increase digestibility or to improve solubility or sweetness of various dairy products. The main sources of lactase are the yeasts *Kluyveromyces fragilis, K.lactis* and the fungi *A.niger* and *A.oryzae*.

Some of the applications of the hydrolysis of lactose in dairy products are:

a) Lactose removal from milk powder or liquid allows it to be ingested by disabled lactase. Moreover, in the case of flavored milk increases the rate of sweetness and enhances the flavor of this product;

b) Reduce the time of fermentation of fermented dairy products;

c) Prevent the crystallization of lactose in ice cream and condensed milk. Remember that lactose concentration in the order of 12% compared to the total milk solids (For example, skimmed) crystallizes easily. When 50% reduction in the level of lactose in milk, it can be used in larger amounts in formulations, which reduces the amount of stabilizer to be added, since this function can be performed by proteins own milk. Furthermore, since the lactose hydrolyzed milk is the number of free molecules increased, which contributes to the freezing point depression solution, when it is used in formulations of creams;

d) The removal of lactose whey allows the manufacture of syrups with 70–75% of solid and can be used in food formulations [11].

4.4   ENZYMES FOR BAKING INDUSTRY

Wheat flour is one of the main ingredients of bakery products. Starch and proteins are the major constituents of wheat flour. Alpha amylase is a special type of amylase which modifies starch during baking to give a significant anti-staling effect.

Besides the traditional enzymes such as malt and alpha-amylases, the advances in biotechnology have made interesting enzymes available for bread making. Hemicellulases, xylanases, lipases, and oxidases can directly or indirectly improve the strength of the gluten network and so improve the quality of finished bread [11].

Within the baking industry, phospholipases can be supplied as effective substitute or supplement traditional emulsifiers, as the enzymes degrade polar wheat lipids to produce emulsifying lipids *in situ*. Lipases from *A. niger, R. oryzae, C. cylindracea* are used in bakery products to increase the volume in bread and bakery [1].

Also, efforts are currently devoted towards the further understanding of bread staling and the mechanisms behind the enzymatic prevention of staling when using α-amylases and xylanases. Studies have confirmed previous findings showing that water-binding capacity and retention in the starch and hemicellulose fractions of the bread, being the substrates of α -amylases and xylanases, respectively, to be critical for maintaining softness and elasticity. This amylase is probably capable of degrading amylopectin to a degree that prevents re-crystallization after gelatinization, without completely degrading the amylopectin network which provides the bread with elasticity [4].

The addition of enzymes to sourdough sponges can also enhance bread volatile compounds. Lipid oxidations through addition of enzymes in the form of active soya flour increase concentrations of hexanal, 1-hexanol, 1-penten-3-ol, 1-pentanol and 2-heptanone, and 2-heptenal and 1-octen-3-ol, only in breads containing soya flour [22]. Addition of lipase, endo-xylanase and α-amylases enhanced acetic acid production by *Lb. hilgardii* 51B. Textural analyses suggest that such sourdoughs with single enzyme have greater stability and crumb softening than breads produced with only exogenous enzymes [22].

The uses of enzymes enable bakeries to extend shelf-life of breads, enhance and control nonenzymatic browning, increase loaf volume and improve crumb structure. There are many industrial manufactures of a wide and continuously evolving range of bakery enzymes to fulfill the needs of the bakery trades.

4.5   OTHER APPLICATIONS

Besides the advances observed, a few new applications of enzymes within the food industry could be mentioned.

Enzymes are used also in drinks and fruit juice manufacturing. Addition of pectinase, xylanase, and cellulase improve the liberation of the juice from the pulp. Pectinases and amylases are used in juice clarification. Almost all the commercial preparations of pectinases are produced from fungal sources. *Aspergillus niger* is the most commonly used fungal species for industrial production of enzymes [23]. The use of laccase for clarification of juice (laccases catalyze the cross-linking of polyphenols, resulting in an easy removal of polyphenols by filtration) and for flavor enhancement in beer are recently established applications within the beverage industry. Brewing is an enzymatic process. Malting is a process, which increases the enzyme levels in the grain. In the mashing process the enzymes are liberated and they hydrolyse the starch into soluble fermentable sugars like maltose, which is a glucose disaccharide [24].

In tea processing, the quality of black tea is dependent great extent on the dehydration, mechanical breaking, and enzymatic fermentation to which tea shoots are subjected. During manufacture of black tea enzymatic breakdown of membrane lipids initiate the formation of volatile products with characteristic flavor properties emphasize the importance of lipid in flavor development and lipase produced by *Rhizomucor miehei* enhanced the level of polyunsaturated fatty acid observed by reduction in total lipid content [1].

Transglutaminase has been applied as a texturing agent in the processing of sausages, noodles, and yoghurt, where cross-linking of proteins provides improved viscoelastic properties of the products. Their wider use is currently limited due to availability of the enzyme in industrial scale. At present only the transglutaminase from *Streptoverticillium* sp. is commercially available at a reasonable scale, and work is ongoing to increase the availability of the enzyme by recombinant production in *Escherichia coli* [4].

4.6 CONCLUSION

The field of industrial enzymes has received great efforts in research and development, resulting in both the development of a number of new products and improvement in the process and performance of several existing products. Microbial enzymes, looking at the increased awareness of environment and cost issues, is gaining ground rapidly due to the various advantages of biotechnology over traditional technologies.

New applications of enzymes within food industry will depend of the functional understanding of different enzyme classes. Furthermore, the scientific advances in genome research and their exploitation via biotechnology is leading to a technology driven revolution that will have advantages for the consumer and food industry alike.

KEYWORDS

- Amylase
- Esterases
- Lipases
- Microbial Enzymes
- Oleochemical

REFERENCES

1. Hasan, F., Shah, A. A, and Hameed, A., Industrial Applications of Microbial Lipases, *Enzyme and Microbial Technology*, **39**, 235–251 (2005).
2. Thakore, Y. Report ID: BIO030E. Analyst: Business Communications Company Inc., Published (January, 2008).
3. Pandey, A. and Ramachandran, S.. In: Enzyme Technology, New Delhi: Asiatech Publishers Inc., pp: 1–10 (2005).
4. Kirk, O., Borchert, T.V.,and Fuglsang, C.C., Industrial enzyme applications. *Current Opinion in Biotechnology*, **13**, 345–351 (2002).
5. Salah, R. B., Ghamghui, H., Miled, N., Mejdoub, H. and Gargouri, Y., Production of Butyl Acetate Ester by Lipase from Novel Strain of *Rhizopus oryzae*, Journal of Biosciense and Bioenginering, **103**, 368–372 (2007).
6. Aguedo, M., Ly, M. H., Belo, I., Teixeira, J. A., Belin, J. M., and Waché, Y. The Use of Enzymes and Microorganisms for the Production of Aroma Compounds from Lipids. *Food Technology Biotechnology*, **42**, 327–336 (2004).
7. Couto, S. R. and Sanromán, M. A. Application of solid-state fermentation to food industry—a review. *Journal of Food Engineering*, **76**, 291–302 (2006).
8. Rossi, S. C., Vandenberghe, L. P. S., Pereira, B. M. P., Gago, F. D., Rizzolo, J. A., Pandey , A., Soccol , C. R., and Medeiros , A.B.P., Improving fruity aroma production by fungi in SSF using citric pulp, *Food Research International*, **42**, 484–486 (2009).
9. Soccol, C. R., Medeiros, A. B. P., Vandenberghe, L. P. S., Soares, M. ,and Pandey, A., Production of aroma compounds. *In: Current developments in solid-state fermentation*, Springer Asiatech Publishers Inc. New Delhi, pp. 357–372 (2008).
10. Soccol, C. R., Medeiros, A. B. P., Vandenberghe, L. P. S., and Woiciechowski, A. L. Flavor Compounds produced by Fungi, Yeasts, and Bacteria. Y. H. Hui (Ed.), *Flavor Technology*. John Wiley & Sons Inc., Hoboken, New Jersey, pp. 179–191 (2007)

11. Soccol, C. R., Vandenberghe, L. P. S, Woiciechowski, A. L., and Babitha, S. Applications of Industrial Enzymes. *In: Enzyme Technology*, Asiatech Publishers, Inc, New Delhi, **742** (2005).

12. Olempska-Beer, Z. S., Merker, R. I., Ditto, M. D., and DiNovi, M. J., Food-processing enzymes from recombinant microorganisms - a review, *Regulatory Toxicology and Pharmacology*, **45**, 144–158 (2006).

13. Carvalho, P. O., Calafatti, S.A., Marassi, M., Silva, D. M., Contesini, F. J., Bizaco, R., and Macedo, G. A. Potencial de Biocatálise Enantioseletiva de Lipases Microbianas. *Quimica Nova*, **28**, 614–621 (2005).

14. Xu, Y., Wang, D., Mu, X. Q., Zhao, G. A., and Zhang, K. C. Biosynthesis of ethyl esters of short-chain fatty acids using whole-cell lipase from *Rhizopus chinesis* CCTCC M201021 in non-aqueous phase. *Journal of Molecular Catalylis Enzymatic*, **18**, 29–37. (2002).

15. Christen, P. and López-Munguia, A., Enzymes and Food Flavor A Review. *Food Biotechnology*, **6**, 167–190 (1994).

16. Krishna, S. H., Manohar, B., Divakar, S., Prapulla, S. G., and Karanth, N.G.,Optimization of isoamyl acetate production by using immobilized lipase from *Mucor miehei* by response surface methodology, *Enzyme and Microbial Technology*, **26**, 131–136 (2000).

17. Ewis, H. E., Abdelal, A. T., and Lu, C.-D. Molecular cloning and characterization of two thermostable carboxylesterases from Geobacillus stearothermophilus. *Gene*, **329**, 187–195. (2004).

18. Torres, S., Baigorí, M. D., Swathy, S. L., Pandey, A., and Castro, G. R. Enzymatic synthesis of banana flavour (isoamyl acetate) by *Bacillus licheniformis* S-86 esterase. *Food Research International*, **42**, 454–460 (2009).

19. Kashima, A., Iijima, M., Nakano, T., Tayama, K.,Koizumi, Y., Udaka, S, and Yanagida, F., Role of Intracellular Esterases in the Production of Esters by *Acetobacter pasteurianus*, *Journal of Biosciense and Bioengineering*, **89**, 81-83 (2000).

20. Van Laere, S. D. M., Saerens, S. M. G., Verstrepen, K. J., Van Dijck, P., Thevelein, J. M., and Delvaux, F. R. Flavour formation in fungi: characterisation of KlAtf, the *Kluyveromyces lactis* orthologue of the *Saccharomyces cerevisiae* alcohol acetyltransferases Atf1 and Atf2. *Applied Microbiology Biotechnology*, **78**, 783–792. (2008).

21. Lilly, M., Lambrechts, M. G., and Pretorius, I. S., Effect of Increased Yeast Alcohol Acetyltransferase Activity on Flavor Profiles of Wine and Distillates, *Applied and Enviromental Microbiology*, **66**, 744–753 (2000).

22. Rehman, S., Paterson, A., and Piggott, J. R., Flavour in sourdough breads: a review, *Trends in Food Science & Technology*, **17**, 557–566 (2006).

23. Jayani, R. S., Saxena, S., Gupta, R., Microbial pectinolytic enzymes: A review *Process Biochemistry*, **40**, 2931–2944 (2005).

24. Ghorai, S., Banik, S. P., Verma, D., Chowdhury,S., Mukherjee, S.,and Khowala, S., Fungal biotechnology in food and feed processing, *Food Research International*, **42**, 577–587(2009).

25. Vandamme, E. J. and Soetaert, W. Bioflavours and fragrances via fermentation and biocatalysis. *Journal of Chemical Technology and Biotechnology*, **77**, 1323–1332. (2002).

26. Berger, R. G. Biotechnology of flavours—the next generation. *Biotechnology Letters*, Published on-line (July 16, 2009).

CHAPTER 5

MEMBRANE TECHNOLOGY IN FOOD PROCESSING

CESAR DE MORAIS COUTINHO

CONTENTS

5.1   INTRODUCTION

The membranes are among the most important industrial applications today, and every year, more indications are found for this technology, such as: water purification, industrial wastewater treatment, dehydration solvent recovery of volatile organic compounds, protein concentration and many others [63]. A membrane is a phase permeable or semi-permeable, consisting of polymer, ceramic or metal, which restricts the mobility of certain compounds. The membrane (barrier) controls the relative rate of mass transport of various compounds through it and then, as in all separation, leads to a free product of certain compounds and a second concentrated product in these compounds.

The performance of a membrane is defined in terms of two factors: Flow and retention or selectivity. Flow or permeation rate is the volume of fluid passing through the membrane per unit membrane area per unit time. Selectivity is the measure of the relative permeation rate of different compounds through the membrane. Retention is the fraction of solute in power retained by the membrane. Ideally, a membrane with high selectivity or retention and high flow or permeability is desirable. Normally, however, procedures to maximize one factor are compromised by the reduction of the other [63].

Membranes are used in several separations, such as mixtures of gases and vapors, liquids miscible, dispersions, solid/liquid and liquid/liquid, dissolved solids, and solutes of liquids. The main feature that distinguishes the membrane separation of other separation techniques is the use of another phase, the membrane. This phase, solid, liquid, or gaseous, is an interface between the two phases involved in the separation and may provide advantages in terms of efficiency and selectivity. The membrane may be neutral or electrically charged, and porous or non-porous, and acts as a selective barrier. The transport of certain compounds through the membrane is achieved by applying a driving force directed through the membrane. This leads to a classification of membrane separation according to the mode, mechanism, and materials transported. The permeate flux through a membrane is kinetically controlled by the application of mechanical force, chemical, electrical, or thermal [63].

5.2   INTRODUCTION TO MEMBRANE PROCESSES

The evolution and expansion of the use of ultrafiltration on an industrial scale became possible after the development of asymmetric polymeric membranes, especially of cellulose acetate and aromatic polysulphones. These membranes were initially developed for desalination of seawater by reverse osmosis and then were used in various applications from other polymeric materials. Until the development of asymmetric membranes, considered as second generation, were available symmetric polymeric membranes, so-called first generation, thicker to obtain mechanical strength and hence poor performance. The development of so-called third-generation membranes, mineral, or inorganic asymmetric membranes that are homogeneous or composite, came to match the needs of high mechanical resistance to pressure, high chemical resistance and high thermal resistance.

The processes of membrane separation most important controlled by mechanical energy, hydrostatic pressure, are listed in Table 1, while the characteristics of membrane processes in terms of the substances distributed between retentate compounds retained by membrane, and permeate, compounds that permeate the membrane or are filtered, are listed in Table 2.

The main processes of membrane separation include microfiltration, ultrafiltration, nanofiltration, reverse osmosis, pervaporation, and gas separation. These processes can be distinguished by the size of the particles or molecules to which the membrane is able to retain or let permeate. This property is related to the size of the pores of the membrane (Figure 1). Obviously, all the particles or molecules that a larger pore diameter, will be retained [58].

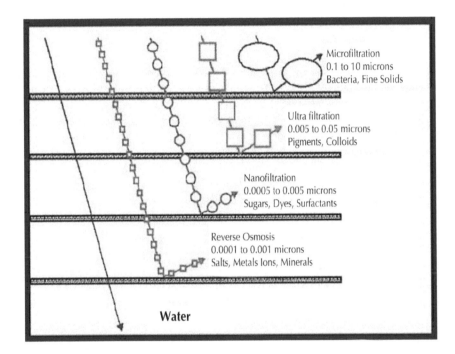

FIGURE 1   Processes of separation by membranes.

Nakao, cited by Snape and Nakajima, summarized the classification of membrane processes based on pressure applied and the particle size or molecular weight [50,66]. The microfiltration uses pressure less than 0.2 MPa and is applied primarily in particle size in the range 0.025 to 10 μm. The ultrafiltration uses pressures above 1 MPa and separates particles of molecular weight between 1 and 300 KDa. The nanofiltration uses pressures between 1 and 4 MPa and separates solutes of molecular weight between 350 and 1000 Da, while the reverse osmosis uses pressure between 4 and 10 MPa and is used to concentrate solutes with molecular weight less than 350 Da

The distinction between nanofiltration and reverse osmosis is not always clear and sometimes manufacturers membranes classified as reverse osmosis, while the above classification, could be nanofiltration. Therefore, in many cases, we adopt the nomenclature of OR/NF to the membrane [66].

TABLE 1   Membrane separations by application of hydrostatic pressure

Separation	Membrane	Major applications
Microfiltration	Micro pore symmetric or asymmetric	Clarification, sterile filtration
Ultrafiltration	Micro pore asymmetric	Separation of macromolecular solutions
Nanofiltration	Asymmetric	Separation of organic compounds of low molecular weight
Reverse osmosis	Asymmetric	Separation of salts solutions

Source: Scott (1995)

In some cases, refers to the pore size of the membrane molecular weight cut off, "molecular weight cut off (MWCO), which represents the molecular weight of less retained component in at least 95% of cases.

In membranes, the mechanism of separation is more complex than the retention or flux of particles or molecules determined by the diameter of the pores of the membrane. Many other variables such as membrane composition, manufacturing method, 3-D configuration or morphology of molecules, their interactions with each other and the membrane surface, the dynamics of fluid in the filtration unit, pressure, and temperature, influence the process of separation [36].

TABLE 2   Characteristics of the processes of separation by membranes.

Process	Driven Force	Retentate	Permeate
Osmosis	Chemical potential	Solutes, water	Water
Dialisis	Difference of concentration	Macromolecules, water	Small molecules, water
Microfiltration	Pressure	Suspended particles, water	Dissolved solutes, water
Ultrafiltration	Pressure	Macromolecules, water	Small molecules, water
Nanofiltration	Pressure	Small molecules, divalent salts, dissociated acids, water	Monovalent ions, non-dissociated acids, water
Reverse osmosis	Pressure	All solutes, water	Water
Electro dialysis	Electric potential	Nonionic solutes, water	Ionized solutes, water
Pervaporation	Pressure	Non-volatile molecules, water	Small volatile molecules, water

Source: Cheryan (1998)

Membranes for commercial use in large-scale processes were not developed until the invention of asymmetric membranes in the late 50's, in the Twentieth Century. With this type of membrane, high flow of permeate, essential for commercial applications, have been established [18].

In asymmetric membranes, the microstructure of the membrane is heterogeneous, determining a different behavior of the membrane, depending on the side in contact with the feed solution [15].

The first benefit for the food industry, appeared quickly when reverse osmosis membranes have been developed for the purification of water, a process known as desalination. After this application, membranes were introduced in various conventional processes, such as concentration by ultrafiltration instead of evaporation. Membranes allow the development of processes and entirely new products. Among the main benefits of using membranes in the food industry, can be mentioned: the separation of molecules and microorganisms, the absence of thermal damage to the products and microorganisms and low power consumption [18]. Examples of applications of membrane technology in food processing are shown in Table 3.

TABLE 3   Examples of applications of membranes in food processing.

Process or Application	Typical product	Membrane process	Industry
Cold sterilization	Beer, wine, milk	Microfiltration	Dairy, beverages
Clarification	Beer, wine, fruit juice	Microfiltration, Ultrafiltration	Beverages
Concentration/clarification	Protein (whey), fruit juice	Ultrafiltration, reverse osmosis	Beverages, dairy
Removal of alcohol	Beer, wine	Pervaporation	Beverages
Fractionation	Proteins (egg, whey, blood), carbohydrates	Ultrafiltration	Dairy, meat, eggs, sugar
Recovery	Citric acid, lactic acid	Ultrafiltration, electrodialysis	Biotechnology
Sensorial improve	Flavors	Pervaporation, reverse osmosis	Beverages
Desalination	Drinking water, reduced hardness water, unsalted cheese	Reverse osmosis, electrodialysis, nanofiltration	Beverages, dairy

Source: Cuperus; Nijhuis (1993)

In principle, almost all separation processes commonly applied in food technology can be achieved using conventional unit operations. For example, sterilization is traditionally achieved by heat treatment. In cold sterilization, microfiltration is used to remove bacterial contamination, at low temperatures. The introduction of cold sterilization by membrane process began with products of high added value in the pharmaceutical industry, to prevent irreversible damage caused by heat treatment. Cold sterilization has become commonplace in the food industry, since it preserves the quality of the product and at the same time, it avoids the appearance of off-flavors [18].

According Cheryan, one of the advantages of membrane technology compared to traditional methods of water removal, is the absence of phase change or state solvent during the process [15]. Evaporation and concentration by freezing are two techniques commonly used in liquid foods.

The evaporation requires energy equivalent to 540 kcal/kg of water evaporated, while the freeze requires 80 kcal/kg of frozen water, only to effect change in the state of water from liquid to vapor and liquid to solid, respectively. Since, the separations by membranes require no change of state for the concentration, resulting in considerable savings, which can be used 65% less energy than the traditional methods. Another advantage is the reduced need for heat transfer or heat generation equipment.

Membrane processes require only electricity needed to pump, as well as the facilities may be located far from the main unit's power plant. Additionally, membrane processes can operate at room temperature, low temperature to prevent problems with growth of microorganisms or degradation of heat sensitive compounds and high temperature to minimize microbial growth, reducing the viscosity of the retentate to reduce pumping costs, and increase the mass transfer and flow. The problems of thermal degradation and oxidative processes common to evaporation can be avoided.

Finally, since small molecules pass freely through membranes of microfiltration and ultrafiltration, the concentration on each side of the membrane will be the same at trial and equal to the original feed solution. Therefore, there is little change in the micro-environment processes during microfiltration and ultrafiltration, without causing changes in pH or ionic strength, a particular advantage when isolating and purifying proteins.

There are some limitations in the processes by membranes. None of them can be applied to the total drying of a product. In reverse osmosis, is often the osmotic pressure of concentrated solutions that limit its. In the case of microfiltration and ultrafiltration, rarely is the osmotic pressure of the macromolecules retained, but the low rates of mass transfer obtained with macromolecules concentrated and high viscosity that makes it difficult to pump the retentate and thereby limit the process. Other problems are plugging of the membranes, poor cleaning and restrict conditions of operation. However, the development of superior materials and better designs of filter modules, have minimized these problems [15].

One can also cite the high costs in general, as referred to membranes, facilities, modules or operational units, and technical support. However, due to the advantages represented by membrane technology compared to conventional processes in the food industry, these costs have been minimized consistently.

According to Cuperus and Nijhuis nanofiltration, ultrafiltration, and microfiltration involve separation mechanisms in porous membranes, while reverse osmosis and pervaporation make use of dense membranes [18]. Membrane ultrafiltration and microfiltration effect the separation by simple mechanisms of exclusion by molecular size. In such cases, the particle size in relation to the average size of the pores of the membrane, determines which particles can permeate through the membrane or not.

In ultrafiltration membranes, the spatial configuration of molecules and electrical charges represent only a minor role in the mechanisms of separation. The effects are small and the osmotic pressure applied to the filtration process, in order of 1 to 7 bar, acts primarily to overcome the resistance caused by the viscosity of the fluid supply through the porous membrane network [63].

Reverse osmosis and pervaporation are able to separate molecules of similar size, such as sodium chloride and water. In such cases, the affinity between the membrane and the target component is important, as the speed of permeation through the membrane. Components that have a greater affinity for the membrane material dissolve in the membrane more easily than other components, causing the membrane material acts as an extraction phase. Differences in diffusion coefficients of components through the membrane allow the separation. In according to the theory of "solution-diffusion", solubility and diffusivity, together, will to control the membrane selectivity. The mechanism by which nanofiltration membranes act is not entirely clear. Probably, both mechanisms size exclusion and the effect of "solution-diffusion" work in the separation of components [18].

5.3 CHARACTERISTICS AND TYPES OF MEMBRANES

Depth filters consist of an array of fibers oriented randomly and that filter or retain the particles inside the filter material. Depth filters are constructed of materials such as cotton, fiberglass, sintered metals and diatomaceous earth. Insoluble or colloidal particles are removed from the fluid by trapping or adsorption to the matrix filter. Particles of 0.01 μm can be retained by these filters. Frequently, various stages of materials are combined in one filter, where the feed comes into contact with the matrix initially more open. Depth filters are operated in perpendicular mode [15].

In contrast, membrane filters effecting the separation of compounds by retention of particles on its surface. Usually, the structure is more rigid, uniform and continuous, with pore size more precisely controlled during manufacture. Unlike depth filters, the membranes are rigid, with little risk of material migration and the growth of microorganisms is often not a problem. Because the membranes are filters with defined pore size, this feature allows for their classification. The advantage of this type of filter is that the particles retained are not lost on the inside, and a high recovery of material removed is possible. This can be important if the target of the process is to maximize the recovery of solid retentate [15].

In relation to its microstructure, the membranes can be classified as symmetric (microporous) or asymmetric (Figure 2), also referred to as membranes with "skin" filter. Microporous membranes are sometimes classified as isotropic, with uniform pore size through the body of membrane or anisotropic, where the pores change in size of a membrane surface to another. In microporous membranes, occur despite retention of all particles larger than the pore size of the membrane, the particles of the same pore size can pass through them and block them, irreversibly sealing the membrane [15].

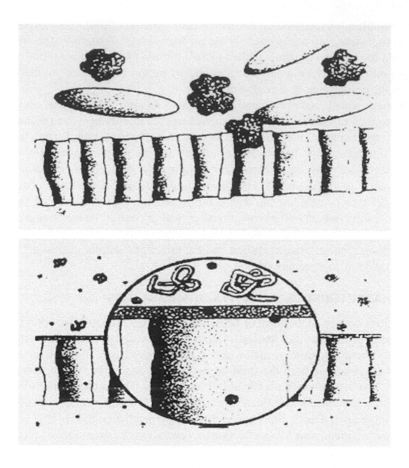

FIGURE 2   Membranes structures: symmetric (top) and asymmetric (bottom).

Asymmetric membranes, on the other hand, are characterized by the presence of a thin "skin" filter on its surface. The layers situated below the skin filter are used to support it. Rejection occurs only on the surface, but due to its unique microstructure, particles or macromolecules retained above the molecular weight cut-off rating not enter the body of the membrane. These asymmetric membranes are rarely "blocked" the way it occurs in microporous membranes, although, like all filters, asymmetric membranes are susceptible to falling of phenomena such as flow concentration polarization and / or fouling. Most membranes used in ultrafiltration, nanofiltration, and reverse osmosis are asymmetric, while the majority of polymeric microfiltration membranes are microporous [15].

According to Porter, the membranes with asymmetric structure, which are the most important currently used in separation processes, the two basic properties required for any membrane, i.e., high rates of mass transport for certain components and mechanical strength are physically separated [58]. An asymmetric membrane consists of a very thin layer of skin selective filter, 0.1 to 1.0 μm, prepared on a thick highly porous substructure, 100 to 200 μm

The skin membrane filter is the real membrane, and their separation characteristics are determined by the nature of the polymer and the pore size, while the rate of mass transport is determined by the thickness of the membrane, since the transport rate mass is inversely proportional to the thickness of the actual filter, the "skin". The sub porous layer acts only as support for thin skin filter, having little effect on the separation characteristics and the rate of mass transfer of the membrane. Asymmetric membranes are primarily used in membrane processes controlled by pressure, such as reverse osmosis and ultrafiltration, where the unique properties of high rates of mass transfer and good mechanical stability can best be used [58].

In addition to high rates of filtration, asymmetric membranes are more resistant to fouling. Membranes act as conventional symmetrical filters and traps particles deep within its internal structure. These particles trapped "blocked" the pores of the membrane and thus the transmembrane flow decreases during the process. Asymmetric membranes behave as filters for surface and retain all rejected particles on the surface, where they can be removed by shearing forces applied by the feed solution during its passage parallel to the membrane surface, the tangential filtration mode [58].

Membranes can be natural or synthetic. Regarding the type of material, the synthetic membranes can be divided into: organic, made of various polymers and inorganic, composed of ceramic or metal (Figure 3).

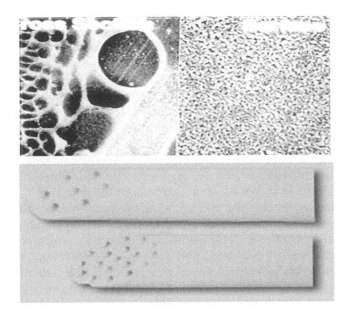

FIGURE 3   Membranes materials: organic (polymeric) (top) and inorganic (ceramic) (bottom).

The literature shows that over 130 different materials have been used in the manufacture of membranes, however, few have achieved commercial importance, and fewer still have passed for regular use in food products [15].

The commercial synthetic membranes are produced from two distinct classes of materials: polymers, made up of organic material such as cellulose acetate, polyamides, polysulphones, polyvinylidene fluoride (PVDF), among others, and the inorganic material, such as metals and ceramics [18].

According to the technological evolution, the membranes can be classified into three distinct classes:

- **First Generation**: Membranes derived from cellulose acetate which had been originally developed for desalination of seawater. They are sensitive to pH (3-8) and temperature (maximum 50°C), and they are susceptible to microorganisms and disinfectants [15].

- **Second Generation**: Membranes prepared with polymers, mainly derived from polyolefins or polyethersulfone. They were introduced in 1975, with different chemical compositions and functional properties, such as polyamide and polyethersulphone membranes. They are resistant to hydrolysis (cleavage of internal links of the polymer), and strong acids and bases and high temperatures. However, they have a low resistance to mechanical compaction. These membranes are commonly used today.

- **Third Generation**: are membranes consisting of ceramic material based on zirconium oxide or alumina deposited on a surface of graphite or other support materials. They have great mechanical strength, supporting high pressures. Moreover, tolerate all pH range (0 to 14) and temperatures above 400°C, are chemically inert but of much higher cost [18].

- **Fourth Generation**: It is representing a combined process recently developed and that consist in an association of conventional electrodialysis and several pore sizes membranes to microfiltration, ultrafiltration, and nanofiltration [1]. This combination is called "continuous electrophoresis with porous membranes" (CEPM), and can be defined as an electrochemical process for the separation of charged organic molecules. In this process, the separation driving force is based on the direct application of an electric field. By the effect of an electric potential, the ions are transported from one solution to another through one or more semi-permeable porous membrane [74].

In ultrafiltration, for example, the majority of the membranes are polymeric however the ceramic membranes also have an important role. Typical polymers for the manufacture of membranes are: polyvinylidene fluoride, polysulfone, polietersulfone, polyacrylonitrile, aliphatic polyamide, polyimide and cellulose acetate, while the typical inorganic materials are alumina, zirconia, and stainless steel sintered [63].

According Cheryan, inorganic ceramic membranes are extremely versatile, since it does not have the disadvantages of common polymeric membranes [15]. However, one should consider that in spite of ceramic membranes are very resistant to extreme operating parameters, the organic materials of the filtration modules are the components that limit the performance of the process, such as seals, gaskets and o-rings.

Among the properties of ceramic membranes, can be mentioned: they are inert to most solvents and chemical agents, have high heat resistance and can easily withstand steam sterilization, 125°C; some membranes can operate up to 350° C; use in a wide range pH, 0.5 to 13 for most of the membranes, and many can operate at pH 0 to 14; long operating life, with some filtration systems for ceramic membranes can operate for more than 10 years after installation with the same set of membranes. Unlike polymeric membranes, whose useful lives are affected primarily by the nature and frequency of cleaning, the inorganic membranes are able to tolerate a regime of frequent aggressive cleaning, in some cases above 2000 mg.kg⁻¹ of chlorine in alkaline solution. Regarding the resistance to pressure, surprisingly, many manufacturers recommend the use of ceramic membranes in static pressures up to 100bar. This is mainly due to the limitations of the seals and the configuration of the "housing" (membrane case) and not because of the membranes [15].

Some limitations of the inorganic ceramic membranes are sensitive to mechanical impact, vibration, temperature and pressure variations, pore sizes limited mainly to the ultrafiltration and microfiltration applications only, need high-capacity pump to obtain the recommended speed flow of 2 to 6 m/s, and finally, high cost, which probably represents the main limitation to the application of inorganic membranes [15].

According to Scott, in organic or polymeric membranes, the mechanical properties, chemical, thermal and permeation is influenced by the state of the molecular structure of the polymer [63]. Important parameters that directly influence the performance of a polymer membrane are the degree of crystallization and the glass transition temperature, ("Tg"). A polymer splits into two states: the glassy state and a gummy state (flexible). In the glassy state, the mobility of polymer chains is very restricted. In the gummy state, there is a high mobility of polymer chains. The degree of crystallization of a polymer represents a great influence on their mechanical properties and transport phenomena. Generally, it is desirable to use a polymer that is chemically and thermally stable. Materials with high glass transition temperature, high melting point and high degree of crystallization, are desirable. Crystalline polymers exhibit high thermal and chemical stability, and are generally used in microfiltration membranes. Non-crystalline polymers, amorphous, with high glass transition temperature are usually recommended to ultrafiltration membranes such as polysulphone (PS) and polietersulphone (PES).

According Cheryan, polietersulphone is a polymer widely used in membrane microfiltration and ultrafiltration, due to several favorable characteristics, such as high heat resistance, up to 125°C, in some cases, they are tolerant to pH of 1 to 13, and this is a great advantage for cleaning purposes, good resistance to chlorine, and some manufacturers provide solutions up to 200 mg.kg⁻¹ of chlorine for cleaning and 50 mg.kg⁻¹ for storage of the membrane; versatility of the polymer for the manufacture of membranes in a variety of settings and modules; wide range of pore sizes available for use in microfiltration and ultrafiltration, ranging from 1 kDa to 0.2 μm in commercial membranes, good resistance to hydrocarbons, alcohols and acids [15].

5.4   CHARACTERISTIC ASPECTS OF MEMBRANE PROCESSES

In the first minutes of an ultrafiltration process occurs a very sharp initial drop in the permeate flux, which is mainly due to the polarization of concentration. This polarization, which is basically a function of hydrodynamic conditions of the process, grows fairly rapidly to the subsequent formation of the gel layer polarization. In a second step, the flow continues to decline due to the effect of interactions between the membrane material and solute, a process called "fouling". In a third step, when the process reaches a state almost stationary, there is a phase where the flow decreases quite slow due to particle deposition and consolidation of the fouling.

The polarization of concentration occurs when hydrocolloids, macromolecules (proteins) or other large particles, compared to the pore membrane are rejected and they begin to accumulate on the surface thereof. This buildup can cause an increase in resistance to the solvent permeation and lead to higher local osmotic pressure. In the case of transmembrane pressure and solute concentration in the feed stream are large enough, the particle concentration can increase up to form a layer of "gel", known as the gel polarization layer, which will act as a "second membrane". The flow of solute to the membrane surface due to convection is balanced by the flow of solute passing through the same plus the diffusive flow originated on the surface due to concentration gradient formed with the current supply. The steady-state conditions are reached when the convective transport of solute to the membrane is equal to the sum of the permeate flow over the diffusive transport of solute in the opposite direction to the permeate flux. Many attempts to explain the mass transfer phenomena in the gel layer polarization were made, the most popular based on the "theory of film" that assumes the presence of a boundary layer of certain thickness, which is the diffusion of solute from the membrane surface for supply current [15,47].

"Fouling" is a phenomenon in which the membrane adsorbs or interacts in some way with the solute of the supply current, resulting in decreased flow and/or increase the rejection of solute by the membrane. Unlike the polarized gel layer, fouling is generally not reversible or time-dependent. As a result of the interaction between the membrane and various solutes in the solution, or even between solutes adsorbed on the membrane and solutes present in the feed stream, the fouling can result from three main factors, or interaction between them: the properties the constituent material of the membrane, the properties of the solute and the operational parameters.

The polarization of concentration can facilitate the irreversible membrane fouling by altering interactions between the solvent and the solute. The key to understanding the phenomenon of fouling is in the chemical composition of the membrane structure; the interaction between solute and membrane and the interaction between molecules of solute present. In particular the interaction between membrane and solute will determine the fouling formed by adsorption of solute molecules on the membrane surface.

The ways in which the pores are blocked presented as a function of size and shape of the solute for the distribution of pore size membrane. The complete blockage of the pores occurs when the particles are deposited on the membrane surface are larger than the pore size of it, completely blocking the pores, and the effect on mass transfer due to reduced area of membrane permeation depends on the tangential velocity of flow that can lead to increased transmembrane pressure applied. The partial blockage of the pores occurs when macromolecules alone or grouped cause a partial seal the pores and can then to former a layer ("cake") on the membrane surface, increasing resistance to permeation. The internal pore blocking occurs when chemical species are adsorbed or deposited inside the pores of the membrane, reducing the amount available to the passage of permeate.

5.5 PHENOMENA THAT AFFECT THE PERMEATE FLUX

Among the main factors extrinsic to the membrane, i.e., that occurs or acts independent of their type and that cause interference with the permeate flux, leading to its decline during the process are: the phenomenon of polarization near the surface of the membrane, gel layer formation - polarized and/or phenomenon of impregnation in the membrane (fouling).

In principle, all the membranes can be considered as selective filters, which are more permeable to certain compounds than others. This means that, at the molecular level, there is an increase in the concentration of retained molecules toward the membrane surface. This phenomenon is known as the concentration polarization. Molecules accumulated retained on the membrane surface, but also dissolved, form an additional resistance to permeation of the solvent, leading to a decline in the flow. However, the concentration of solutes near the membrane interface can reach such high values that promote the formation of a gel layer, this phenomenon occurring with great intensity in proteins. The formation of the gel layer is often referred to as fouling of the membrane and is irreversible or, at best, only partially reversible. Fouling of the membrane can also be caused by other phenomena, such as the adsorption of molecules and occlusion or closing the pores of the membrane. In adsorption, molecules are smaller than the diameter of the pores, but instead of moving freely, get stuck to the wall thereof, while in occlusion, the molecules have very close to the size of the pores of the membrane. These molecules are forced to pass through them, by applying pressure during the process, leading thus to block them. An immediate and reversible decline in the flow through the membrane can be defined as due to the concentration polarization, while a long-term decline is due to irreversible fouling in the membrane, which in turn is attributed to the formation of gel layer by the evolution of the concentration polarization, adsorption and occlusion of its pores. Both the concentration polarization and the fouling usually occur in all processes of filtration membranes, but its effects are more important in microfiltration, ultrafiltration and reverse osmosis [18].

If the fouling is present in the membrane as a result of the filtration process, the deposited material can, in some cases, be removed by aggressive cleaning agents such as detergents, acids, bases, or to organic solvents. The advantage of using a chemically resistant membrane is that powerful cleaning agents can be used. However, even with regular cleaning, the permeate flux cannot always be restored to baseline, resulting in fouling in the long term [58]. This phenomenon is a residual fouling, limiting the operational life of the membrane.

5.6 MAIN PHYSICAL PARAMETERS THAT AFFECT PERMEATE FLUX IN MEMBRANES

The main operating physical parameters that affect permeate flux are: pressure, temperature, viscosity and density (related to the concentration of solutes) of the feed, tangential velocity and filtration mode (perpendicular/tangential) (Figure 4). For pure water or other solvents, there is a linear relationship between permeate flux and transmembrane pressure up to a certain limit. With a real solution as feed, there is the tendency of the flow curve reach an asymptotic behavior with increasing pressure. This is because of several phenomena, including the concentration polarization, gel layer formation, fouling and osmotic effects. The flow values are also dependent on temperature, i.e., these values generally increase with increasing temperature as a result of the decrease in viscosity and diffusivity of the fluid supply [63].

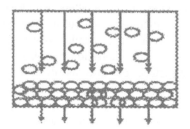 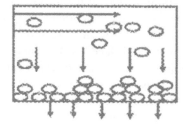

FIGURE 4 Permeation mode: perpendicular ("dead-end") (left) and tangential ("cross-flow") (right).

The permeate flux is directly proportional to the applied pressure and inversely proportional to the viscosity of the fluid supply. Viscosity is controlled primarily by two factors: concentration of solids from the feed solution and temperature. The increase in temperature or pressure, increase the permeate flux. This is partially true, that depends on certain conditions, such as: low pressure, low concentration of solids in the fluid feed and high permeation speed. When the process deviates from any of these conditions, the flow becomes independent of applied pressure.

The permeate flux is directly proportional to the applied pressure and inversely proportional to viscosity. The viscosity can be controlled by two factors: the concentration of solids in the feed and temperature. Under strict conditions, the increase in pressure or temperature causes an increase of permeate flux. However, the use of high pressure can cause compression of the gel layer, increasing the pore plugging of the membrane. The increase in the concentration of solutes affects the viscosity, density and diffusivity of the feed solution, causing a decrease in the permeate flux. In general, higher temperatures provide higher permeate flux in regions where there is pressure control and mass transfer. The effect of temperature is caused due to the reduction in fluid viscosity and the mobility of molecules, i.e. the diffusivity. The increase in tangential velocity increases the rate of permeation to cause more turbulence. The turbulence from agitation or pumping flow, promotes a disruption in the concentration of solutes on the membrane surface, reducing the thickness of the boundary layer. This is one of the most simple and effective methods to control the effect of concentration polarization [68].

There is a linear relationship between flow and the inverse of the viscosity of the solvent. In membranes for nanofiltration and ultrafiltration, the main mechanism of mass transport in these systems is by convection. In nanofiltration membranes, differences in concentration of organic solvents in aqueous solutions (water-miscible such as ethanol) to influence the primary mechanism of mass transport, and concentrations of ethanol up to 50%, the viscosity was the main factor, while above this percentage, other physical properties such as molar volume of solvent and surface tension, ruled the behavior of permeation. In contrast, most of the ultrafiltration membranes, the linear relation between the flow of solvent and the inverse of viscosity, remains independent of the concentrations of organic solvents miscible in aqueous solutions [75].

Similarly, Shukla and Cheryan, studying the behavior of permeation of 18 ultrafiltration membranes, found that 15 of these membranes agree Darcy model, in which the permeate flux decreased in linear correlation with increasing viscosity of the permeation solvent, indicating that in these 15 ultrafiltration membranes, the transport phenomenon of solvent was affected by viscosity [63].

The setting also affects the performance of membranes. A high performance of a membrane on the permeate flux and solute retention desired, must be balanced on the characteristics of it, as more tendency to pore plugging, cost, easy of cleaning and replacement [58].

Assuming that in the passage of fluid through the membrane is not occurring fouling, among other factors due to precipitation of salts, starch gelatinization or proteins denaturation, increased temperature leads to increased permeate flux through the membrane due to the effect this factor on the density and viscosity of the fluid, and increases the diffusivity of the same through the membrane [15].

The increase in concentration of the solution leads to increased osmotic pressure resulting in a decrease in the flow of solvent through the membrane towards to permeate. In addition, high solute concentration causes an increase in the viscosity of solution, which leads to a decrease in flow through the membrane. Solute size and very low molecular weight, when present in a highly concentrated solution they may plug the pores of the membrane, thereby decreasing the flow through the same [62].

The increased pressure in processes using membranes, resulting in a higher rate of convective transport of solute to the membrane surface, increasing the concentration of the interface, which has lead to increased diffusivity of solvent in the opposite direction to which the process pressure acts by decreasing the permeate flux [58].

In a pressure regime high enough, the permeate flux becomes independent of applied pressure, which is the critical flow of the process. The presence of a layer of particles trapped and compressed on the surface of the membrane, leading to maintenance of a constant pressure drop in the gel layer polarization, and this pressure is the critical pressure of the system. Considering that the thickness of the layer retained on the membrane surface is very small, relative to the diameter of the pore channel, we can neglect its effect in relation to the hydraulic conditions of flow, and thus the flow on the surface can be given as zero, thus characterizing the critical flow.

When applying a pressure that results in a flow less than the critical flux, the transmembrane pressure quickly stabilized at a new level, and the flow rate increased or decreased. On the other hand, is achieved when in a flow greater than the critical, the transmembrane pressure decreases quite slow when the flow is decreased again. Even when the flow is reduced in critical levels the transmembrane pressure does not re-stabilize. This behavior of transmembrane pressure with the flow shows that the gel polarization layer formed on values above the critical flow begins to be consolidated into a structure resistant to re-dispersion [13].

The time required for the ultrafiltration process reaches a steady state decreases with increasing flow rate, while the permeate flux, with the process already in steady state increases with increasing fluid velocity. The increase in tangential velocity may also decrease the accumulation of solute on the membrane surface and thereby increase the permeate flux [14].

A process of membrane filtration can be conducted for a long time without being observed a significant change in the permeate flux, provided it is performed at a low pressure so as not to reach the critical flow. By increasing the pressure of work can increase the formation of fouling, causing a rapid decline in permeate flux.

The asymptotic relationship between pressure and flow is due to the effects of the concentration polarization. At low pressure, low solute concentration of the feed and high fluid velocity, the effects of the concentration polarization are minimal, and the permeate flux is affected by transmembrane pressure. Deviations from the linear relationship between flow and pressure are observed at high pressures, independent of other operating parameters, due to the consolidation of the polarized gel layer of the solute.

The gel layer is dynamic, so changing operating conditions, such as reducing the pressure or the concentration of feed solutes or increase the speed of the feed, you can revert the system to an operating system controlled by pressure. If, however, the fouling is consolidated, the permeate flux does not return to original levels, except after cleaning procedures of the membrane.The flow in the region regardless of the pressure can be controlled by the efficiency in reducing the thickness of the boundary layer and increasing the rate of transfer of polarized molecules back to the center of feed; transport by diffusion, which opposes the flow of solutes by convection toward surface of the membrane [15]. Regarding the effects of temperature, the rate of permeation is inversely proportional to the viscosity of the fluid, and since the viscosity of the solvent decreases with increasing temperature, have been found experimentally in many cases filtration membranes that permeate flux increases 3% on average, per °C increase in temperature, in aqueous systems [58]. The initial applications in filtration by membranes were performed by a designed mode in which the fluid permeated perpendicular to the membrane, ("dead-end mode"). This classic mode of filtration is marked by the concentration polarization, fouling and formation of "cake" on the membrane surface, leading to a sharp decline in the flow and hence the inefficiency of the process.

While this mode of filtration is very simple, in virtually all cases have been used the principle of "cross-flow" filtration, ("cross-flow mode"). In this technique, the fluid is pumped in parallel to the membrane surface, thus reducing the hydraulic stagnant layer, where there is turbulence, and reducing the intensity of the concentration polarization and fouling.

The cross-flow filtration is often used in combination with the procedure known as "backflushing", whereby the filtration flow is reversed for a short period of time, so that particles or molecules very attached are removed from the membrane surface. The speed of cross-flow filtration, transmembrane pressure and the frequency of "backflushing" are important process parameters, which are optimized to obtain low fouling, high permeate flux and low energy costs. In many such cases where we use the tangential filtration mode, the application of low pressure can minimize fouling and concentration polarization. The use of larger area of membrane permeation is more efficient than the increased pressure of the fluid supply [18].

In the process of ultrafiltration, high-speed tangential flow tends to prevent fouling, and facilitate cleaning of the membrane, while low pressures are used to avoid compression of the gels on the surface [58].

However, despite the permeate flux can be increased by the increase in tangential velocity the pressure drop in the supply channel can become very high and result in large reduction in transmembrane pressure, causing the process failure. It is necessary to optimize the speed and configuration of the membrane, in order to obtain higher filtration efficiency [63].

5.7 MEMBRANE - SOLVENT - SOLUTE INTERACTION

The particles brought into the surface of the membrane by convection can interact with the membrane by adsorption or by physically blocking the pores, acting on a surface or inside the pores and then stay connected to other particles in the gel layer polarization. At sufficiently high concentrations, these particles can form a "cake" on the membrane surface. If the repulsive forces are weak and/or convective forces are strong, the particles can bind to a layer structure in a reversible or irreversible state [13].

The microfiltration process is used to reject particles in the range of 0.05 to 10μm, while the ultrafiltration rejects particle of order of 1 to 50nm. The transport of solvent, carried out by convection is directly proportional to the applied transmembrane pressure, which can be described, among other models, by Hagen-Poiseuille model.

The adsorption occurs as soon as the membrane surface is in contact with the solution (macromolecular), when the solute molecules will adsorb on the membrane surface due to chemical and physical interactions, e.g. hydrophobic interactions (dispersion forces), polar interactions (dipole-dipole and induced dipole). The nature of the membrane material, the type of solute, the solute concentration, ionic strength and pH are parameters that determine the extent of adsorption [47].

The relationship between the type of solvent, polar or non-polar, and the type of membrane, hydrophilic or hydrophobic, used in separation processes must be carefully analyzed. Polar solvents have significantly higher flow (8 to 10) than non-polar solvents in processes carried out in hydrophilic membranes. In turn, the non-polar solvents have a flow of 2 to 4 times greater than the polar solvents in hydrophobic membranes. As examples can be cited the values obtained for the passage of methanol (polar) and hexane (non-polar); in a pressure of 13bar, through a hydrophilic membrane, resulting in flows, respectively, $18L/m^2h$ and $2.52 L/m^2h$. On the other hand was observed for the same compound and same pressure, when the process is conducted in a hydrophobic membrane, the flows obtained were, respectively, 21.6 and $10.8 L/m^2h$ to methanol and hexane [8].

In the processes of membrane separation is quite clear the importance of the effect of possible associations between the solute and solvent and the membrane material. It can be observed in the case of retention of Sudan IV, an organic compound of molecular weight equal to 384; in this molecule are present 4 aromatic rings, and its rejection by a hydrophobic membrane was 25% and 10%, respectively, when dissolved in n-hexane, at a pressure of 15bar, and methanol, 20bar. Under the same conditions, was performed a crossing of two solutions in a hydrophilic membrane, resulting in the values of retention of 86% for the compound in the solution of methanol and 43% to the same compound in the solution of n-hexane [8,9].

The interaction between the factors of molecular weight and solute rejection by the membrane can be observed by the results obtained in the analysis of the rejection of triacylglycerols present in a solution of n-hexane, where for higher molecular weights were obtained higher percentages of rejection. One can cite as an example, that the same pressure and temperature, the values of retention were 92 and 72% respectively for tri-molecular weight equal to 807 and 554 [8].

Studying the interaction of three different membranes, two synthetic polymers and cellulose, and solutions containing different concentrations of dextrose, was evaluated the relative reduction of water flow, given in percentage, the difference between the flow before and after solute adsorption divided by flow prior to adsorption. In the case of polymeric membranes, SG-PES and PES-GR, was observed reductions in flow of about 4 to 12% and 11 to 15% respectively, while for the cellulose membrane was not observed a significant reduction in the flow, indicating a significant adsorption of dextrose on the surface of membranes made of synthetic polymers (more hydrophobic). The significant difference in the reduction of the flow in the membranes of synthetic polymers is due to differences in the characteristics of the membrane surface [73].

In the process of membrane separation performed in a mixture of soybean oil and hexane can be observed that the decrease of pore diameter membrane leads to greater efficiency. Hexane having a lower viscosity, lower molecular weight, a lower molar volume and a smaller molecular diameter in relation to soybean oil, it has a diffusivity much larger than the oil. The decrease in the pore of the membrane leads to an increase in the difference between the diffusivities of the two compounds is up to the smallest pore size analysis, reaching a ratio of diffusivity respectively between hexane and oil of 26.8 [78].

5.8 ASPECTS OF PHYSICAL-CHEMISTRY AND PROCEDURES IN MEMBRANE CONDITIONING TO IMPROVE PERMEATE FLUX, STABILITY TO SOLVENTS AND SOLUTE RETENTION

The extraction of vegetable oils is usually carried out by hexane. The miscella formed in this process is evaporated to separate the oil from this solvent. This practice requires considerable amount of energy (around 530 kJ per kg of oil), since it involves the phase change of the solvent. Additionally, the explosive vapors generated in the industrial plant represent a security problem. These limitations can be partly solved by membrane technology, i.e., the miscella can be pumped through a suitable membrane which is permeable to solvent and hold the oil. Permeate (containing essentially pure solvent) would be recycled, while the retentate pass through successive stages of membrane units to maximize the recovery of solvent and minimize the amount of solvent to be evaporated. Both reverse osmosis membranes and nanofiltration membranes are suitable for separating the oil (triacylglycerols) of the solvent. However, few commercially available membranes are stable to hexane.

Polymeric membranes are resistant to solvents, but do not cover all potential applications. Experimental observations and semi-empirical models have shown that the permeation of organic solvents through polymeric membranes is not only based on the viscosity or molecular diffusion, but depend on additional parameters caused by phenomena of interaction between the solvent and the membrane (surface tension, sorption, hydrophilicity or hydrophobicity interface). Ceramic membranes, on the other hand, present some differences, the rigidity of the porous network and the absence of the phenomenon of sorption, in addition to being resistant to organic solvents. However, the modeling of permeation of organic solvents in porous ceramic membranes is very incipient [26].

Typical membranes are made of polymers such as aromatic polyamides, polyssulfone, of methyl cellulose acetate, among others. These materials have both hydrophilic sites and hydrophobic sites in their structures. In dense membranes (non-porous), the process of permeation of water is typically activated by a diffusion of many hydrophilic sites. Charged residual groups present in polymer chains (e.g. carbonyl) facilitate the transport of water by increasing the sorption capacity of solute. Additionally, the presence of other solutes, the water is preferentially transported through these hydrophilic sites, then leading to high rates of permeation. However, in most nonaqueous systems, the permeate flux through such membranes containing hydrophilic sites would be considerably less than that of water, because of links type hydrogen bond, which are limited in alcohol and absent in hydrocarbons [8].

Hydrophilic membranes have a higher difference in flows due to pure solvent polarity when compared to hydrophobic membranes. According Bhanushali et al. permeation of pure solvents in hydrophilic membrane (composed of aromatic polyamide) show that polar solvents (methanol, ethanol, iso propanol) have significantly higher flow (8 to 10 times) than the non-polar solvents (pentane, hexane, octane) [8]. In contrast, the flow of non-polar solvents was two to four times greater than the flow of polar solvents on hydrophobic membranes (consisting of di-methyl silicone).

Not only the flow but also the stability of polymeric membranes used in organic solvents is critical. Many commercially available membranes have low stability in non-polar hydrocarbons such as hexane. Some membranes composed of polyvinylidene fluoride (PVDF) are also destroyed by hexane, probably due to the incompatibility of the material used for membrane support in this solvent. The understanding of interactions between membrane material and solvent is then essential to the development of materials and optimization of filtration [8].

In a study of conditioning of polymeric membranes in organic solvents, Van der Bruggen et al. submitted hydrophilic membranes (composed by polyetherssulfone) and hydrophobic membranes (polymeric composition not reported) at a pre-treatment consisted of 24 hours of immersion of the membranes in ethanol (for flows with ethanol) or hexane (for flows with hexane) [77]. These authors found that pretreatment of the membranes with ethanol increased the flow of solvent in the hydrophobic membrane, and decreased flow to the membrane hydrophilic. In the first case, this effect is explained by the decrease of the difference in surface energy between the solvent and the membrane, whereas in the latter case, the resistance to permeation of ethanol increases by increasing the difference in surface energy between the solvent and membrane. This is achieved by reorganizing the structure of the membrane, i.e., immersion in ethanol causes an effect of "clustering" in the hydrophobic and hydrophilic groups in the surface layer of the membrane, thus hydrophilic membranes become more hydrophobic (in addition to increased pore size), while hydrophobic membranes become more hydrophilic, these statements were supported by images of scanning electron microscopy (SEM).

In pre-treatment with hexane, the hydrophilic membrane showed no flow of hexane when untreated, have become permeable to this solvent. This shows that the membrane structure changed to a less hydrophilic conformation, allowing the passage of hexane (reduced resistance to mass transport). These changes were not uniform and showed differences between samples due to the small membrane area (0.00024 m²) used in the experiments, leading to variability in the pattern of hexane flux in these membranes after pretreatment with this solvent [77].

According to Van der Bruggen et al., parameters that reflect on hydrophilicity/hydrophobicity can also have a considerable influence on the rejection of solutes [77]. Components in aqueous solution can be wrapped with a layer of water (hydration) which increases the effective size of this component, as this effect does not occur for components in organic solutions, the effective size of the particle will be smaller, reducing the rejection or retention of solute. Additionally, there is increased mobility of polymer chains due to contact with organic solvents, increasing the average size of pores and, thus, further decreasing the rejection of the components. So, the concept of pore size is ambiguous in organic solvents, considering that the polymer chains and, consequently, the pores are not rigid. The rejection can drop dramatically due to this phenomenon. Therefore, the molecular weight cut off, determined in aqueous solutions is not fully applicable in determining rejection properties of organic solvents. Consequently, the flow of solvents through polymeric membranes is highly dependent on the solvent properties, among them the most important is the polarity, and an increase in solvent polarity decreases the flow in hydrophobic membranes and increases flow in hydrophilic membranes, due to fact that the polarity of organic solvents is strongly related to surface tension. These findings confirm the experimental observations obtained by Yang et al., under which both the viscosity and surface tension are the major parameters that influence the flow of solvents in nanofiltration membranes [79].

For the retention of solutes, even when the rejection in aqueous solution is high, the rejection of the same compound in organic solvents can be very low due to the absence of the formation of hydration shell around the solute (resulting in a smaller size) in the latter case. Finally, the structure of the surface layer of a membrane can be altered by exposure to organic solvents. As a result, flows (and rejection) are modified as a function of time of exposure to solvents. Because of these changes in the structure of the polymer chains are dependent on the interaction between solvent and exposure time, the concept of stability to organic solvents becomes ambiguous [77].

According to Yang et al., the MWCO rating specified by the manufacturer in polymeric membranes, is only valid for predicting rejection in aqueous solutions, and that the retention of solutes in organic solvents, molecules of the same molecular weight, or even for the same molecule is unpredictable and dependent on a specific solvent [79].

Despite the viscosity and polarity are the main parameters to influence the flow of solvents in polymer membranes (the latter related to surface tension), in addition, there is an interaction between membrane and solvent, which influences the mechanism of mass transport that depends on the type of membrane material and of the specific properties of the solvents and thus is important in determining the flow of the solvent. However, this is another mechanism of mass transport in organic solvents has not yet been elucidated and needs is a better understanding of the complex mechanisms of specific interaction between solvent-solute and polymer [79].

The stability of membranes in organic solvents depends on the physicochemical characteristics of solvents and membranes. Interactions of solvent with the membranes can result in dilation (swelling), plasticization, or dissolution of the membrane material and subsequent change of membrane structure, leading to changes in the properties of separation and loss of mechanical resistance to pressure. The characteristics of solvents such as molar volume, solubility, viscosity, surface tension and dielectric constant, and properties of membranes as hydrophilicity/hydrophobicity and solubility play an important role in determining the stability of membranes [75].

Among the factors that determine the permeation of organic solvents, Bhanushali et al. included the solvent molar volume, surface energy, and sorption [8]. Moreover, Machado et al. found no correlation between flow and molecular volume of organic solvents, and only the viscosity and surface tension (polarity) determined the permeation of the solvent [42].

There is a linear relationship between flow and the inverse of the viscosity of the solvent in membranes for nanofiltration and ultrafiltration, indicating that the main mechanism of mass transport in these systems is through convection. In nanofiltration membranes, differences in concentration of organic solvents in aqueous solutions (water-miscible such as ethanol) to influence the primary mechanism of mass transport, and concentrations of ethanol up to 50%, the viscosity was the main factor while above this threshold, other physical properties such as molar volume of solvent and surface tension, ruled the behavior of permeation. In contrast, most of the ultrafiltration membranes, the linear relation between the flow of solvent and the inverse of viscosity, remains independent of the solvent concentrations in aqueous solutions [75].

Membrane manufacturers usually provide them semi-dry or immersed with water. It is important condition in most polymeric membranes before use with organic solvents. Conditioning comprises rinsing the membrane and permeation of solvent under appropriate pressure, in order to remove impurities and wetting the surface membrane and inside of pores. The conditioning ensures complete contact of the membrane with the solvent, facilitating the permeation and increasing the performance of the membrane. The method of conditioning has marked effect on the flow, structural integrity of the membrane and improves the levels of work pressure in polymeric membranes [75].

According to Shukla and Cheryan, in many cases, there were breaks in the polymeric matrix of ultrafiltration membranes that were not properly conditioned [63]. The conditioning promotes a gradual change in the membrane, minimizing the degradation of the pores.

Shukla and Cheryan studied the effect of conditioning on the performance of ultrafiltration polymeric membranes in solutions with various concentrations of ethanol in water [63]. Were analyzed membranes made of materials such as polysulfone, polieterssulfone, polyacrylonitrile, polyvinylidene fluoride, and methyl cellulose acetate. The methods of conditioning consisted of gradual change in the levels of ethanol in aqueous solution at 0 to 70%, with increases of 10 to 10% (Method 1), a direct change of 0 to 70% ethanol (Method 2), exposure to direct solution of 70% ethanol (Method 3), and reduced from 100 to 70% ethanol (Method 4). After each conditioning method, the membranes were tested for flow and rejection of zein (protein present in corn, soluble in ethanol and with a molecular weight of 22,000 Da, approximately). The conditioning data were expressed in terms of a model of mass transport by convection ($J = LPPT/\mu$), where "J" is the flow, "PT" transmembrane pressure "μ" the viscosity of permeate and "LP "represents the coefficient of permeability of the membrane. Membranes not affected by the solvent showed a linear relationship between flow and the inverse of viscosity, as in cases where a significant effect of solvent on the membrane (swelling of the polymer and pore dilation), this relationship was not linear. Since the viscosity of solutions of ethanol is maximum at 50% concentration of this solvent in water, the flow should show a corresponding minimum at this concentration of ethanol, assuming that the membrane is not affected by the solvent in any other way. This phenomenon is governed by Darcy's law and Hagen-Poiseuille, according to the equation mentioned above. From 18 ultrafiltration membranes tested, 15 were submitted to the Darcy model, in which the flow decreases in linear correlation with the increased viscosity of the permeation solvent. These results indicated that in these 15 ultrafiltration membranes, the transport phenomenon of solvent was affected by the viscosity parameter. Regarding the effect of conditioning methods on the permeability and in the rejection of zein, the results indicated that in some cases (combination of method/membrane), where the solvent flows were high (as in the membrane composed of PVDF, conditioned by 2, 3, and 4 methods), the probable reason was attributed to an expansion of the pores, which also resulted, therefore, low rejection of protein (zein). Among the methods, the membranes conditioned by method 1 (gradual change of 0 to 70% of ethanol solution in water), maintained its integrity (as shown by the increased flow of solvent in relation to the pressure) and had high rejection of zein. Membranes conditioned by other methods resulted in low rejection of the protein, in almost all cases. Some membranes that showed high rejection of zein at low pressures (138 kPa) have become permeable to this protein at high pressures (> 275 kPa). This occurred because the solvents tend to decrease the glass transition temperature of polymers, acting as plasticizers, which in turn reduces the ability of polymer to withstand high pressures. There are also evidence of solvents capable of forming hydrogen bonds are usually much less destructive to the polymer matrix that highly non-polar solvents such as hexane. Therefore, under exposure to organic solvents, the full working pressure was significantly reduced in many cases, except when the membranes were conditioned gradually, using Method "1".

The efficiency of the above-mentioned membrane conditioning done gradually with different concentrations of miscible solvents enter into agreement with the results obtained by Giorno et al., which studied the effects of organic solvents on ultrafiltration membranes of polyamide with molecular weight cut off of 10 and 50 kDa, in order to prepare emulsions of oil in water by an emulsification technique, and using tangential filtration [25]. These authors reported that the permeation of polar solvent isooctane through the polar membranes (hydrophilic) was possible by pre-treatment (conditioning) of the membranes in a miscible solvent gradient of decreasing polarity, with the purpose of removing water from the pores and fill them with isooctane.

Four different procedures were evaluated based on the percentage of the solvents and the time of contact with the membranes. The influence of solvents on polyamide membranes was studied by analysis of scanning electron microscopy (SEM), which showed pronounced change in the structure and morphology of the thin surface layer (selective) of the membranes. When the membranes were subjected to pretreatment with solvents, there was an inverse relationship between the molecular weight of the original cut and the final size of the pores.

For example, the original membrane of 10 kDa presented average pore size of 0.038 μm, and after pre-treatment was to 0.075 μm, while the original membrane of 50 kDa was 0.02 μm to 0.04 μm of medium pore size. This effect was probably due to different properties of mass transfer and the swelling capacity of the structure of the thin surface layer on the membrane of 10 kDa (dense) and 50 kDa (porous). Isopropanol caused due to the effect of swelling, a more open structure of pores, "spongy".

Additionally, in the membrane of 10 kDa, the isooctane led to partial melting of the dense layer of polymer surface, with the formation of tortuous pore channels. In general, increasing the contact time of the membranes with solvents (isooctane and isopropanol) caused an increase in average pore diameters and the range of size distribution. Although, these effects were significant in both membranes, were more pronounced in the 10 kDa membrane.

The authors concluded that despite the pre-treatments have caused the changes described previously in the structure and morphology of the membranes, it was possible to obtain stable and reproducible performance in terms of permeate flux of isooctane and the formation of emulsions of oil in water, which were stable for at least three months. Then, the methodology of pre-treatment allowed obtain high permeate flux (up to 70 L/h/m² in the membrane of 50 kDa) and low transmembrane pressure (0.5 bar).

García et al. studied the effect of pretreatment on the flux of hexane through polymeric and ceramics membranes. The pretreatment consisted of immersing the membranes in solvent mixtures of decreasing polarity, starting with 50% water/isopropanol for 24 hours, 50% isopropanol/hexane for 24 hours and finally 100% hexane for 24 hours. The flow of hexane in the polymer membrane (polieterssulfone-PES) was virtually "0", which was attributed to low surface tension of this solvent (18 mN/m at 25°C) when compared to water (73 mN/m at 25°C), considering that the membranes of PES are slightly hydrophilic.

According to Van der Bruggen et al., cited by García et al., PES membranes that had zero flow in hexane, were permeable to this solvent, after pre-treatment which consisted of immersing the membranes in hexane for 24 hrs [77]. According to these authors, this indicated a change in the membrane structure to a conformation less hydrophilic, which allowed the permeation of hexane.

Using the pre-treatment recommended by García et al., the flow of hexane obtained in the membranes of PES was high and stable over time in each transmembrane pressure used. These authors reported then that the pre-treatment used was highly effective for these polymeric membranes. Two explanations have been suggested for this behavior. The pre-treatment may have helped to remove the pore water, initially by replacement with isopropanol and, after, with hexane. Moreover, the pretreatment also promote a gradual change in the polarity of the membrane, facilitating the permeation of hexane.

In ceramic membranes (zirconia), however, the pre-treatment was ineffective. In these membranes, the permeate flux (hexane) decreased over time and between successive tests, indicating a possible interaction between the solvent and the membrane material, or adsorption of solvent on the membrane surface. This effect of adsorption between the hexane and the surface of zirconia may decrease the effective size of the pores and, consequently, lead to a decrease in the flow of hexane. Additionally, the surface of zirconia may be able to catalyze the isomerization of cyclohexane and decomposition reactions, which could also explain the reduction in permeate flux observed. These authors concluded then that the membranes of polietherssulfone submitted to pre-treatment showed high flows with hexane, stable over time and increases linearly with pressure, following the usual behavior for pure solvents. The stability of these membranes pre-treated to hexane allows a full potential of applications in the edible oils. In ceramic membranes, however, the decline in the flow of hexane observed as a function of time indicated that the solvent interacted with the membrane, being adsorbed on its surface. As a result, the flow of hexane through the ceramic membranes was much lower than those obtained with polyetherssulfone membrane with similar values of molecular weight cut off. The results are not explained by simple models used to predict conventional solvent flow, as the equation of Hagen - Poiseuille therefore parameters other than viscosity must be taken into account, such as surface tension and hydrophilicity of the membrane.

5.9 TERMS COMMONLY USED IN MEMBRANE TECHNOLOGY

The following terms or parameters are essential for the control of the filtration process mediated by membranes:

a) Permeate flux (F) or (J): measures the volume (liters) of permeate obtained during the process, the area (m^2) of the membrane versus time (hours), whose equation is $F = L/m^2h$. There is a correlation between pore size and flow of the experimental water. Flow of solvent through the pores of a membrane is represented as a function of average pore diameter (d p), number of pores (N), porosity (ε), applied pressure (PT), the solvent viscosity (μ) and membrane thickness (Δ x). The model most often used to describe the flow through the pores of the membrane is based on the model of Hagen-Poiseuille law for laminar flow through the channels, whose equation is: $J = (\varepsilon \ dp2PT) / 32 \ x\mu$. In this equation, the flux (J) is given in terms of speed (g/cm/sec), which can be converted to $L/h/m^2$, since the density of water is 1 g/cm^3. This model assumes that all pores of the membrane are perfect cylinders, so in some cases, the discrepancy between the estimated flow and the flow test, may be attributed to a factor of "tortuousness" related to the actual pore configuration of the membrane [15].

b) Concentration factor (Fc) measures the reduction of fluid amount during the filtration process, that is, the ratio of the mass of feed (M_A) at the start of filtration and the mass of retentate (M_R) at the end of process, whose equation is: $F_C = M_A/M_R$

c) Retention coefficient (R): measuring the retention rate (%) of a particular solute, retained by the membrane during the filtration process. Retention (R) can be defined by the following equation: $R = 100 \ (1-(C_P / C_R))$, where C_P = concentration of solute in the ultrafiltrate (permeate) and C_R = concentration of solute in the retentate [58].

d) Permeability of the membrane (J_V) is represented by the following equation: $J_V = V/A \ t \ P$ where: "V" is the volume of permeate, "A" refers to the area of membrane permeation, "t" is the time of filtration and "P" represents the operational pressure [15].

e) Transmembrane pressure (PTM) is the average between the entrance pressure (PE) (exerted by the fluid feed in the retentate side) and exit pressure (PS) (by permeate side, usually at ambient pressure), i.e. PTM = (PE + PS)/2.

5.10 MEMBRANE TECHNOLOGY APPLIED TO VEGETABLE OILS

The membrane technology is still developing, with more and more applications are being found in food processing. The applications of conventional membranes, such as microfiltration, ultrafiltration, and reverse osmosis, can currently be considered standard unit operations, which are being implemented in many cases. Currently, the focus of this technology for the food industry is the production of membranes more suitable to a particular process, product, or even to improve the quality of existing products [18].

Recent research on the application of membrane technology in vegetable oils, aimed mainly dessolventization and acidification [61]; degumming [44]; synthesis and purification of structured lipids [44,45]; separation of compounds present in small quantities in oil, as antioxidants, and clarification or bleaching.

5.10.1 GENERAL ASPECTS ON REFINING OF VEGETABLE OILS BY MEMBRANE TECHNOLOGY

Typically, in refining of vegetable oils, the seeds are first cleaned and, depending on the type of seed on the oil content, can be processed in two different ways. Seeds containing high oil content are initially subjected to pre-pressing process, before extraction by solvent. Moreover, seeds containing low oil are processed to flakes or expanded and subjected to direct extraction by organic solvent (hexane). The resulting mixture of oil and solvent, which is called miscella leaves the extractor with 70 to 75% solvent by weight and, sequentially, solvent is recovered by distillation. Triacylglycerols make up over 95% of the composition of crude vegetable oils, and the minor components consisting of phospholipids, free fatty acids, pigments, sterols, carbohydrates, proteins, and their degradation products. These substances can give the oil modified color and taste, and shorten its shelf life. Therefore, crude vegetable oils need to be subjected to the refining processes to achieve the desired quality. Because of this process consists of several stages, a large amount of energy is consumed.

Engineering processes and design of equipments for refining has been improved considerably in recent years, mainly by reducing the energy requirements and losses of neutral oil. However, the basic principles of the processing of edible oils have remained virtually unchanged for nearly sixty years [36].

Membrane technology can be applied to the oil industry to simplify the process of refining, reducing energy consumption and waste water [41].

Membranes are semi-permeable barriers that separate different compounds in a solution, by restriction or not of the passage of components of the mixture [15].

Theoretically, triacylglycerols and phospholipids have similar molecular weight, about 900 Daltons, which makes difficult to separate them by a membrane. However, phospholipids are surfactants molecules, i.e., they have hydrophilic and hydrophobic regions, forming reverse micelles with globular structure, in a non-aqueous system. The micelles formed have molecular weight of 20,000 Daltons or more. Thus, phospholipids can be separated from triglycerides, by using appropriate membranes. Most of the pigments and some of the free fatty acids and other impurities are adsorbed on the micelles of phospholipids and removed [41].

Until the mid-90s, the literature did not contain sufficient information on membrane degumming and for problems with their stability, fouling and cleaning, and most of the membranes evaluated did not had reasonable selectivity and satisfactory permeate flux, necessary to make this technology commercially viable. This situation began to change with the appearance of stable membranes to hexane, creating great opportunities for the application of membrane technology in the refining of crude vegetable oils [41].

In degumming by membranes, typically the permeate flux is inversely proportional to the retention of phospholipids, and is desirable in a practical application, to obtain greater flow and maximum retention possible. According to Hance et al., two important factors controlling the efficiency of membrane separations are: the molecular size exclusion and chemical interaction between the membrane surface and the components of the fluid feed. The formation of micelles of phospholipids influences the separation in a positive way since monomers pass through the pores of the membrane more easily than micelles. Moreover, the adsorption of compounds on the membrane surface is often cited as the primary cause of fouling, resulting in blocking of pores. This phenomenon of adsorption is influenced by the polarity of the membrane surface and the polarity of the liquid medium in which are dissolved solutes to be separated. In organic medium, as miscella of soybean oil in hexane, these authors reported that hydrophobic membranes showed greater tendency to adsorption of phospholipids and triacylglycerols of that hydrophilic membranes.

In the traditional process of degumming, using water, phospholipids are precipitated by hydration, and then agitation and centrifugation. The phosphorus content of soybean oil is reduced to a range of 60 to 200 mg/kg, this reduction being due to the hydratable phospholipids. The remaining non-hydratable phospholipids not need to be removed through the steps of acidification and clarification. In the process of degumming mediated by membranes, the almost total removal of phospholipids occurs in one step, indicating that the membranes are able to reject both phospholipids hydratable and non-hydratable. In degumming of soybean oil by polymer membranes, more than 99% of the content of phospholipids present in the crude oil decreased in the permeate [69].

The main dynamic phenomena limiting the use of membrane technology in the oil industry, as well as other applications in food, are represented by the polarization of concentration, formation of gel layer and the fouling.

"Fouling" has been described as the result of complex mechanisms which occur during the processes of filtration, i.e., adsorption of substances of low molecular weight on the walls of the pores, "cake" formation on the membrane surface and mechanical seal the pores. Some molecules such as proteins, polysaccharides and fatty acids are known to cause drastic reduction in the permeate flow during the filtration process, despite its size, often much smaller than the diameter of the pores. This fouling is attributed to the internal adsorption of these compounds on the surface of membranes. It is postulated that the fouling is initiated by interactions between the solutes and the membrane material. Chemical bonds, forces of Van der Waals and electrostatic and acid-base interactions are the main phenomena involved in the interactions that occur between surfaces and solutes on a molecular level. Consequently, the physicochemical properties of the surfaces of membranes are altered and this facilitates the deposition of other molecules and aggregates, leading to a dramatic fouling, which added to the permeate low flow due to high fluids viscosity, limiting, in some cases, the industrial application of membrane technology in large scale [27].

Studies suggest that the fouling appears mainly due to the adsorption of phospholipids on the surface of membranes, and, therefore, this is more relevant at the stage of degumming of the crude oil.

In a study on fouling by adsorption on inorganic membranes for filtration of vegetable oils, Hafidi, Pioch and Ajana reported that there is a drastic decline in permeate flux during filtration of crude sunflower oil in different membranes composed of alumina [27]. Purified sunflower oil, with or without free fatty acids did not show fouling relevant during cross-flow filtration, indicating that both triglycerides and free fatty acids are not involved in this phenomenon to a large extent. In contrast, the presence of phospholipids caused a drastic drop in the permeate flux. The main reason for the occurrence of this phenomenon was attributed to rapid adsorption of phospholipids on the surface of membranes, which occurred in the first moments during the filtration process. These authors concluded that the phospholipids formed supra molecular structures on the surface of membrane, causing fouling.

Prior to the consolidation of the fouling, which represents the penetration of the solute molecules of the feed fluid in the pores of the membrane, blocking them, there is an increased concentration of solutes on the membrane surface due to the concentration of solutes in solution, resulting from transport by convection, which is known as concentration polarization. This polarization leads to precipitation of solute molecules on the surface of the membrane, a phenomenon known as the formation of gel layer. Finally, the adsorption of small molecules on the inner wall of the pores, and a complete occlusion by the molecules of solute is the fouling consolidated. These phenomena determine a rapid reduction in the permeate flux.

According to Wu and Lee, in an experiment using ultrafiltration of miscella of crude soybean oil by a porous ceramic membrane, it was found that the polarization of concentration profoundly influences the efficiency of ultrafiltration, solute deposition occurs on the membrane surface during the filtration process [78].

The main operational parameters that influence the performance of a process mediated by membrane separation are: tangential velocity, solute concentration in the feed fluid, pressure and temperature [61].

Concentration polarization can be minimized by increasing the turbulence of the fluid, tangentially to the membrane. Turbulence in the retentate side is an effective way to reduce the concentration polarization and thus increase the efficiency of separation [78].

According to Kim et al., agitation in the retentate side causes an increase in the rate of permeate flux [30]. Turbulence in the feed fluid derived both for pumping and agitation has great effect on the flow in the region controlled by mass transfer. Turbulent regime of fluid near the membrane surface dissolves the accumulation of solutes and reduces the hydraulic resistance of the "cake", reducing the boundary layer thickness. It is also believed that high shear force obtained by intense agitation in the fluid, effectively reduces the thickness of the gel. Anyway, this is one of the most simple and efficient method of controlling the effects of concentration polarization. However, when the system is in the region controlled by the pressure, the effect of turbulence is negligible. This occurs because the concentration of solutes in the fluid supply is not high enough, the pressure is very low and the speed is high enough to minimize the formation of boundary layer or gel. When these conditions do not prevail, the effect of speed or turbulence begins to become significant.

Pioch et al. evaluated the effect of tangential velocity on the permeate flux in the filtration of crude sunflower oil by a polymeric membrane [57]. The speed of 3.5 m / s get a significant increase on the permeate flux, compared to the speed of 2.5 m / s. However, at 4.5 m / s, the flow curve showed no increase compared to the intermediate speed. These authors interpreted these results to the fact that the polarized layer of solutes has reached a threshold thickness at 3.5 m / s. In their study, this demonstrated that a limited increase in the flow can be obtained by increasing of the tangential velocity.

Wu and Lee reported the effects of transmembrane pressure and stirring rate on the retention of phospholipids in the ultrafiltration of crude soybean oil, by a ceramic membrane [78]. The rejection decreased with increasing transmembrane pressure without agitation. However, the influence of transmembrane pressure decreased with agitation, particularly at high speeds. The rejection was significantly increased with the agitation. The permeate flux also increased with increasing agitation speed. According to these authors, of course, the agitation facilitated the minimization of the concentration polarization, thus increasing the rejection and permeation. After the fall of permeate flux during the first two hours of filtration, the increase in transmembrane pressure did not affect the permeate flux, indicating that there was accumulation of solutes, triacylglycerols and phospholipids of soybean oil, which formed a gel layer on the membrane surface. These authors concluded that the polarization of concentration is an important phenomenon in the process of ultrafiltration and can be reduced by agitation on the retentate side. The results also showed that smaller pore size, down 0.02 μm, is essential for an efficient separation of phospholipids in ultrafiltration by ceramic membranes. However, the permeate flux becomes lower when reduces the size of pores, implying that a larger area surface of membrane is needed.

The decrease in the effect of pressure on the permeate flux after a certain time of filtration previously reported by Wu and Lee enter into agreement with the results obtained by Kim et al. in degumming of crude soybean oil by a polyimide ultrafiltration membrane [30,78]. According to these authors, the gel layer represented by the precipitation of solute on the membrane surface, makes the permeate flux independent of pressure, and an increase in pressure resulted in a layer of solute denser and thicker. Initially, with the increase in the work pressure, the rate of permeation of oil miscella in hexane increased. However, up to 3 kg/cm^2, the rate of permeation of miscella tended to become constant, which is a behavior usually observed in ultrafiltration membranes. This trend was observed in three concentrations of miscella analyzed (20, 30 and 40%, m / m). This is attributed to the fact that the phospholipid layer due to the polarization of concentration retained on the membrane surface can control the permeation rate above the critical pressure. From this stage, therefore, the permeation rate is not more related to a function of the pressure.

Considering the effects of the parameters: solute concentration in the fluid supply and temperature, a higher concentration of solutes leads to high viscosity of the fluid, reducing the permeate flux. On the other hand, increased temperature reduces the viscosity, thus leading to an increase in the permeate flux.

Kim et al. reported that the increase in the concentration of the miscella, lead to a drop in permeate flux [30]. These authors attributed this result to the fact that higher concentrations of solutes promote fouling. The transport of solutes to the membrane by convection, cause a steep concentration gradient within the boundary layer and leads to a transport of solutes back into the interior due to diffusion phenomena. No more solute molecules can be accommodated due to the close arrangement and increased mobility of smaller molecules. This increase in concentration leads to precipitation and therefore the plugging or fouling in the membrane.

According to Kim et al., generally, high temperatures lead to higher permeate flux, both in the region controlled by the pressure and in the region controlled by mass transfer [30]. One can assume from this statement that other unusual effects do not occur simultaneously, such as fouling of the membrane due to precipitation of insoluble salts and denaturing or gelatinization of solids at high temperatures. In the region controlled by the pressure, the effect of temperature on the flow is due to its effects on density and viscosity of the fluid supply. The diffusivity also increases with increasing temperature. Therefore, the temperature determines a significant effect on the increased flow.

Pioch et al. studied the effect of temperature on the permeate flux and retention of phosphorus in the cross-flow filtration of crude sunflower oil, by a ceramic membrane [57]. These authors reported that an increase of 20°C in the oil temperature, of 25°C to 45°C, allowed a small increase in the rate of permeate flux, from 230 to 280 L/h.m^{-2}. However, the efficiency of the process regarding the retention of phosphorus had a sharp decline, from 88% to 24%, with 70 mg / kg of phosphorus in the permeate at 45°C instead of 11 mg / kg to 25°C. Due to the size of the molecules to be retained and disposed of oil, compared to the size of the pores, it is assumed that this stage of filtration is based on the retention of the micelles. In this case, the loss of efficiency in the retention due to increase in temperature can be attributed to well-known tendency of the molecules solve in rather than join together with the increase in temperature, or also by a change in the interactions between oil components and the membrane. For that reason, despite the limitation in the rate of flow, processing temperatures lower would be more appropriate for edible oils due to the preservation of minor components.

Stafie, Stamatialis and Wessling investigated the filtration of refined miscella of sunflower oil, in hexane, by polyacrylonitrile composite membranes as a support and polydimethylsiloxane as a selective layer, with 30 and 50 kDa molecular weight cut off [67]. These authors reported that with increasing pressure, the increase in the flow of hexane was much larger than the relative increase in the flow of oil, leading to an increased retention at high pressures. Both the flow and retention of the solute (oil) were dependent on the applied transmembrane pressure and feed concentration. Increased pressure was favorable in terms of flow and retention. Regarding the effect of concentration of the feed, besides the higher flows obtained at low concentrations, the oil retention ranged from 80% to the fluid supply with 30% concentration of sunflower oil in hexane (m / m), to 90% retention, in 8% concentration of oil in the feed, of which 19% of oil concentration, the retention of the same stood at intermediate values between 80 to 90%.

The magnitude of the effects of different filtration parameters (temperature, pressure, tangential velocity and solute concentration in the fluid supply), varies according to type of membrane used in the permeation of vegetable oils, namely at the level of ultrafiltration porous membranes, or level of non-porous or dense nanofiltration membranes. Subramanian et al. investigated the effect of temperature on the permeate flux during filtration of oil (high oleic sunflower oil), for non-porous polymeric membranes [72]. The results showed that the increase in temperature led to increase in the total flow of permeate due to the cumulative effect of the increase in solubility and diffusivity. However, the change in viscosity of the fluid due to increased temperature did not exert significant role on the permeate flow. These authors concluded that the behavior observed in these filtration systems suggests that "solution-diffusion" is the predominant mechanism of transport of the components of vegetable oils by non-porous membranes. The effect of viscosity, which is related to temperature, suggesting that the flow through convection exists in these membranes, but its extent is not significant.

However, in porous membranes, which are used in ultrafiltration, the viscosity factor of the fluid supply, which in turn is related to temperature and solute concentration, represents major role in determining the permeate flux. According to Kim et al., except for the anomalous behavior of water, all other solvents show a clear correlation between the flow increased and a reduction of viscosity [30]. According to these authors, this indicates that the viscosity is the main factor to influence the flow of solvent through ultrafiltration membranes.

The performance of ultrafiltration membranes is directly related to the phenomena of interaction between membrane and solvent, which may vary according to changes in solvent properties such as viscosity, molecular size, surface tension (related to the polarity) and dielectric constant Kim et al. analyzed the performance of an ultrafiltration membrane of polyimide with 20 kDa molecular weight cut off in degumming of crude soybean oil [30]. They performed also the filtration of various solvents (water, methanol, ethanol, acetone and hexane), where the flow was measured, with the aim of evaluating the resistance of the membrane to a specific solvent. The permeability, measured permeate flux as a function of pressure, was higher for the solvent with lower viscosity (hexane, 0.31 cP) while the lowest flow was determined for the solvent of higher viscosity (ethanol, 1.08 cP). Regarding the effect of polarity on the permeate flux, these authors reported that due to the hydrophobicity of the membrane of polyimide, the lowest flows were recorded in the case of polar solvents such as water and alcohols, while higher flows were found with non-polar solvents such as acetone and hexane.

An important factor that influences the performance of membrane filtration is its mode of operation. According to Gekas, Baralla and Flores, the operating mode for most unit operations with membranes, among which the nanofiltration and ultrafiltration are in "cross flow" mode, in which the feed stream is parallel the membrane surface, while the permeate flux is transverse to it [23]. Thus, the flow of feed and permeate are intertwined, justifying the terminology "cross flow". Moreover, in the conventional mode of filtration, the feed stream are perpendicular to the membrane or filter media, and is known as "dead end" or "flow through".

The de-acidification of soybean oil was done through polymeric membranes of reverse osmosis, the rate of permeate flux was higher at lower transmembrane pressures than at high values, which can be attributed to the polarization concentration and compression of the membrane, as pressure increased on the surface of it, this phenomenon is particularly intense, according to these authors, when using a "dead end" filtration design, as used in that experiment [5].

5.10.2   RECOVERY OF SOLVENT

The recovery of solvent in the processing of vegetable oils is done mainly by evaporation, with vapors recovered from the non-condensable gases by absorption in mineral oil. For economic, environmental and security make the recovery of a solvent of the most critical steps in the processing of edible oils. The first reverse osmosis membranes were composed of cellulose acetate and various polymers, which become damaged quickly by exposure to hexane. In the past, these membranes have limited its application to aqueous systems and were not suitable for processes such as recovery of hexane from the miscella / oil obtained by solvent extraction [35].

In a process model combining membrane / distillation proposed for solvent recovery from oil miscella in hexane, a unit of reverse osmosis membrane permeates the miscella from the extractor. The section of miscella through the membrane, the permeate resulting in a hexane permeate rich stream is recycled back to the extractor, while the retentate, oil-rich stream is processed by distillation, to recover the remaining solvent [35].

The de-solventization membrane, however, has some limitations. Results have shown that membranes can be stored in organic solvents for several weeks, no change in their initial flows, however, the combination of organic solvents and high pressure, leads to considerable changes in membrane performance [66].

Koseoglu et al. reported that in many reverse osmosis membranes used to recover solvent from miscella oil, cotton seed oil in hexane, 25% m/m, only one membrane was not damaged by hexane, but had low flow and low oil retention [36].

The chemical composition of the membranes has high impact with regard to the transport of organic solvents. Non-polar solvents have higher flows in the membranes of hydrophobic nature, whose separation mechanisms involve interactions polymer-solvent, solvent-solute and solute-polymer [8].

The application of polymeric membranes for organic use requires physical and chemical stability against the hexane and temperatures around 50°C to 60C. Materials such as aromatic polyamide, aromatic polyimide, PVDF (polyvinylidene fluoride) and polytetrafluoroethylene may be appropriate [29].

Pagliero et al. evaluated the recovery of solvent in miscella degummed sunflower oil/hexane with concentrations between 25 - 45% (w / w) of oil. The membranes are synthesized from PVDF and prepared by the process of phase inversion, were evaluated for their flow and selectivity towards the oil. The tests were performed in a 400 mL filtration unit, with an effective area of membrane equal to $31.66.10^{-4}m^2$ and agitation of 750 rpm [53]. The best separation was achieved at pressures between 4 and 6 bar, temperature of 50°C and 25% oil in miscella (w/w), corresponding to a flow of 30 L/m²h.

Koseoglu, Lawton and Lusas reported the use of nano and ultrafiltration membranes, aimed at separation of crude cotton seed oil (25% w / w), in miscella containing hexane, ethanol and isopropanol as solvents [37]. Were tested membranes made of polysulfone, fluorinated polymers, polyamide and cellulose acetate, with an average molecular weight cut-off (MWCO) ranging between 150 and 1000Da. Only polyamide membranes (OSMO Sepa O, OSMO 192T-89, OSMO 192T-O, Osmonics) were stable in hexane. The best performance was obtained with the membrane OSMO 192T-89, with a molecular weight cut off between 300 - 400 Da, a permeated flux of 7.14 L/m²h flow, 21.6% of oil in the permeate, resulting in retention of oil equals 13,6%.

Wu and Lee used porous ceramic membranes for ultrafiltration of miscella of soybean oil / hexane, based on the different diffusivities between hexane and oil. They used a crude extract of soybean oil / hexane with 33% oil, without any pretreatment [78]. They performed a cross-flow filtration with a membrane disk with pore diameter of 0.02 μm and thickness of about 0.1 μm. The best separation conditions were found using a pressure equal to 4 kg/cm² and agitation speed of 120 rpm. The concentration of soybean oil decreased from 33% in the feed to 27%, in permeate, this representing 20% rejection. The results suggest that smaller pores are essential for an efficient separation. However, the flow becomes even lower with the reduction of pores, implying the need for a large surface area of membrane. It should be noted that the pore size is not considered effective for separating oil / hexane, as is much higher, about 20,000 Da, compared to the molecules in question.

Raman, Rajagopalan and Cheryan examined nanofiltration membranes (provided by Kiryat Weizmann, Israel), resistant to hexane in the recovery process the solvent miscella constituted 20% of refined soybean oil [59]. In a first stage, with average flow of 9 L/m²h, 2.76 MPa and 24°C, was obtained a retentate with 45% oil. This was again concentrated through nine successive filtrations on a membrane similar, with an average flow of 20 L/m²h, in the same pressure and temperature. The separation of oil in the combined system was approximately 99%.

Geng, Lin and Tan investigated the separation of the solvent from a miscella of crude soybean oil/hexane (30%v/v) [24,41]. We used commercial ceramic membranes with different pore sizes (1kDa, 3kDa, and 5kDa) and maximum pressure of 6bar. The membrane pore equal to 1kDa provided satisfactory results, with rejection of 70% oil, while low permeate flows. The authors suggest that higher flows can be achieved from the application of higher pressure.

Kuk, Hron and Abraham applied a polyamide membrane with molecular weight cut off of 1000 Da, in a mixture containing ethanol and crude cotton seed oil [39]. The recovery of solvent were 99% with an operational pressure of 2 to 4 bar, temperature of 25°C and permeate flow of L/m²h ranging between 1 and 4 L/m²h. The authors conclude that small pore diameters result in low permeate flows.

5.10.3   DE-ACIDIFICATION

The conventional step of de-acidification, chemical refining, is the major economic impact in the processing of vegetable oils. This is due to several factors, such as loss of oil due to the hydrolysis of triacylglycerols by the alkali, and the soap itself has a low commercial value, however, since fatty acids have many applications, the soap is usually treated with concentrated sulfuric acid, this process resulted in the generation of highly polluting effluents, the water used to wash the oil after treatment with alkali needs to be treated before being released to the environment. In theory, the membrane technology would solve most of these problems. The ideal process would use hydro-phobic membranes. Appropriate nanofiltration membranes allow a partial separation of fatty acids [15].

In an experiment simulating refining crude soybean and canola oils, without addi-tion of organic solvents and using polymeric membranes, Subramanian et al., reported that a filtration process in a single step, is able to remove phospholipids, pigments and oxidation products, as well as increased from 12 to 26% depending on the type of oil, the content of tocopherols in the permeate compared with the original levels of tocopherols in the crude oils, due to the preferential permeation (negative rejection) of these compounds, in relation to the triglycerides. However, the process did not separate the free fatty acids from crude oils and also the permeate flux was low [70]. These authors concluded that this single step process using membranes was suitable as an alternative to chemical refining only for degumming and clarification processes, not for de-acidification.

In another study, Snape and Nakajima, using the membrane separation for lip-ids classes in hydrolyzed sunflower oil, observed that free fatty acids permeated the membrane and preferentially concentrated in the permeate, while triglycerides were retained [66]. Mono and diglycerides showed intermediate behavior, i.e., were equally distributed between permeate and retentate.

Koike et al. using a commercial dense reverse osmosis membrane, consisting of cellulose acetate and silicone-polyimide, could efficiently separate free fatty acids, monoacylglycerols, diacylglycerols and triacylglycerols of high oleic sunflower oil, hydrolyzed by lipases, diluted with solvents (ethanol and hexane) [32].

Lai, Soheili and Artz, in a combined process of permeation of soybean oil with reverse osmosis nanofiltration membranes and in combination with extraction by sub-critical liquid pressurized carbon dioxide, they obtained a preferential permeation of oleic acid in relation to the triglycerides [40]. From a system model of 40% of oleic acid and 60% triglycerides (soybean oil), the permeation through the reverse osmosis membrane (BW 30) resulted in a permeate above 80%w/w oleic acid , while the per-meation through the membrane for nanofiltration (NF 90, MWCO = 200Da) resulted in a permeate with approximately 50%w/w oleic acid. However, the last membrane showed a significantly high permeate flux compared to that obtained with the BW 30.

5.10.4   REMOVAL OF PIGMENTS (BLEACHING)

Vegetable oils contain various pigments, including chlorophyll, carotenoids, xantho-phylls and its derivatives, which need to be removed to obtain oil color acceptable to the consumer. Additionally, the oxidation of vegetable oils is favored by the presence of some pigments, and chlorophyll has been implicated in the "poisoning" of nickel used as a catalyst in hydrogenation reactions. The high cost of land clarifier, the oil losses and subsequent problems associated with disposal of waste generated has led to interest in the application of membrane technology to replace the traditional process of removal of pigments [66].

Koseoglu et al. examined fifteen different ultrafiltration membranes in the removal capacity of pigments in various vegetable oils, including soybean, canola, cotton seed oil, and peanuts [37]. Only five membranes were resistant to hexane, however, their identifications were not disclosed. Chlorophyll and β-carotene were retained by membranes, but the efficiency was variable between membranes and oils. In general, the color readings in the permeated oils were in the order of one tenth of those obtained in crude oils.

Reddy et al., cited by Snape and Nakajima investigated the use of membrane NTGS 2100, for the removal of chlorophyll and β-carotene from crude sunflower oil, getting above 95% rejection to chlorophyll, however, the flow of permeate was extremely low, 0.1 to 0.2 kg/m²/h [60,66]. Addition of hexane mixed to the crude oil, up 50%, increased the permeate flux, but reduced the rejection of chlorophyll around 70%.

5.10.5 REMOVAL OF WAXES

Waxes can be defined as a mixture of long chain non-polar compounds and are constituted of a variety of different chemical groups including hydrocarbons, esters, aldehydes, ketones, sterol esters, fatty alcohols, fatty acids, and sterols [56].

The total waxes amount present in oils depends on the type of oil, temperature, type of extraction and solvent used. Wax content in vegetable oils can cause turbidity when the oil is stored at low temperature. In this condition, the waxes present will to crystallize causing the oil turbidity or forming a precipitate. Due this reason oils need to submit to a process called "winterization" that consists in keep the oil at low temperature to crystallize the waxes content and remove them by filtration or centrifugation [33].

Waxes content in crude vegetable oils are extremely variable. Sunflower, rice, and corn oils have waxes in their composition, and crude sunflower oil can have 1.0%w/w of wax [43].

The waxes must be entirely removed by refining processes, since refined oil containing up to 8 mg.kg⁻¹ of these compounds can to promote turbidity in 48 h, at low temperatures [76].

Removal of wax is critical in sunflower oil, since this oil, after refining processes, contains around of 6 mg.kg⁻¹ [12].

Waxes present in vegetable oils can be removed by microfiltration membranes with average pore diameter from 0.05 to 1 μm. Depending on the type of oil its temperature is adjusted to -10 to 20°C to crystallize the wax before of the microfiltration process. Most waxes are retained with a minimal amount in permeate. For example, sunflower oil containing 2600 mg.kg⁻¹ wax can be cooled to 5°C and processed through a membrane made of polyethylene, in "hollow fiber "configuration with 0.12 μm average pore diameter. At a pressure of 2 bar and temperature of the fluid supply of 10°C, the average permeate flow was 10 L/h/m². The content of wax in permeate was 30 mg.kg⁻¹ and during the "cold test", it showed no turbidity. The fouling in the membrane caused by the wax was offset by "backflushing" procedure with nitrogen under high pressure [48].

Removal of waxes from sunflower oil by microfiltration membranes has been carried out on an industrial scale for several years by a company in Japan [66].

In permeation, the effect of the wax in the formation of fouling can be minimized or canceled by the temperature of the feed. In experiments with ultrafiltration of miscella of crude sunflower and soybean oils, Pagliero et al. reported that high temperatures (50°C) led to complete dissolution of the waxes present in crude sunflower oil, equaling the fouling obtained in this oil to the crude soybean oil and, consequently, leading to similar permeated flows in both sunflower and soybean oils [54].

5.10.6 DEGUMMING

The step of degumming is usually done after the separation of oil from the miscella. The "gum" are phospholipids that can be removed by the addition of water to crude oil, causing them to be hydrated and can be removed. In contrast, the degumming by membranes can be done directly with the miscella. Being amphoteric molecules, phospholipids form reverse micelles in the miscella, with molecular weight above 20 kDa and molecular size from 20 to 200 nm. In ultrafiltration, degumming by membranes can be used to separate them "micelles" from the oil-hexane mixture ("miscella"). The permeate consists of hexane, triglycerides, free fatty acids and other small molecules, while almost all the phospholipids are retained by the membrane. Membranes suitable for this application should be resistant to hexane, made, for example, in: polyamide (PA), polyethersulphone (PS), polyvinylidenefluoride (PVDF), polyimide (PI), polyacrylonitrile (PAN), or inorganic [15].

Reversed micelles, to be formed, trapping pigments in their structure, some amount of free fatty acids and other impurities, and then removed in the retentate fraction with most of the phospholipids [53].

Both ultrafiltration polymeric membranes and polymeric membranes for nanofiltration may be suitable for the degumming of vegetable oils. While in ultrafiltration membranes, the rejection mechanism of phospholipids is mediated by the molecular size exclusion, in non-porous membranes, occurs also rejection by the "solution-diffusion effect", which is related to the interactions between these solutes and the membrane active layer. Additionally, membranes retain hydratable phospholipids and non-hydratable, resulting in almost total rejection of these compounds [69].

Membrane degumming can be performed either with the miscella, generally 20 to 30% of crude oil in hexane, m/m, and with the crude oil without the addition or removal of solvent.

In the case of miscella degumming, the greatest difficulty has been the low stability of the membranes to organic solvents, whereas in degumming of crude oils, the main problem is the low permeate flow as a result of high oil viscosity, resulting in sharp fouling, dropping this technology for application in industrial large scale [51].

According to Subramanian et al., in a study of degumming of crude sunflower oil extracted by cold pressing, using polymeric composite membranes, they obtained not only retention of phospholipid by 100%, depending on the type of membrane, but also pigments and oxidation products were retained [70]. However, the permeate flux would need to be increased for an industrial application.

In a study of the degumming of crude sunflower and soybean oils, with no added solvent, using polymeric ultrafiltration membrane with molecular weight cut-off of 15 kDa. Koris and Vatai, obtained at an operational pressure of 5 bar, 60°C and a flow rate of 0.3 m³/h, 77%, and 73.5% of retention of phospholipids in the sunflower and soybean oils, respectively [34]. Permeate flows of both oils were not compared. With addition of 1% water to sunflower oil, the retention of phospholipids increased to 97%, probably, according to these authors, due to increased formation of reverse micelles.

In an experiment with miscella degumming of soybean oil in hexane, 25% m / m, using ultrafiltration membranes made from PVDF, PES and PS, Ochoa et al. found that the stability of the membranes in organic liquids is influenced both by the type of polymer and by the average pore size [51]. Small pores membranes become more stable, while that for the type of polymer, the results showed that PVDF is more stable in hexane than PES and PS membranes.

During the degumming, the authors reported a sharp decline in the permeate flux at the beginning of the process of permeation, and attributed this behavior to the concentration polarization and to membrane fouling, and this initial sharp drop in the flow was more intense in PES and PS membranes. The decline in the permeate flux was much less pronounced in long term, 250 min, that initially, 50 min, this trend implies that a system of consolidation of fouling caused by the "cake" may have affected the membrane after the initial filtration. Additionally, the PVDF membranes achieved the highest retention of phospholipids, up to 98%.

In miscella degumming of sunflower and soybean oil in hexane, 25%m/m, using ultrafiltration membranes made of PVDF with polyvinylpyrrolidone, PVP as an additive, Pagliero et al., under experimental conditions of 2 to 6 bar and 30°C to 50°C, reported that the permeate flux versus time indicates that the degumming of sunflower oil produced more fouling in the membrane that soybean oil degumming, and attributed this phenomenon to the presence of waxes in the sunflower oil, and the effect of these waxes on the formation of greater fouling in the lower temperature of filtration [53]. However, the retention of phospholipids was slightly higher in sunflower oil, up 100% in some cases. The process of degumming resulted also in the reduction of red color in both oils, but only the yellow color was reduced in the sunflower oil.

On the effect of pressure and temperature parameters on permeate flux during degumming, Pagliero et al., in an experiment with degumming of crude soybean oil miscella in hexane, 25%m/m, found that under high pressure, 6 bar, the phenomenon of the concentration polarization begins to exert an important effect on the process of permeation, making the permeate flux less sensitive to the transmembrane pressure applied [55]. At high pressure, a layer of rejected molecules, gel polarized layer, was deposited on the membrane surface and permeate flux was highly dependent on the consolidation of this layer, being little affected by pressure. About the influence of temperature, as it increases, the permeate flux increases in direct proportion, because the increase in the temperature of the feed reduces its viscosity.

5.10.7 SEPARATION OF MINOR COMPONENTS: TOCOPHEROLS AND CAROTENOIDS

SEPARATION OF TOCOPHEROLS

Tocopherols are natural antioxidants present in vegetable oils, with beneficial effect on their quality. Therefore, recent studies have focused on the application of membrane technology in the preservation or concentration of these compounds [72].

The preferential permeation of tocopherols in relation to triacylglycerols in membrane processes was first observed by Subramanian et al., that performed a filtration of peanut oil and crude sunflower oil, which produced permeated with high levels of tocopherols [70]. They used polymeric membrane prepared in laboratory, with silicon as active layer and polysulphone and polyimide as support layer, without description of characteristics of pore. The experiments were conducted at 30bar, 40°C and 400 rpm. The peanut oil and sunflower oil showed negative retentions of tocopherols of -60% and -96%, respectively. According to the authors, tocopherols do not seem to have affinity with the micelles of phospholipids and low molecular weight of these compounds, together with possible interactions with the membrane material, result in a higher permeation or, similarly, in negative values of retention.

Subramanian, Nakajima and Kawakatsu showed negative retention of tocopherols (-18 to -24%) for crude soybean oil through filtration in commercial membranes containing silicon as the active layer and polysulphone and polyimide as support layers (no description of pore size), operational conditions of 30 bar, 40°C and 800 rpm [70]. This resulted in an increase of 12-15% of these compounds in permeate. Experiments with canola oil performed under the same conditions increased the content of tocopherols in permeate at 25%.

Subramanian et al. reported the application of dense membranes (not porous) in the study of permeation of different tocopherols. Membrane was used with an active layer of silicon and polyimide as support layer (NTGS-2200, Nitto Denko, Japan) [72]. The experiments were carried out with high oleic sunflower oil with enriched-tocopherol acid, with pressure and temperature ranging from 20 to 50bar and 20°C to 50°C, respectively. Tocopherols showed preferential permeation when compared to triacylglycerols, corresponding to negative values of rejection (-30% to -52%). However, the concentration of tocopherols in the feed led to reduction in the rate of permeation of the same, but without significant effect on the total flow. The authors do not mention the characteristics of the membrane pore.

Nagesha, Sankar and Subramanian evaluated the selectivity of membranes with dense silica and polyimide as the active layer and support, respectively, on the permeation of tocopherols [49]. They used samples of deodorized distillate of soybean oil (DDOS), diluted or not in hexane and deodorized distillate of soybean oil ester, without dilution. For comparison were selected model systems consisting of mixtures of oleic acid and tocopherols in different proportions (80:20 and 50:50), diluted or not in hexane. By model system free of hexane was added to a mixture of fatty acids methyl esters (FAMEs) from soybean oil and tocopherols (90:10). The experiments were conducted on a pressurized filtration unit with nitrogen, at 30 bar, 30°C and 800 rpm. The study revealed that occur preferential permeation of tocopherols in relation to other constituents of low molecular weight (fatty acids and FAMEs). The selectivity of the membrane to the tocopherols increased with the esterified DDOS. The presence of hexane positively influenced the permeation of tocopherols, by providing greater solubility of these compounds in relation to other constituents of the feed. The authors conclude that the membrane processes have good prospects in the enrichment or concentration of tocopherols of DDOS. However, pilot scale experiments would be needed for the complete development of the process.

SEPARATION OF CAROTENOIDS

The carotenoids are generally tetraterpenoids of 40 carbon atoms, consisting of color pigments of red, yellow or orange present in fruits and vegetables, and are classified as carotenes or xanthophylls. Carotenoids are polyene hydrocarbons with varying degrees of unsaturation, and the xanthophylls are synthesized from carotenes, by reactions of hydroxylation and epoxidation. The β-carotene and lycopene are examples of carotenes, whereas lutein and zeaxanthin are xanthophylls. Carotenoids are composed of 8 isoprene units and methyl groups. Have a central long carbon chain with conjugated double bonds. These molecules are removed during the process of refining oils and fats. More than 600 types of carotenoids have been isolated from natural sources. Of these, approximately 50 are precursors of "vitamin A". The carotenoid precursor has at least one ring of β-ionone not replaced side chain polyenic with a minimum of 11 carbon atoms. Among the carotenoids, the β-carotene is the most abundant in foods and that has the highest activity of vitamin A [3].

According Darnoko and Cheryan, palm oil contains high concentrations of carotenoids and tocopherols, and that can be removed in its conversion to methyl esters and subsequent application of membrane technology to separate the carotenoid esters [19]. Several solvent stable nanofiltration membranes were studied for this application. The flow, with a standard solution of methyl ester of crude palm oil was in the range 0.5 to 10 L/m²/h, and retention of β-carotene was 60-80% at a pressure of 2,76 MPa at 40°C. A process of multi-stage membrane is designed for continuous production of methyl esters of concentrated carotenoids of crude palm oil. With feed rate of 10 ton per hour of methyl esters of palm oil containing 0.5 g/L of β-carotene, the process might produce 3611 L/h of concentrate of carotenoids containing 1.19 g/L of carotene and 7500 L/h of bleached methyl esters containing less than 0.1 g/L of β-carotene. In conclusion, the authors claim that the economy of this process is promising.

Chiu et al. performed the concentration of carotenoids from crude palm oil using membrane technology [16]. They used a flat sheet polymeric nanofiltration membrane (NP10), composed by polyethersulphone (PES), with a molecular weight cut off of 200Da. In this experiment, the red palm oil ethyl esters flux was 7.5 $Lm^{-2}h^{-1}$, and the rejection of β-carotene was 75%, at the operational pressure of 2.5MPa and temperature of 40°C. According these authors, the results of this study showed that the separation using membranes is effective in the recovery of carotenoids from red palm oil, presenting a high potential regarding the use of this technology for industrial application in the near future.

SEPARATIONS OF EMULSIONS

Emulsions are heterogeneous systems consisting of one immiscible liquid completely diffused in another liquid in droplets of 0.1μm diameter or greater [4].

According to Kocherginsky, Tan and Lu, Polymeric membranes have been shown to be an effective means of separation in the breakage of water-in-oil emulsions [31].

The separation of emulsions is only possible using hydrophilic membranes with a pore size smaller than the diameter of the droplets in the emulsion. In general, the smaller the mean pore size the better the efficiency of separation of the components in the emulsion. The use of low pressures favors this separation. According these authors, the membranes act by facilitating coalescence of the droplets present in the dispersed phase simultaneously with permeation through the pores of the membrane or even through non-porous membranes.

According to Coutinho et al. ceramic membranes can also be used to efficiently separate emulsions [17]. Fontes et al., using a compound microporous membrane made from alumina and silica with a tubular structure, sintered in the laboratory at a final temperature of 1450°C and with a mean pore size of 0.3-20μm, suitable to separate emulsions of 2%v/v of oil (sunflower and soybean) in water [21]. They achieved demulsify this emulsion efficiently by cross flow microfiltration mode in a turbulent regime ($Re > 10,000$) with transmembrane pressures of 1.5 to 3.0bar, obtained a carbon retention over 90%. In another study, Del Colle et al. performed a de-emulsification of a mixture of sunflower oil and water (1%v/v) at 25°C with microfiltration by tubular alumina membranes sintered in laboratory at 1450°C and with 0.5μm of pore diameter [20]. These tubular ceramic membranes were impregnated with a zirconium citrate solution and submitted to 600°C and 900°C. The membrane that was treated at 600°C showed better separation performance, achieved 99% retention of the oily phase. However, the membrane submitted to 900°C had a low permeate carbon concentration and additionally, presenting a better flow rate. According these authors, these differences in the flow rate showed that the heat treatment (600°C or 900°C) used after the impregnation of the ceramic ultra-structure of the membrane, was the component that interfered with the behavior of the two membranes studied, probably due to changes in the microcrystalline structure of the zirconium nano-particles after the heat treatment.

STUDY OF THE MEMBRANE CLEANING PROCESSES

Due to the limiting effects imposed by fouling on the permeate flux, it is necessary to adopt efficient methods of cleaning of the membranes, in order to bring the values of flow to baseline and thus prolong the life of the membranes.

Basso, Viotto and Gonçalves, performed an efficient cleaning process for alumina ceramic membrane, used in degumming by ultrafiltration of crude soybean oil on a pilot scale in order to restore the levels of permeate flux. In this cleaning method, which used only hexane, the best results were obtained by the combination of high tangential velocity of the feed fluid (5.0 m/s), and at low transmembrane pressure (0.45 bar) [6].

Smith et al., studying the effect of "backwash" in maintaining the flow in membrane systems, and concluded that the performance of this procedure, during the process of membrane filtration, can effectively remove most of the compounds responsible for the reversible fouling of the membrane, reducing the pressure of work and increasing the permeate flux [66].

Cakl et al., analyzed the effects of "backflushing" to restoration of the permeate flux in microfiltration process of oil emulsions, performed in a ceramic membrane. The results showed that the "backflushing" can maintain the permeate flux close to an appropriate value for a long time. It was observed that the effect of "backflushing" was more significant with a procedure of "backpulsing" shorter and a higher speed of permeation [11]. The magnitude of the effect of transmembrane pressure difference in the reverse flow caused no significant effect.

Gan et al. studied the optimization of the cleaning process in ceramic membranes used previously in the filtration of beer, now with the fouling out. The cleaning experiments were performed at a transmembrane pressure of 0.2bar and a flow rate of 2m/s. When using the "backwashing", and also periodically conducted a "backpulsing", has achieved a substantial increase in flow after 8 minutes by rinsing, suggesting a partial removal of membrane material. The increase in temperature had a significant impact on cleaning of the membrane as well as recuperation of the initial flow of it, and the value of restoring the stream to a temperature of 80°C was 20% higher than to a temperature of 22°C, while increasing the temperature of 40°C to 80°C led to a reduction in cleaning time of 26 to 11 minutes.

Cabero, Riera and Alvarez evaluated the influence of the rinse to remove protein waste from the ceramic membrane fouling has been established [10]. They compared the process with and without rinsing water recycling, and obtained a more efficient removal of waste to the process carried out without the water circulation, a fact explained by re-deposition of particles in the water re-circulated. The efficiency of the rinsing performed at 50°C was higher than in the 20°C because water has a higher power of solubilization at higher temperatures. By varying the transmembrane pressure of 0.05 MPa to 0.3 MPa, and keeping the temperature constant and the speed it was found that the higher efficiency of rinsing occurred at a pressure of 0.15 MPa whereas the lowest efficiencies were obtained for the extreme pressure values.

KEYWORDS

- **Evaporation**
- **Cold Sterilization**
- **Pervaporation**
- **Food Processing**
- **Ultrafiltration**

REFERENCES

1. Aider, M., De Halleux, D., and Bazinet, L. Potential of continuous electrophoresis without and with porous membranes (CEPM) in the bio-food industry: Review. Trends in Food Science and Technology, 19, 351–362 (2008).
2. Alicieo, T. V. R., Mendes, E. S., Pereira, N. C., and Motta Lima, O. C. Membrane ultrafiltration of crude soybean oil. Desalination, 148, 99–102 (2002).

3. Ambrósio, C. L. B., Campos, F., and Faro, Z. Carotenóides como alternativa contra a hipovitaminose A. Revista de Nutrição, 19, 233–243 (2006).

4. Araújo, J. M. A. Química de Alimentos: Teoria e Prática. Viçosa: Universidade Federal de Viçosa (1999).

5. Artz, W. E., Kinyanjui, T., and Cheryan, M. Deacidification of Soybean Oil Using Supercritical Fluid and Membrane Technology. Journal of American Oil Chemists' Society, 82, 803–808 (2005).

6. Basso, R. C., Viotto, L. A., and Gonçalves, L. A. G. Cleaning process in ceramic membrane used for the ultrafiltration of crude soybean oil. Desalination, 200, 85–86 (2006).

7. Bazinet, L., Lamarche, F., and Ippersiel, D. Bipolar membrane electrodialysis: Applications of electrodialysis in the food industry. Trends in Food Science and Technology, 9, 107–113.

8. Bhanushali, D., Kloos, S., and Bhattacharyya, D. Solute transport in solvent resistant nanofiltration membranes for non-aqueous systems: Experimental results and the role of solute-solvent coupling. Journal of Membrane Science, 208, 343–359 (2002).

9. Bhanushali, D., Kloos, S., Kurth, C., and Bhattacharyya, D. Performance of solvent-resistant membranes for non-aqueous systems: Solvent permeation results and modeling. Journal of Membrane Science, 189, 1–21 (2001).

10. Cabero, M. L., Riera, F. A., and Álvarez, R. Rinsing of ultrafiltration ceramic membranes fouled with whey proteins: Effects on cleaning procedures. Journal of Membrane Science, 154, 239–250 (1999).

11. Cakl, J., Bauer, I., Dolecek, P., and Mikulasek, P. Effects of backflushing conditions on permeate flux in membrane cross-flow microfiltration of oil emulsion. Desalination, 127, 189–198 (2000).

12. Carelli, A. A., Frizzera, F. M., Forbito, P. R., and Crapiste, G. H. Wax composition of sunflower seed oils. Journal of American Oil Chemists' Society, 79, 763–768 (2002).

13. Chen, V., Fane, A. G., Madaeni, S., and Wenten, I. G. Particle deposition during membranes filtration of colloids: Transition between concentration polarization and cake formation. Journal of Membrane Science, 125, 109–122 (1997).

14. Cheng, T., and Lin, C. T. A study on cross flow ultrafiltration with various membrane orientations. Separation and Purification Technology, 39, 13–22 (2004).

15. Cheryan, M. Ultrafiltration and microfiltration handbook. Chicago: Technomic Publ.. (1998)

16. Chiu, M. C., Coutinho, C. M., and Gonçalves, L. A. G. Carotenoids concentration of palm oil using membrane technology. Desalination, 1–4 (2009).

17. Coutinho, C. M., Chiu, M. C., Basso, R. C., Ribeiro, A. P. B., Gonçalves, L. A. G., and Viotto, L. A. State of art of the application of membrane technology to vegetable oils: A review. Food Research International, 42, 536–550 (2009).

18. Cuperus, F. P., and Nijhuis. H. H. Applications of membrane technology to food processing. Trends in Food Science and Technology, 7, 277–282 (1993).

19. Darnoko, D., and Cheryan, M. Carotenoids from red palm methyl esters by nanofiltration. Journal of American Oil Chemists' Society, 83, 365–370 (2006).

20. Del Colle, R., Longo, E., and Fontes, S. R. Demulsification of water / sunflower oil emulsions by a tangential filtration process using chemically impregnated ceramic tubes. Journal of Membrane Science, 289, 58–66 (2007).

21. Fontes, S. R., Queiroz, V. M. S., Longo, E., and Antunes, M. V. (2005). Tubular microporous alumina structure for demulsifying vegetable oil / water emulsions and concentrating macromolecular suspensions. Separation and Purification Technology, 44, 235–241.

22. Garcia, A., Álvarez, S., Riera, F., Álvarez, R., and Coca, J. Water and hexane permeate flux through organic and ceramic membranes. Effect of pretreatment on hexane permeate flux. Journal of Membrane Science, 253, 139–147 (2005).

23. Gekas, V., Baralla, G., and Flores, V. Applications of membrane technology in the food industry. Food Science and Technology International, 4(5), 311–328 (1998).

24. Geng, A., Lin, H. T., and Tam, Y. Solvent recovery from edible oil extract using nano-filtration ceramic membranes. In World Conference and Exhibition on oilseed and edible, industrial, and specialty oils (pp. 17). Istanbul: abstracts (2002).

25. Giorno, L., Mazzei, R., Oriolo, M., De Luca, G., Davoli, M., and Drioli, E. Effects of organic solvents on ultrafiltration polyamide membranes for the preparation of oil-in-water emulsions. Journal of Colloid and Interface Science, 287, 612–623 (2005).

26. Guizard, C., Ayral, A., and Julbe, A. Potentiality of organic solvents filtration with ceramic membranes. A comparison with polymer membranes. Desalination, 147, 275–280 (2002).

27. Hafidi, A., Pioch, D., and Ajana, H. Adsorptive fouling of inorganic membranes during microfiltration of vegetable oils. European Journal of Lipid Science and Technology, 105, 138–148 (2003).

28. Hancer, M., Patist, A., Kean, R. T., and Muralidhara, H. S. Micellization and adsorption of phospholipids and soybean oil onto hydrophilic and hydrophobic surfaces in non aqueous media. Colloids and Surfaces, 204, 31–41 (2002).

29. Iwama, A. New process for purifying soybean oil by membrane separation and an economical evaluation of the process. Journal of American Oil Chemists' Society, 64, 244–250 (1987).

30. Kim, I., Kim, J., Lee, K. and Tak, T. Phospholipids separation (degumming) from crude vegetable oil by polyimide ultrafiltration membrane. Journal of Membrane Science, 205, 113–123.

31. Kocherginsky, N. M., Tan, C. L., and Lu, W. F. Demulsification of water-in-oil emulsions via filtration through a hydrophilic polymer membrane. Journal of Membrane Science, 220, 117–128 (2003).

32. Koike, S., Subramanian, R., Nabetani, H., and Nakajima, M. Separation of oil constituents in organic solvents using polymeric membranes. Journal of American Oil Chemists' Society, 79, 937–942 (2002).

33. Kolodziejczyk, P. P. POS pilot plant corporation – Canola Council of Canada: identification of waxes in canola waxes (2000).

34. Koris, A., and Vatai, G. Dry degumming of vegetable oils by membrane filtration. Desalination, 148, 149–153 (2002).

35. Koseoglu, S. S. Membrane technology for edible oil refining. Oils and Fats International, 5, 16–21 (1991).

36. Koseoglu, S. S., and Engelgau, D. E. Membrane applications and research in edible oil industry: An assessment. Journal of American Oil Chemists' Society, 67, 239–249 (1990).

37. Koseoglu, S. S., Lawhon, J. T., and Lusas, E. W. Membrane processing of crude vegetable oils: Pilot plant scale removal of solvent from oil miscellas. Journal of American Oil Chemists' Society, 67, 315–322 (1990).

38. Koseoglu, S. S., Rhee, K. C., and Lusas, E. W. Membrane processing of crude vegetable oils: Laboratory scale membrane degumming, refining and bleaching. In D. R. Erickson (Ed.), Proceedings of world conference on edible fats and oils processing: Basic principles and modern practices. Champaign: American Oil Chemists' Society (pp. 182–188) (1990).

39. Kuk, M. S., Hron, S., and Abraham, G. Reverse osmosis membranes characteristics for partitioning triglyceride – solvent mixtures. Journal of American Oil Chemists' Society, 66, 1374–1380 (1989).

40. Lai, L. L., Soheilli, K. C., and Artz, W. E. Deacidification of soybean oil using membrane processing and subcritical carbon dioxide. Journal of American Oil Chemists' Society, 85, 189–196 (2008).

41. Lin, L., Rhee, K. C., and Koseoglu, S. S. Bench-scale membrane degumming of crude vegetable oil: Process optimization. Journal of Membrane Science, 134, 101–108 (1997).

42. Machado, D. R., Hasson, D., and Semiat, R. Effect of solvent properties on permeate flow through nanofiltration membranes. Part I: Investigation of parameters affecting solvent flux. Journal of Membrane Science, 163, 93–102 (1999).

43. Martini, S., and Añón, M. C. Storage of sunflower seeds: Variation on the wax content of the oil. European Journal of Lipid Science and Technology, 107, 74–79 (2005).

44. Moura, J. M. L. N., Gonçalves, L. A. G., Petrus, J. J. C., and Viotto L. A. Degumming of vegetable oil by microporous membrane. Journal of Food Engineering, 70, 473–478 (2005).

45. Moura, J. M. L. N., Gonçalves, L. A. G., Sarmento L. A. V., and Petrus, J. J. C. Purification of structured lipids using SCCO2 and membrane process. Journal of Membrane Science, 299, 138–145 (2007).

46. Moura, J. M. L. N., Ribeiro, A. P. B., Grimaldi, R., and Gonçalves, L. A. G. Reator de membrana enzimático e fluidos supercríticos: Associação de processos. Química Nova, 30, 965–969 (2007).

47. Mulder, M. H. V. (1995). Polarization phenomena and membrane fouling. In R. D. Noble & S. A. Stern (Eds.), Membrane separations technology: Principles and applications (pp. 45 – 84). Amsterdam: Elsevier.

48. Mutoh, Y., Matsuda, K., Ohshima, M., and Ohushi, H. US Patent n. 4545940 (1985).

49. Nagesha, G. K., Subramanian, R., and Sankar, K. U. Processing of tocopherol and FA systems using a non-porous dense polymeric membrane. Journal of American Oil Chemists' Society, 80, 397–402 (2003).

50. Nakao, S. Determination of pore size and pore size distribution. 3. Filtration membranes. Journal of Membrane Science, 96, 131–165 (1994).

51. Ochoa, N., Pagliero, C., Marchese, J., and Mattea, M. Ultrafiltration of vegetable oils. Degumming by polymeric membranes. Separation and Purification Technology, 417–422 (2001).

52. Pagliero, C., Mattea, M., Ochoa, N. A., and Marchese, J. Aplicación de membranas para el desgomado de aceite crudo de soja y girasol. In J. C. C. Petrus, L. M. Porto, and J. B. Laurindo (Eds.), 4o Ibero-american congress in membrane science and technology. Florianópolis: anals, pp. 292–296 (2003).

53. Pagliero, C., Mattea, M., Ochoa, N. A., and Marchese, J. Recuperación de solvente a partir de miscelas de aceite de girasol / hexano usando tecnologia de membranas. In J. C. C. Petrus, L. M. Porto, J. B. Laurindo (Eds.), 4o Ibero-american congress in membrane science and technology. Florianópolis: abstracts, p. 51 (2003).

54. Pagliero, C., Mattea, M., Ochoa, N. A., and Marchese, J. Fouling of polymeric membranes during degumming of crude sunflower and soybean oil. Journal of Food Engineering, 78, 194–197 (2007).

55. Pagliero, C., Ochoa, N. A., Marchese, J. and Mattea, M. Degumming of crude soybean oil by ultrafiltration using polymeric membranes. Journal of American Oil Chemists' Society, 78, 793–796 (2001).

56. Peris-Vicent, J., Adelantado, J. V. G., Carbó, M. T. D., Castro, R. M., and Reig, F. B. Characterization of waxes used in pictorial artworks according to their relative amount of fatty acids and hydrocarbons by gas chromatography. Journal of Chromatography A, 1101, 254–260 (2006).

57. Pioch, D., Largueze, C., Graille, J., Ajana, H., and Rouviere, J. Towards an efficient membrane based vegetable oils refining. Industrial Crops and Products, 7, 83–89 (1998).
58. Porter, M. C. Handbook of industrial membrane technology. New Jersey: Noyes Publications (1990).
59. Raman, L. P., Cheryan, M., and Rajagopalan, N. Solvent recovery and partial deacidification of vegetable oils by membrane technology. Fett. / Lipid., 98, 10–14 (1996).
60. Reddy, K. K., and Nakajima, M. Membrane decolorization of crude oil. In International Congress of Membranes Processes. Yokohama, p. 894 (1996)
61. Ribeiro, A. P. B., Moura, J. M. L. N., Gonçalves, L. A. G., Petrus, J. C. C., and Viotto, L. A. Solvent recovery from soybean oil / hexane miscella by polymeric membranes. Journal of Membrane Science, 282, 328–336 (2006).
62. Satyanarayana, S. V., Bhattacharya, P. K., and De, S. Flux decline during ultrafiltration of kraft black liquor using differents flow modules: A comparative study. Separation and Purification Technology, 20, 155–167 (2000).
63. Scott, K. Handbook of Industrial Membranes. Oxford: Elsevier (1995).
64. Shukla, R., and Cheryan, M. Performance of ultrafiltration membranes in ethanol – water solutions: effect of membrane conditioning. Journal of Membrane Science, 198, 75–85 (2002).
65. Smith, P. J., Vigneswaran, S., Ngo, H. H., Bem-Aim, R., and Nguyen, H. A new approach to backwash initiation in membrane system. Journal of Membrane Science, 278, 381–389 (2006).
66. Snape, J. B., and Nakajima, M. Processing of agricultural fats and oils using membrane technology. Journal of Food Engineering, 30, 1–41 (1996).
67. Stafie, N., Stamatialis, D. F., and Wessling, M. Insight into the transport of hexane-solute systems through tailor-made composite membranes. Journal of Membrane Science, 228, 103–116 (2004).
68. Strathmann, H. Synthetic membranes and their preparation. Handbook of Industrial Membrane Technology. M. C. Porter (Ed.), Noyes Publications, New Jersey (1990).
69. Subramanian, R., Ichikawa, S., Nakajima, M., Kimura, T., and Maekawa, T. Characterization of phospholipid reverse micelles in relation to membrane processing of vegetable oils. European Journal of Lipid Science and Technology, 103, 93–97 (2001).
70. Subramanian, R., Nakajima, M., and Kawakatsu, T. Processing of vegetable oils using polymeric composite membranes. Journal of Food Engineering, 38, 41–56 (1998).
71. Subramanian, R., Nakajima, M., Kimura, T., and Maekawa, T. Membrane process for premium quality expeller-pressed vegetable oils. Food Research International, 31, 587–593 (1998).
72. Subramanian, R., Raghavarao, K. S. M. S., Nakajima, M., Nabetani, H., Yamagushi, T., and Kimura, T. Application of dense membrane theory for differential permeation of vegetable oil constituents. Journal of Food Engineering, 60, 249–256 (2003).
73. Susanto, H., and Ulbricht, M. Influence of ultrafiltration membrane characteristics on adsorptive fouling with dextrans. Journal of Membrane Science, 266, 132–142 (2005).
74. Tanaka, Y. Irreversible thermodynamics and overall mass transport in ion-exchange membrane electrodialysis. Journal of Membrane Science, 281, 517–531 (2006).
75. Tsui, E. M., and Cheryan, M. Characteristics of nanofiltration membranes in aqueous ethanol. Journal of Membrane Science, 237, 61–69 (2004).
76. Turkulov, J., Dimic, E., Karlovic, D., and Vuska, V. The effect of temperature and wax content on the appearance of turbidity in sunflower seed oil. Journal of American Oil Chemists' Society, 63, 1360–1363 (1986).

77. Van der Bruggen, B., Geens, J., and Vandecasteele, C. Fluxes and rejections for nanofiltra-
     tion with solvent stable polymeric membranes in water, ethanol and n-hexane. Chemical
     Engineering Science, 57, 2511–2518 (2002).
78. Wu, J. C. S., and Lee, H. Ultrafiltration of soybean oil / hexane extract by porous ceramic
     membranes. Journal of Membrane Science, 154, 251–259 (1999).
79. Yang, X. J., Livingston, A. G., and Santos, L. F. Experimental observations of nanofiltra-
     tion with organic solvents. Journal of Membrane Science, 190, 45–55 (2001).

CHAPTER 6

TENDERIZATION OF MEAT AND MEAT PRODUCTS: A DETAILED REVIEW

B.G. MANE, S.K. MENDIRATTA, and HIMANI DHANZE

CONTENTS

6.1   INTRODUCTION

The meat tenderness is defined as a sum total of mechanical strength of skeletal muscle tissue [1]. It is rated as one of the most important quality attribute by the consumers [2] and the most significant factor affecting the consumer satisfaction and perception of taste [3]. The eating satisfaction is the result of the interaction between tenderness, juiciness and flavor. However, the variation in meat tenderness might be due to genetic make of animal or created pre-slaughter and/or post-slaughter or combination of all [4-6] . The variation in the meat tenderness results in the consumer dissatisfaction. It is challenge to the scientific community and to the meat industry to solve the problem of variation in meat tenderness to assure and/or to give guarantee about the meat and meat products to the consumer [7]. In this chapter, the various aspects of meat tenderness such as process of tenderness of meat, practices of meat tenderness, influences of various conditions on meat tenderness, methods of tenderization of meat, and meat products and physico-chemical determinants of meat tenderness are discussed.

6.2   TENDERIZATION PROCESS OF FRESH MEAT AND/OR CARCASSES

It is also called as ageing/conditioning/ripening/setting-up of meat or carcasses. It is defined as holding the fresh carcasses or meat at refrigeration temperature for extended periods following initial chilling. Resolution of rigor and tenderization of muscle (ageing) is indicated by softening of muscle that takes place in a progressive manner after a variable period in each species. Conditioning is the term applied to the natural process of tenderization when meat is stored or aged post-rigor (pork 4-10, lamb 7-14 and beef 10–21 days at 1°C [8].

The muscle tissue is composed of three proteins:
a.   Sarcoplasmic,
b.   Myofibrillar, and
c.   Connective tissue.

However, myofibrillar and connective tissues protein are important in deciding the meat tenderness. The tenderness of meat is determined by the background toughness due to the connective tissue component of muscle tissue, the toughening phase due to the state of myofibrils and sarcomere length during and after rigor mortis and tenderization/proteolysis phase of meat [5,9-11]. The tenderization process begins just after slaughter and it measured subjectively by consumer panel or objectively by using shear force machines [7]. It is well established that the post-mortem proteolysis of muscle proteins, particularly the myofibrillar proteins are responsible for the tenderization of meat and three proteolytic enzyme systems in muscle play the very important role in post-mortem proteolysis that is the calpain system, the lysosomal cathepsins, and multicatalytic proteinase complex [11]. The degradation of structural, cytoskeletal, and regulatory proteins of myofibrillar proteins by these enzyme systems leads to the meat tenderization.

The cathepsins enzyme systems were initially considered as an important enzyme play role in muscle tenderness. However, the discovery of calpain enzyme system and their importance in meat tenderness can be due to their ability to alter Z-disk [12]. The calpains are calcium-activated enzymes and they are of three types of proteases that is μ-calpain, m-calpain and skeletal muscle-specific calpain, (p94 or calpain 3), and the inhibitor of μ- and m-calpain that is the calpastatin [11]. The activity of μ- and m-calpains is a large extent dependent on calcium ions. Takahashi *et al.* (1995) pointed out a special role of calcium ions in muscle proteolysis and of para-tropomyosin protein in relaxing the interaction between actin and myosin. Recently, the role of the 20S proteasome in the meat tenderization has been reported by [13,17] More recently, the role of apoptosis in the process of meat tenderization were reported by researchers. They reported that after the slaughter of animal to maintain the homeostasis under the very stressful condition, muscle cell has no alternative except of initiating the apoptosis process [14]. The meat tenderization is strongly correlated with apoptotic processes due to inversion of membrane polarity [15] and the post-mortem changes showing evidences for a significant contribution of caspases in both flesh and meat [14,16].

6.3 FACTORS INFLUENCING THE MEAT TENDERNESS

6.3.1 PRE-SLAUGHTER FACTORS INFLUENCING MEAT TENDERNESS

Tenderness of meat is highly variable, it is affected by pre-slaughter factors such as species, breed, sex, age, feeding and nutrition, exercise, environmental, and managemental conditions. Animal age has more influence on tenderness attributes than the sex of the animal. However, increased post-mortem ageing time improved myofibrillar tenderness attributes regardless of sex or age of animal [18]. Age and sex are the most notable factors that determine the tenderness of meat. Cross *et al.* (1973) reported that connective tissue components were not significantly associated with sensory panel evaluations of tenderness. However, Huff and Parrish, (1993) reported that as animal age increases, the amount of connective tissue left after chewing of cooked meat also increases.

6.3.2 POST-SLAUGHTER HANDLING CONDITIONS INFLUENCING MEAT TENDERNESS

Meat is exposed to different handling conditions and processing steps before final preparation and consumption. There is great need to evaluate newer methods of cooking along with traditional methods for recommending suitable methods for tough meat cuts/muscles [19]. Several post-mortem factors such as holding time after slaughter, time of boning, and chilling before cooking play a very important role in determining quality of meat products [6]. Identification of appropriate processing conditions that manage quality in a cost effective manner are also important for commercial meat processing operations. After slaughter, contraction and stiffening of muscle take place. This causes decrease in tenderness. After completion of rigor stage again increase of tenderness take place. Any handling practices which causes shortening of muscle fibers lead to toughness of meat. If the meat is frozen before completion of rigor mortis, this led to toughness due to "thaw rigor". The production of tender meat is also greatly influenced by the conditions prevailing during the period between slaughter and full development of rigor mortis. Thus the meat processors by using unsuitable processing methods can make potentially tender meat tough. It has been well established that tenderness and other meat quality parameters are related to post-mortem shortening and processing method [20].

Most of the biochemical and structural change in the muscle tissue takes place during the first 24 hrs of post-mortem. It was reported that handling condition affects the quality of fresh meat and were found to have a significant effect in protein percentage, pH, WHC, cooking loss and extractable proteins percentage [21]. Pre-rigor meat has high pH, responds to electrical stimulation [22] and has higher emulsifying capacity and greater extractability of salt soluble proteins (Myosin, Actin, Tropomyosin) than post-rigor meat [23]. Mendiratta *et al.* evaluated the quality characteristics of spent sheep meat curry prepared from meat subjected with different handling conditions and reported that significantly higher pH, moisture, cooking yield, and sensory attributes score were observed for meat cooked within 1–2 hr of slaughter.

6.3.3 POST-SLAUGHTER TEMPERATURE INFLUENCING MEAT TENDERNESS

The temperature immediate after post-slaughter has great influence on meat tenderness. The various researchers also pointed that the post-slaughter temperature during the ageing is crucial for the tenderness. The activity of muscle proteinase such as the calpains [24], cathepsins [40], and their inhibitors calpastatin immediately after post-mortem are influenced by temperature. Dransfield *et al.* (1994a) found an intermediate temperature region of approximately $10-25^\circ$C, where preferential activity of μ calpain occurs, thereby tenderization is found to be more at this region. Tornberg (1996) investigated that at 35°C, lack of tenderness was due to denaturation of calpains rather than by calpastain inhibition. A high degree of tenderness is associated with holding temperature around $10-15^\circ$C, while temperature below and above this range causes heat and cold shortening [25]. Hannula and Puolanne (2004) also reported that temperature at the onset of rigor mortis has major influence on meat tenderness and meat should be held above $7-10^\circ$C before the onset of rigor. Rate of pH fall depends on temperature, but the effect of temperature is not same with all the muscles [26]. White *et al.* (2006) have also reported that rate of pH decline increased with increasing temperature.

- **High Temperature**: The conditioning is defined as the holding of freshly slaughtered carcasses, or excised muscles at temperature above normal chilling temperature (0° to 2°C) for a period of time before subsequent chilling at low temperatures (West, 1979). The optimum temperature for high temperature conditioning has been reported to be between 14°C and 20°C as this temperature represented minimal shortening [27]. Holding of carcass at an elevated temperature not only stimulate the release of lysosomal enzymes into the soluble cell fraction but also increases the activity of these enzymes during the conditioning period because the combined effect of high temperatures and low pH approaches optimal condition for these enzymes to act [28]. Henderson *et al.* (1970) observed that Z-line degradation in excised bovine muscle occurred much faster at temperatures above 25°C than at lower temperatures.

Parrish *et al.* (1973) also reported similar findings for intact beef muscle conditioned at 16°C as compared to those chilled at 20°C.

- *Chilling (Low Temperature)*: The two main determinants in defining the pre-rigor condition are the rate of pH fall and rate of chilling [29]. These two variables are obviously not independent, as temperature will affect the pH fall. The pH fall can also be manipulated independently by other factors. The enhanced tenderness of slowly chilled meat can also be attributed to the earlier onset of proteolytic breakdown of the myofibrillar structure [30] and thereby exhibited a faster pH fall than rapidly chilled muscles. It has been observed that delayed chilling resulted in faster pH fall, and improvement in textural quality of excised muscle compared to direct chilling [31,32] White *et al.*, 2006). Locker *et al.* (1975) reported that cold shortening produced more adverse effects on the tenderness of meat than age, sex and plan of nutrition. Pre-rigor freezing and subsequent thawing also caused myofibrilar fragmentation but did not produce detrimental effects on tenderness. Rathina-Raj *et al.* (2000) suggested that the pre-rigor buffalo muscles should be held at ambient temperature ($26\pm2^{\circ}$C) for the first 6 hr followed by chilling in order to improve their tenderness quality.

6.3.4　POST-SLAUGHTER PROCESSING INFLUENCING MEAT TENDERNESS

The post-slaughter processing such as hot boning, electrical stimulation, and so on also greatly influences the tenderness of meat and meat products. The meat and meat products prepared from the pre-rigor meat have higher water holding capacity which results in higher juiciness and tenderness. Application of some integrated technologies like electrical stimulation, immersion chilling, and muscle stretching and forming to enhance meat quality, also need hot boning as a prime option [33].

- *Hot Boning*: Hot boned meat is meat that has been removed from the carcass before the chilling process [34]. Hot boned meat generally results in higher processing yield than cold boned meat because higher pH above the isoelectric point of meat proteins resulted in higher WHC [41] (Hamm, 1960). Hot boning is economical since considerable amount of energy is required for chilling of carcass [35] (Sharma *et al.*, 1988). Cooking of hot boned meat utilizes the inherent heat (38°C) of the muscle and hence reduces total cooking time. Cooking yield was also reported to be higher for hot boned roasts than percentage yield of cold boned roasts [36] (Loucks *et al.*, 1984). Other advantage considered for hot boning is an improved yield of meat because evaporation during cooling can be avoided [35] (Sharma *et al.*, 1988). Hot boned meat generally results in higher processing yield than cold-boned meat [34] (Claus *et al.*, 2006). Lan *et al.* (1995) observed that contractile state of the myofibrils affected gel strength viscosity and cooking loss. Sharma *et al.* (1988) investigated that tenderness of goat meat was significantly adversely affected by hot boning and this trait was further exacerbated by freezing.

- **Electrical Stimulation**: Electrical stimulations (ES) cause intense muscle contractions and speeds up the normal post-mortem processes that is the post-
  mortem pH decline is rapid and sarcomere length is longer resulting in tender
  meat [37] (Savell *et al.*, 1978). In addition to prevention of the cold shortening,
  the ES also appears to tenderize the meat per se and improves appearance and
  possibly flavor [38] (Kondaiah and Mendiratta, 2002). The time, duration and,
  voltage for the ES have been standardized for chicken [42] (Kannan *et al.*,
  1990) and mutton [43] (Vijaya-Kumar *et al.*, 1990), while post exsanguination
  ES has also been carried out [44] (Mahajan and Panda, 1991) to increase the
  tenderness of meat. In the process of ES, the intense contractions generated
  and that might be physically disrupt and weaken the muscle structure. Calcium
  released during contraction might stimulate the calpains and cause greater proteolytic break down as the muscle temperature and pH are high. Low (up to
  100 V) and high (500 V) voltage systems are used for ES. Low voltage systems applied within 5 min of stunning for up to 20 seconds in pulses of 1–2
  seconds duration. High voltage systems applied up to 60 min PM for up to 90
  seconds in pulses of 1–2 seconds' duration. Safety is more important with high
  voltage. Low voltage systems work by stimulating the musculature via the
  still living nervous system, while high voltage systems stimulate the muscles
  directly and can be applied with some delay after stunning.

6.3.5   METHODS OF COOKING

As cooking progresses, the contractile proteins in meat become less tender and the
major connective tissue protein (collagen) becomes tenderer. Thus, for the cuts that
are low in connective tissue, the preferred methods of cooking are dry heat, broiling,
roasting, and barbecuing. Dry heat raises the temperature very quickly and the flavor
of meat will develop before the contractile proteins become significantly less tender.
For cuts with a high amount of connective tissue, the recommended method of cooking is long and slow at low temperatures using moist heat such as braising and stewing.
The application of moist heat for a long time at low temperatures results in conversion
of tough collagen into tender gelatin. The changes in meat toughness during cooking
may be determined by the mechanical strength of perimysium at the perimysium-
endomysium interface, while endomysium shrinkage results in loss of water from the
muscle [45] (Bernal and Stanley, 1987). Both these contribute to the texture of cooked
meat. Degree of doneness also affects tenderness. Tender cuts of meat cooked to a rare
degree of doneness are tenderer than when cooked to medium or well-done. If tender
meat is cooked up to well-done stage that may lead to improvement in flavor but tenderness is greatly reduced.

Low temperature and slow cooking times cause connective tissues to become gela-
tinized, soft and tender but high temperature, and faster cooking periods cause muscle
proteins to coagulate and shrink quickly and moisture is squeezed out of the meat.
Davey and Gilbert (1968) demonstrated that at high temperature of 60–70°C, there
was increase in toughness of meat. It was attributed to shrinkage of collagen fibers.
The cooking of meat changes, the structure of the intramuscular connective tissues,

and its mechanical properties due to denaturation of the collagen. Thus, tenderization treatment has become essential for the meat cuts that are subjected to fast cooking methods like frying. Cyril *et al.* (1996) reported decline in tenderness of rabbit meat by frying and roasting. Carpenter *et al.* (1968) reported that pork chops cooked by microwave were less tender than oven broiling. Similar results were also reported by Swenson *et al.* (1994) and El-Shimi (1992) in microwave cooking.

The highly tenderization effects of pressure cooking was observed by various researchers on meat and meat products. High pressure is one of the new technologies that have been earmarked for meat tenderization through liberation of lysosomal enzyme [46] (Elgasim *et al.*, 1983; [47] Cheftel and Culioli, 1997) or by enhancement of the activity of other enzymatic systems such as calpains [48] (Homma *et al.*, 1995). Prasad *et al.* (1988) reported that for Chevon pressure cooking resulted in better lean retention and water holding capacity. Moist heat (stewing) suited better for mutton but pressure cooking caused toughness, whereas, for chevon pressure cooking resulted in better lean retention and WHC [49] (Prasad *et al.*, 1988). It was reported that pressure cooking has greater tenderization effect in beef and buffalo muscles [50] (Robertson *et al.*, 1984; [51] Syed-Ziauddin *et al.*, 1994). Rathina-Raj *et al.* (2000) reported that pressure cooked muscles were significantly tenderer than normal cooked ones. However, they found that cooking loss values were lower, that is to say 41–43% for normal cooked compared to 46–47% for pressure cooked muscles. These findings of improvement in tenderness by pressure cooking are in agreement with Bouton *et al.* (1974) for veal muscles, Mahendrakar *et al.* (1988) for ovine muscles, and Syed-Ziauddin *et al.* (1994) for buffalo muscles. This could be due to greater degree of muscle fibers and protein coagulation in muscles cooked at higher temperatures in pressure cooking than in normal cooking [52] (Asghar and Pearson, 1980). Dry heat cooking methods have been reported to cause less lean retention and higher cooking losses in chevon than mutton [49] (Prasad *et al.*, 1988).

6.3.6   PRESSURE TREATMENT

The consequences of high pressure treatment on meat tenderness or meat ageing are not yet clear. Beilken *et al.* (1990) reported that the application of pressure/heat treatment led to a decrease in meat toughness, but this treatment condition induced discoloration. For post-rigor meat, Macfarlane *et al.* (1980) found no difference or a slight increase in shear force measurements between treated and untreated meat (150 MPa, 3 hr $0°C$), although there were important modifications in myofibrillar structures. Suzuki *et al.* (1992) also reported a decrease in hardness of post-rigor muscle pressurized below 300 MPa at $10°C$ for five minute. A 'Hydrodyne' process was reported for effective tenderization of beef muscle wherein the meat is subjected to a shock wave generated while it is submerged under water [53]. Placing meat in sealed water filled chamber and setting off an explosion destroys most of Z-lines and improve tenderness. Pressure treatment that is subjecting meat to a very high pressure accelerates post-mortem glycolysis and cause improvement in tenderness. Pressure treatment can also be used in reducing toughness at post-rigor stage. However, additional application of heat is required to raise the temperature up to 60°C during pressure treatment.

A hydrostatic pressure of 1.05×10^7 Kg/ m^2 at 30–35°C for 2 min duration was reported effective at increasing meat tenderness [54]. In developing countries, pressure cookers where pressure up to 15 psi can be achieved are used for cooking of tough meats. Mahendrakar *et al.* (1989) reported that greater tenderization of pressure-cooked muscle occurs due to thermal shrinkage of muscle and greater solubilization of collagen at higher temperature.

6.4   METHODS OF MEAT TENDERIZATION

6.4.1   MECHANICAL METHODS OF MEAT TENDERIZATION

The various mechanical methods are employed for the tenderization of meat. These methods include hanging by achilles/pelvic suspension to blade tenderization, tumbling, massaging, and many more. The tough meat particularly from older animals can be better utilized for products preparation by employing mechanical comminution such as grinding, mixing, chopping, and so on. Reducing particle size through simple mincing or by grinding/ bowl chopping and making the minced meat products or restructured products by incorporating appropriate ingredient has been the most-practical and effective way of utilizing tough meat beneficially. Modern technologies such as tumbling, massaging, flaking, blade tenderization, and so on could also be tried to reduce toughness and to improve texture properties of the meat product. Singh *et al.* (2002) also demonstrated improved marination in chicken breast fillets by mechanical tenderization in a blade tenderizer and rotary marination in a meat tumbler.

Various methods of carcass hanging are used to improve tenderness. Stretching muscles when they pass into rigor would increase sarcomere length and result in more tender meat. In particular, pelvic suspension is gaining popularity as it significantly improves meat tenderness due to muscle stretching. When carcass suspended from achilles tendon some muscle groups stretch, while others freely contract. Suspension of the carcasses during chilling by pelvic hanging (passing the hook through obturator foramen) than the conventional hind leg (passing hook behind the achilles tendon) and producing tenderness in the Longissimus dorsi, Semimembranosus, and *Semitendinosus* is described as tender stretch method of tenderization [38] (Kondaiah and Mendiratta, 2002). Pelvic suspension has been reported to be more effective in enhancing tenderness than by using delayed or slow chilling [30]. However, it requires more labor. Other method used for meat tenderization is tender cut. It involves cutting the backbone in certain places on the carcass, without cutting the loin muscle [39].

6.4.2   CHEMICAL METHODS (ACID MARINADE/CALCIUM CHLORIDE/ PHOSPHATES) OF MEAT TENDERIZATION

Many methods of altering meat tenderness have been evaluated, including the use of marinades[55] (Gault, 1985). Marinades of tenderizing capacity are particularly important in applications involving muscle rich in connective tissue. These muscles make up the cheaper carcass cuts and the tenderizing effects of marinating offers a commercially important means of upgrading them [56] (Gault, 1991; [57] Lewis and

Purslow, 1991). While marinating is recognized as a means of enhancing the quality and versatility of meat, commercial marinades often have little influence on the tenderness of meat but they improve palatability by enhancing or complementing the flavor of meat ([58] Rao, 1989; [56] Gault, 1991). Acidic marination involves the immersion of meat in an acidic solution of vinegar, wine or fruit juice [59] (Stanton and Light, 1990; [57] Lewis and Purslow, 1991). The mechanism of the tenderizing action of acidic marinades is believed to involve several factors including weakening of structure due to swelling of the meat, increased proteolysis by cathepsins, and increased conversion of collagen to gelatin at low pH during cooking [60] (Berge *et al.*, 2001). Marination to improve palatability of meat has been traditionally practiced. Salt solutions or acids (acetic acid) improve tenderness of meat. Acids may encourage collagenases and cathepsins, which work at low pH to break down muscle structure. Also they may increase myofibril swelling with increasing water holding contributing to tenderness and juiciness. Genaro *et al.* (1989) have also reported that the incorporation of acetic, citric, and lactic acid into restructured beef steaks alters the properties of collagen and other meat proteins. Infusion or injection of solutions has been a further development to marinading. Injection of calcium chloride solution (0.3M) to tenderize lamb [61] (Koohmaraie *et al.*, 1988), beef [62] (Koohmaraie *et al.*, 1990), pork, and buffalo meat [63] (Mendiratta *et al.*, 1997) has been reported. Marination of spent hen cuts in 0.3 M $CaCl_2$ and 0.5% lactic acid for 4 or 20 hr followed by tumbling (intermittent tumbling in a cycle of 10 min working and 5 min pause for 1 hr cuts in sealed bags) decreased shear force values and improves sensory attributes for better utilization as fried product [64] (Kanimozhi, 1998). Calcium chloride may be used directly after slaughter. The high temperature of a carcass directly after slaughter tends to increase the reaction speed and intensifies tenderization processes. However, the palatability of such meat is slightly poorer than that of meat maturing traditionally. Activation of the proteolytic calpain system and possibility of direct action on muscle proteins as that of salt has been considered as the mechanism of action. Undesirable effects on taste and flavor at higher concentrations were reported. Faster oxidation of the haem pigments and surface browning could be overcome by combined use of calcium chloride with ascorbic acid [65] (Wheeler *et al.*, 1996). Injection of calcium chloride is practiced 1 or 2 days, post-mortem to avoid stimulating muscle contractions. By injecting intravenous or intraperitoneal at antemortem (before slaughter) or within half an hour after slaughter into the muscle, either phosphates alone or in combination salt or enzymes improves tenderness. Cooking losses in meat products were reduced considerably (even up to less than half the values of control) with addition of phosphates. Cooking losses reduced significantly when phosphate is added to goat meat in chunks form [66] (Kondaiah and Sharma, 1989).

6.4.3 ENZYMATIC METHODS (PLANT ENZYME) OF MEAT TENDERIZATION

Uses of plant proteolytic enzymes are the traditional, cheaper, and convenient methods for tenderization of meat and meat products. These proteolytic enzymes already exist in the tissues or obtained from the diverse sources like plants and microbes.

Endopeptidases which catalyses hydrolysis of internal peptide bonds in protein molecules are important in enzymatic tenderization of meat. Introduction of the enzyme in to the meat is very important. There are three different methods of introducing the proteolytic enzymes into meat cuts postmortem, such as dipping in solution containing proteolytic enzymes, pumping enzyme solution, and rehydration of freeze dried meat in a solution containing proteolytic enzymes [67] (Mendiratta, 2000). Sinha and Panda (1985) reported that activity of proteolytic enzymes depends on concentration of active enzyme, pH and temperature, stability of the enzyme, ability to come in contact with its substrate and its ability to digest native proteins as compared to heat denatured proteins. Most of enzymes (papain, ficin, and bromelin) often degrade the texture of meat, due to their broad substrate specificity, and this leads to unfavorable taste due to over tenderization [68] (Kang and Rice, 1970). Therefore, the ideal meat tenderizer would be proteolytic enzyme with specificity for collagen and elastin protein in connective tissue at the relatively low pH of meat which would act either at the low temperature at which meat is stored or the higher temperature achieved during cooking [69] (Cronlunds and Woychik, 1987).

The three plant enzymes were discussed in detail below having their great potentials for application for tenderization in natural ways:

- **Cucumis** *(Cucumis trigonus Roxb):* The cucumis plant is found growing wild in India and the dried fruits of this plant are being traditionally used during cooking of meat to improve taste that is tenderness and juiciness. Hujjatullah and Baloch have reported the use of dried coarsely ground fruits of C. *trigonus Roxb* as a meat tenderizer. They extracted the protease activity in 10% sodium chloride and ammonium sulfate, and named the proteolytically active principle as *"Cucumin"*. They reported highest enzymatic activity at pH 5.0 and temperature of 40°C. The enzyme activity was retained up to a temperature as high as 70°C. Similar plant Proteolytic enzyme named as *"Ivrin"* was isolated from unripe fruits of *Cucumis pubescens wild* by Yadav (1982). He found its greater proteolytic activity on myofibrillar, sarcoplasmic, and stroma proteins compared to papain and ficin. He also concluded that *Ivrin* was superior to papain and ficin in terms of improving juiciness, texture, and overall acceptability. Kumar and Berwal treated leg and breast cuts of spent hens with 0–5% level cucumis extract at 45°C for 4 hr. They concluded that 4% cucumis extract was optimum for tenderization of spent hen meat and preparation of tandoori chicken. Naveena reported that dried cucumis powder was having higher activity than fresh peels. He also reported more effect of cucumis on myofibrillar than on connective tissue proteins. Sinku *et al.* studied the physico-chemical and lipid status in meat of broiler, desi, and culled birds upon the *In vitro* action of *Ivrin*. He reported significant decrease in pH and WHC of *Ivrin* treated meat. Naveena *et al.* (2004) have reported increase in flavor, tenderness and overall acceptability scores of buffalo meat roasts treated with cucumis. They concluded that cucumis could be effectively used at household or industrial level as a better alternative to papain for tenderization of tough buffalo meat.

*   ***Papain (Carciya papaya)***: It is produced from latex and fruit of *Carciya papaya* and has very powerful hydrolyzing activity on fiber proteins and connective tissue. It has optimum activity at higher temperature (65°C) [70] (Ockerman, 1991) and most tenderization takes place during early stages of cooking. The papain enzymes can be injected just before slaughter/after slaughter/injected just before cooking. Nayga-Marcado (1984) has reported that the incorporation of 1, 1.5 and, 2% papain powder in roasted pork produced significantly (P<0.01) more tender meat compared to no papain treatment and concluded that papain treatment greater than 1% of the fresh weight of meat produces over tenderized meat and had no added advantage. Bawa *et al.* (1981) reported increase in tenderness of spent hens with increasing level of enzyme. Mendiratta (1992) reported synergistic effect of papain enzyme and pressure treatment for tenderization of spent hen meat. Naveena (2002) found that marination of tough buffalo meat chunks in 0.2% (w/w) papain was optimum for desirable tenderness and sensory attributes. Bawa *et al.* (1985) established that 5 min interval between the injection of 25 ppm papain (on the basis of starved live weight) and slaughter is sufficient to improve the tenderness of meat from spent white leghorn hens. Mendiratta (1992) reported that a solution (8 mg/ml) of commercial samples of crude papain, when injected into wing vein @ 6 ml/kg live wt) 30 min prior to slaughter caused desired tenderness in spent hen meat.

*   ***Ginger (Zingiber Officinale Roscae):*** Washed and dried rhizome of a rootstock of ginger is widely used for its spicy properties and medicinal value. Ginger has also been investigated by different workers for its tenderizing [71-73,67,74], antioxidant [75] and antimicrobial [76,77] properties in meat and meat products. Ginger rhizome was investigated as a new source of plant proteolytic enzyme by Japanese researchers [71] and proteolytically active principle was named *zingibain* [78]. Ginger protease is a thiol proteinase and its maximum activity can be achieved at 60°C and rapid denaturation of the enzyme occurred at 70°C. Maximum proteolytic activity of the ginger rhizome can be extracted at pH 6.0. Highest digestion of the meat proteins treated with the ginger proteinases has been reported to occur at pH 4.0 and temperature of 45°C. So the activity of the ginger proteinase was higher under acidic conditions than under alkaline conditions [79]. Proteolytic activity of the ginger extract was found higher at ambient temperature compared to chilled temperature [72]. The antioxidant and antimicrobial actions of ginger extract are unique which no other tenderizing agent has or possesses. Tenderization of beefsteaks using ginger extract at the level of 0.2 ml for 6.5 cm^2 surface area was achieved without detrimental effects of mushiness [80]. Precooked lean beef treated with 0.5% v/w ginger extract in the absence of salt was sufficient to give acceptable tenderness. However, addition of ginger at level greater than 1% v/w resulted in meat that was too soft [81,82]. Treatment of chilled spent hen meat with 3% v/w ginger extract was found more effective than prechilled meat in improving tenderness and sensory attributes [83, 84]. Naveena reported that 5% w/v ginger extract was optimum for desirable tenderization

of buffalo meat. Mendiratta *et al*. suggested that 3% ginger extract can be use for improvement of the sensory quality and shelf life of mutton chunks.

6.5 THE PHYSICO-CHEMICAL DETERMINANTS OF MEAT TENDERNESS

6.5.1 COLLAGEN CONTENT AND THEIR SOLUBILITY

The collagen content was 10–13% of total protein [85]. Connective tissue in the buffalo meat had a bigger contribution to toughness [86]. The total concentration of connective tissue components were not closely related to the scores for muscle fiber tenderness. Chronological age was significantly related to the collagen content in the muscle. The collagen content increased significantly with advancing age of the male Murrah buffaloes [87]. A hydroxyproline content of 0.12% was recorded in high protein diet fed young male buffaloes [88]. The muscles from young buffaloes of 1 to 2 years showed less collagen (0.91 to 1.71 g/100g) than from 12 year old buffaloes (1.16 to 2.23 g/100g) [51]. Some authors did not find any differences in the amount of connective tissue in young and aged animals [89,90]. As animals get older the collagen cross links are stabilized. After cooking the collagen cross links weaken but do not break, so contributing to the toughness of meat from old animals [91]. Greater tenderization of pressure cooked muscles could be explained due to greater thermal shrinkage of muscles and greater solubilization of collagen at higher cooking temperature. Ruantrakool and Chen also reported significantly higher soluble collagen content in pressure cooked chicken gizzard and breast meat compared to boiling. As animals get older the collagen cross links were stabilized and the collagen was much less soluble [90,91]. Soluble collagen percentage was significantly related to the contribution of connective tissue to toughness as assessed by sensory panel. The soluble collagen decreases (13.5 to 3.6) as the age of the animal increases [89,90]. A collagen solubility of 6.58% was observed in meat chunks from spent female Murrah buffaloes [74].

6.5.2 MUSCLE FIBER DIAMETER

A significant relationship was observed between muscle fiber diameter and tenderness. Measurement of muscle fiber diameter could be useful for selection of animals with tender meat [92]. A significant relationship was observed between muscle fiber diameter and tenderness. Large muscle fibres are generally indicative of less tender beef [93]. The meat obtained from high protein diet fed young male buffaloes showed a muscle fiber diameter of 35.32μm [88]. As the age and slaughter weight increased there was an increase in muscle fiber diameter [95]. A much less diameter of muscle fibers were observed in young animals [91,94]. A muscle fiber diameter of 49.2 to 49.7 μm was observed in raw beef from 18 month old bulls [98]. In concentrate fed male Murrah buffaloes muscle fibre diameter of 80 μm was observed. As the age and slaughter weight increased there was an increase in muscle fiber diameter [91,95]. A muscle fiber diameter of 60.76 μm was recorded in meat chunks from spent female Murrah buffaloes [74]. Fiber diameter was positively correlated to shear force values but negatively correlated to tenderness and sarcomere length of the muscle [96,100].

6.5.3 SARCOMERE LENGTH

The sarcomere length was positively correlated to lean texture [93]. There is substantial evidence that supports sarcomere shortening as the cause of muscle toughning [27]. Sarcomere shortening during rigor phase has been reported to be the main cause of muscle toughing during the first 24 hr of slaughter [99]. However, contradictory results were obtained by Smith *et al.* and Koohmaraie *et al.*. A sarcomere length of 2.3 μm was observed in raw beef from 18 month old bulls [98] . Young buffalo bulls showed a sarcomere length of 1.73 to 1.88 μm. The longer the sarcomeres more tender the meat is influenced by the boning. [95] [25]. Males had significantly shorter sarcomere length than females in turkeys [97]. However, the variation in sarcomere length had no impact on tenderness of beef from Charolais sires [101]. Amount and rate of shortening is temperature dependant and the rate is much higher at 37°C than at 15°C rigor temperature [102]. Devine *et al.* reported minimum shortening region of 10–15°C for the *Longismus dorsi* muscle and 7–13°C for *Semi membranous* muscle. *Semi membranous* muscles have more correlation between shortening and ultimate tenderness both in warm and cold shortening region [102,103] but in case of *Longissimus muscle*, only cold shortening region have this relationship. Geesink *et al.* have observed a relationship between sarcomere length and incubation temperature during storage and reported that muscle contraction is minimum at 15°C and above 25°C have detrimental effect. White *et al.* reported that Cold shortening was induced in rapidly chilled muscles, as they had shorter sarcomere length than slowly chilled muscle up to 21 days postmortem.

6.5.4 MYOFIBRILLAR FRAGMENTATION INDEX

The phenomenon of myofibrils breaking into shorter segments at or near Z-line during post-mortem storage of muscle is termed as the myofibrillar fragmentation. Myofibrillar proteins are largely responsible for rigor mortis and tenderness of meat. Goll *et al.* (1983) investigated that calpain proteolytic system has a primary role in myofibrillar protein degradation. Protein solubility increased with storage time, while gel strength decreased in chevon and beef during storage [104]. Myofibrillar fragmentation followed a pattern of chicken > sheep > goat > cattle on post-mortem storage at different temperature [105]. Myofibrillar Fragmentation Index (MFI) is a measure of myofibrillar protein degradation [93]. This is highly related to shear force and sensory tenderness ratings [106] and [18] and negatively correlated to lean color [93]. Longissimus tenderness was highly and positively correlated with MFI and indicates the amount of myofibrillar proteolysis that has occurred [109]. The MFI was observed to be 87.5 in six year old male Murrah buffaloes [108]. Geesink *et al.* have reported that incubation temperature of muscle above 25°C during pre-rigor and early post-rigor period reduces myofibrillar fragmentation. Earlier onset of proteolytic breakdown of the myofibrillar structure has been reported in slowly chilled meat [30]. However, by 21 days post-mortem, both rapidly and slowly chilled muscles underwent proteolysis to the same extent [32].

6.5.5   SHEAR FORCE VALUE

Shear force value provides basic information on tenderness, WHC and texture of meat. Murray *et al*. found that Warner Bratzler (W-B) shear force value and hardness (first bite) were highly dependent on muscle fiber variations, where as overall tenderness (panel) was only slightly dependent on them. It is also differed between the muscles in age and nutrition of animals [88,95,107]. Berry *et al*. also reported that pre-rigor cooked meat had significantly higher shear force values than post rigor cooked meat. Sharma *et al*. (1988) also observed lower shear value of chilled bone meat. The order of tenderness was higher in young bulls followed by steers and then cows [113]. Shear force value was reported to have positive and higher correlation with fiber diameter, hydroxyproline content, and toughness of the meat and negatively correlated with the sarcomere length of the meat [96].

6.5 6   TEXTURE PROFILE

The structure of meat from older animals has coarser appearance than the meat from younger animals (Ognjanovic, 1974). Fine texture of muscles may be due to young age of the animals and stall fed management [107]. Shear force values are closely related to muscle fiber properties than to connective tissue components in cooked muscle [110,112]. Pressure cooked meat from old buffaloes showed a higher shear force value compared with young male buffaloes [51]. Sex of the animal did not affect tenderness from shear force test [113].

6.6   CONCLUSIONS

The meat tenderness is one of the most important quality attributes rated by consumers. The variation in the tenderness affects the consumer satisfaction and it is the challenge to the scientific community to solve the problem in the natural ways without affecting the other quality attributes and safety of products. The various pre-slaughter and post-slaughter factors influencing the tenderness meat and meat products. Until now various techniques were employed to solve the problems ranging from mechanical, chemical and enzymatic to improve the tenderness of tough meat. The results were varying; however, the combination of methods such as enzymatic and mechanical or any other methods combined with enzymatic methods of tenderization has wide future scope for industrial and household application.

KEYWORDS

- **Connective Tissue**
- **Tenderizing**
- **Peptide Bonds**
- **Papain**
- **Carcass**

REFERENCES

1. Bawa, A. S., Orr, H. L., and Usborne, W. R. Enzymatic tenderization of spent white leghorn hens. *Poultry Science*, **60**, 744–74 (1981).
2. Bawa, A. S., Orr, H. L., and Usborne, W. R. Effect of anti-mortem papain injection on the tenderness of spent white leghorn hens. *Journal of Food Science and Technology*, **22**, 254–257 (1985)
3. Beilken, S. L., MacFarlane, J. J., and Jones, P. N. Effect of high pressure during heat treatment on the Warner Bratzler shear force values of selected beef muscles. *Journal of Food Science*, **55**, 15–18 (1990).
4. Berry, B. W., Ray, E. E., and Stiffler, D. M. Effect of electrical stimulation and hot boning on sensory and physical characteristics of prerigor cooked beef roast. *Proceedings of the Annual Meeting of European Meat Research Workers*, Colorado Springs Co., **26**, 1 (1980)
5. Bouton, P. E., Harris, P. V., Shorthose, W. R., and Ratcliff, D. Changes in the mechanical properties of veal muscles produced by myofibrillar contraction state cooking temperature and cooking time. *Journal of Food Science*, **39**, 211–325 (1974).
6. Carpenter, Z. L., Abraham, H. C., and King, G. T. Tenderness, and cooking loss of beef and pork. I. Relative effects of microwave cooking, deep fat frying, and oven-broiling. *J. Am. Diet. Assoc.*, **53**(4), 353–6 (1968).
7. Cyril, H. W., Castellini, C., Bosco, A. D. Comparison of three cooking methods of rabbit meat. *Italian Journal of Food Science*, **8**, 337-340 (1996).
8. Davey, C. L. and Gilbert, K. V. Studies in meat tenderness: 4. Changes in extractability of myofibrillar protein during meat ageing. *Journal of Food Science*, **33**, 2 (1968).
9. Devine, C. E., Wahlgreen, M. A., and Tornberg, E. The effects of rigor temperature on shortening and meat tenderness. In: *Proc. 42nd Int. Congr. of Meat Science and Technology, Lillehammer, Norway*, pp. 396–397 (1996).
10. El-Shimi, N. M. Influence of microwave cooking and reheating on sensory and chemical characteristics of roast beef. *Food Chemistry*, **45**, 11–14 (1992).
11. Geesink, G. H., Bekhit, A. D., and Bickerstaffe R. Rigor temperature and meat quality characteristics of lamb longissimus muscle. *Journal of Animal Science*, **78**, 2842–2848. (2000)
12. Genaro, A. C. and Norman, M. G. Organic acids as tenderizers of collagen in restructured beef. *Journal of Food Science*, **54**, 1173. (1989)
13. Goll, D. E., Otsuka, Y., Nagainis, P. A., Shannon, J. D., Sathe, S.R., and Muguruma, M. Role of muscle proteinases in maintenance of muscle integrity and mass. *Journal of Food Biochemstry*, **7**, 137–177 (1983).
14. IIannula, T. and Puolanne, E. The effect of the cooling rate on beef tenderness the significance of pH at 7^0C. *Meat Science*, **67**, 403–408 (2004).
15. Henderson, D. W., Goll, D. E., and Stromer, M. H. A comparison of shortening and Z-line deradation in post-moterm bovine, porcine and rabbit muscle. *American Journal of Anatomy*, **128**, 117 (1970).
16. Hujjatullah, S. and Baluch, A. K. Proteolytic activity of Cucumis trigonus Roxb, extraction, ectivity, charectristtics. *Journal of Food Science*, **35**, 276–278 (1970).
17. Kim, K. J. and Lee, Y. B. Effect of ginger rhizome extract on tenderness and shelf life of pre-coocked lean beef. *Journal of Korean Society of Food Science*, **11**, 119–121 (1995a)
18. Koohmariae, M. Muscle proteases and meat aging. *Meat Science*, **36**, 93 (1994).
19. Kumar. M. and Berwal. T. S., Tenderization of spent Hen meat with Cucumis Trigonus Roxb (Kachri). *Indian Journal of Poultry Science*, **33**, 67–70 (1998).

20. Lan, Y. H., Novakofski, T. R., Brewer, M. S., Carr, T. R., and McKeith F. K. (1995). Effect of treatment of pre and post rigor porcine muscles with low sodium chloride concentration on subsequent extractability of protein. *Journal of Food Science,* **35,** 268–270.

21. Locker, R. H. and Danies, G. J. Rigor mortis in beef sterno-mandibularis muscle at 37^0C. *Journal Science of Food Agriculture,* **26,** 1721 (1975).

22. Locker, R. H. and Leet, N. G. Histology of highly-stretched beef muscle. The fine structure of grossly stretched fibres. *Journal of Ultrastructure Research,* **52,** 64 (1975).

23. MacFarlane, J. J., McKenzie, I. J. and Turner, R. H. Pressure treatment of meat effects on thermal transitions and shear values. *Meat Science,* **5,** 307–317 (1980).

24. Mahendrakar N. S., Dani N. P., Ramesh B. S., and Amla, B. L. Effect of postmortem conditioning treatments to sheep carcasses on some biophysical characteristics of muscles. *Journal of Food Science and Technology,* **25,** 340–344 (1988).

25. Mahendrakar, N. S., Dani, N. P., Ramesh, B. S., and Amla, B. L. Studies on influence of age of sheep and postmortem carcass conditioning treatments on muscular collagen and its thermolability. *Journal Food Science and Technology,* **26,** 102–105 (1989).

26. Mendiratta, S. K. *Synergistic effect of pressure heat and enzyme treatment for tenderization of spent hen meat.* Ph. D thesis, submitted to CCS Haryana Agricultural University, Hissar-India (1992).

27. Murray, A. C., Jeremiah, L. E, and Martin, A. H. *Journal of Food Technology,* **18,** 607. (1983).

28. Naveena, B. M. *Studies on use of cucumis, ginger and papain for tenderization of buffalo meat.* PhD thesis submitted to IVRI, Izatnagar-243122, India (2002)

29. Nayga-Marcado, L. C. Effect of varying levels of powdered papain and internal meat temperatures on the tenderness, cooking loss, drip loss and cooking time of roasted pork muscle. *CMU Journal of Agriculture, Food and Nutrition,* **6,** 205–215 (1984).

30. Ognjanovic, A. Meat and meat production. In: *The Husbandry and Health of the Domestic Buffalo.* W. Ross Cockrill (ed.) FAO, Rome. pp. 377–410 (1974).

31. Parrish, F. C. J. R., Young, R. B., Miner, B. E., and Anderson, L. D. Effect of post-mortem conditions certain chemical, marcological and organoleptic properties of bovine muscles. *Journal of Food Science,* **38,** 690 (1973).

32. Ruantrakool, B. and Chen, C. Collagen content of chicken gizzard and breast meat tissues as affected by cooking method. *Journal of food science,* **51,** 301–304 (1986).

33. Singh, R P., Yadav, A.S., and Verma, S. S. Effect of filleting and marination processes on the quality of roasred chicken breast fillets. *Journal of Food Science and Technology,* **39,** 155–157 (2002).

34. Sinha, S.K. and Panda, P. C. Tenderization of meat with proteolytic enzymes- part II. *Livestock Advisor,* **18,** 57–64 (1985).

35. Sinku, R. P., Prasad, R. L., Pal, A. K., and Jadhao, S. B. Effect of plant proteolytic enzymes on physico-chemical properties and lipid profile of meat from culled, desi and broiler chicken. *Asian-Australian Journal of Animal Science,* **16,** 884–888 (2003).

36. Smith, M. E., Kastner, C. I., Hant, M. C, Kropf, D. H. and Allen, D. M. Elevated conditioning temperature effect on beef carcass from four nutritional regimes. *Journal of Food Science,* **44,** 158 (1979).

37. Suzuki, A., Kim Homma, N., Ikeuchi, Y., and Satio, M. Acceleration of meat conditioning by high pressure treatment. C. Balny, R. Hayashi, K. Heremans, and P. Masson (Eds.). *High pressure and biotechnology,* INSERM/John Libbey Eurotext Ltd, Montrouge, pp. 219–227 (1992).

38. Swenson, J. A., Marriott, N. G., Claus, J. R., Wang, H., and Graham, P. P. Characteristics of microwavable pork chops. *Journal of Muscle Foods,* **5,** 389–406 (1994).

39. Thomas, R., Anjaneyulu, A. S. R., and Kondaiah, N. Quality and shelf life evaluation of emulsion and restructured buffalo meat nuggets at cold storage (4±1°C). *Meat Science,* **72**, 373–379 (2006).
40. Tornberg, E. Biophysical aspects of meat tenderness. *Meat Science*, **43**, S175–S191. (1996)
41. Totosaus, A., Gurerrero, I. and Gerardo-Montejano, J. Effect of tempreture and storage time on chevon and beef protein gels. *Journal of Food Science*, **36**(5), 487–488 (2001).
42. West R. L. Effect of high temperature conditioning on muscle tissue. *Food Technology*, **33**, 41–46 (1979).
43. Yadav, B. S. *Studies on physico-chemical properties of muscle proteins and use of tenderizing enzymes in buffaloes*. Ph. D. Thesis, Indian Veterinary Research Institute, Izatnagar, (India) (1982).
44. Yadav, B. S. and Singh, L. N. Changes in sarcoplasmic, myofibriller and stroma protein in buffalo skeletal muscle with advancing age. *Indian Journal of Animal Science*, **56**, 20–25 (1986).

CHAPTER 7

BIOLOGICAL PROPERTIES OF MUSHROOMS

CARLOS RICARDO SOCCOL, LEIFA FAN, and SASCHA HABU

CONTENTS

7.1 INTRODUCTION

Mushrooms belong to Fungi Kingdom, although in the past, they were classified as plants. The fruiting bodies receive attention for their forms and colors when they are found in nature. The interest in mushrooms is old, oriental countries use the mushrooms for the treatment and prevention of several diseases such as arthritis, rheumatism, bronchitis, gastritis, cancer, as well as in health and longevity maintenance. Scientists search answers about the action mechanism of mushrooms and about alternative treatments based on natural compounds. Many species are appreciated in culinary around the world because of their good taste and their nutritious potential. There are researches about bioactive compounds such as polysaccharides, proteo-glucans, phenol compounds, nucleotides, and their action mechanism to treatment for several diseases. Studies have shown significant results in the immune system improvement, antimicrobial and anti-angiogenic activity, and in the treatment of the cancer and high cholesterol. The aim of this chapter describes any biological properties of mushrooms to the treatment of several diseases.

Mushrooms are a special group of macroscopic fungi. Macromycetes arranged in the phylum Basidiomycota and some of them in the Ascomycota are known as the higher fungi [70,93]. It is considered that exist in the planet about 140.000 different species of mushrooms, however, only about 10% are known. Half of them present nutritious properties, 2.000 species are safe, and approximately 70 of them are known by presenting some pharmacological property.

Edible mushrooms are attractive because of their flavor, taste, and delicacy. Although many species of edible mushrooms exist in the nature, less than 20 species are used as food, and only 8–10 species are regularly cultivated in significant extent.

Fresh mushrooms can be acquired from grocery stores and markets, including straw mushrooms (*Volvariella volvacea*), oyster mushrooms (*Pleurotus ostreatus*), shiitakes (*Lentinula edodes*), and enokitake (*Flammulina* spp.). There are many other fungi like milk mushrooms, morels, chanterelles, truffles, black trumpets, and porcini mushrooms (*Boletus edulis,* also known as 'king boletes') [23].

Many worldwide cultures, especially in the orient, recognize that extracts from some edible and non-edible mushrooms were recognized for their potential health benefits. In China, the dietary supplements and nutraceuticals made from mushroom extracts are used, along with various combinations of other herbal preparations [4,8].

These mushrooms have attracted attention because they are source of non-starchy carbohydrates, with a high content of dietary fiber (chitimous wall), moderate quantities of proteins (20–30% of dry matter) with most of the essential amino acids, minerals and vitamins (B) [1,23].

Various compounds with important pharmaceutical properties have been isolated from these organisms. Substances that act as anti-aging, in longevity, modulating the immune system, having hypoglycemic activity and to inhibit tumor growth have been isolated from mushrooms, such as polysaccharides. Polysaccharides can interconnect several points forming a wide variety of branched or linear structures, for example, ß-glucans (Ooi and Liu, 2000). Polysaccharides have structural variability for regulatory mechanisms of various interactions in higher organisms [1,8]. Furthermore, other bioactive substances such as triterpenes, lipids, and phenols have also been identified and characterized in mushrooms with medicinal properties [66].

Fungi can be produced technically through fermentative process. The media may be in form of available substrates from valued cheap sources like agro-biomass and industrial waste; transformed into high value added food and pharmaceutical products.

7.2   BIOLOGICAL PROPERTIES OF MUSHROOMS

7.2.1   ANTIOXIDANTS ACTIVITY

Exogenous chemical and endogenous metabolic process in the human body or in the food system might produce highly reactive free radicals, especially oxygen derived radicals, which are capable of oxidizing biomolecules, resulting in cell death and tissue damage [20]. During the reduction of molecular oxygen, reactive oxygen species are formed, and there is a continuous requirement for inactivation of these free radicals. The superoxide and hydroxyl radicals are the two most representative free radicals. In cellular oxidation reactions, superoxide radical is normally formed first, and its effects can be magnified because it produces other kinds of cell-damaging free radicals and oxidizing agents. Damage induced by free radicals can affect many biological molecules, including lipids, proteins, carbohydrates and vitamins present in food. Reactive oxygen species also implicate in the pathogenesis of various human diseases, such as DNA damage, carcinogenesis, rheumatoid arthritis, cirrhosis, arteriosclerosis as well as in degenerative processes associated with ageing. Evidences have been indicating that diet rich in antioxidant reduce risks of some diseases [20,62]. Mushroom contain vitamins A and C of ß-carotene and a great variety of secondary metabolites such as phenolics compounds, polyketides, terpenes, steroids and phenols, all have protective effects because of their antioxidant properties [25,35].

Researchers investigate several edible and non-edible mushrooms with antioxidant properties for applications in food, cosmetics and treatment of diseases (Table 1). Water or ethanolic extracts of fruiting bodies or biomass resulting by fermentation have been studied and tested.

TABLE 1   Potential antioxidant of mushrooms

Mushroom	Biological Activity: Antioxidant		
	Source	Substance	Reference
Agaricus bisporus	Fruit body	Methanolics	Elmastas *et al.*, 2007
Agaricus brasiliensis	Fruit body	Methanolic	Soares *et al.*, 2009
Agaricus silvaticus	Fruit body	Methanolic	Barros *et al.*, 2008b
Agrocybe cylindracea	Fruit body	Ethanolic and hot water extracts	Tsai *et al.*, 2007
Antrodia camphorata	Submerged Fermentation	Methanolic	Shu and Lung, 2008

TABLE 1   *(Continued)*

Boletus edulis	Fruit body	Hot water extracts	Ribeiro *et al.*, 2008
Boletus edulis	Fruit body	Alkaloids	Sarikurkcu *et al.* 2008
Boletus badius	Fruit body	Methanolic extracts	Elmastas, *et al.*, 2007
Cordyceps sinensis	Submerged Fermentation	Polysaccharide	Leung *et al.*, 2009
Geastrum saccatum	Fruit body	Glucans	Dore *et al.*, 2007
Grifola frondosa	Fruit body	Water extracts	Lee *et al*, 2008b
Hypsizigus marmoreus	Fruit body	Cold and Hot water, Ethanolic	Lee *et al.*, 2007
Inonotus obliquus	Fruit body	Methanolic	Lee *et al.*, 2007b
Laetiporus sulphureus	Fruit body	Ethanolic	Turkoglu *et al.*, 2007
Lentinula edodes	Fruit body	Ethanolic	Zheng *et al.*, 2005
Leucopaxillus giganteus	Fruit body	Methanol	Barros *et al.* 2008b
Lepista nuda	Fruit body	Methanolic	Elmastas *et al.*, 2007
Phellinus linteus	Fruit body	Ethanolic	Song *et al.*, 2003
Pleurotus ostreatus	Fruit body	Methanolic	Elmastas *et al.*, 2007
Pleurotus ostreatus	Fruit body	Ethanolic	Jayakumar Thomas and Geraldine, 2009
Polyporus squamosus	Fruit body	Methanolic Extracts	Elmastas *et al.*, 2007
Russula delica	Fruit body	Methanolic	Elmastas *et al.*, 2007
Suillus collitinus	Fruit body	Methanol extracts	Sarikurkcu Tepe and Yamac, 2008
Turbinaria conoids	Fruit body	Fucoidan	Chattopadhyay *et al.*, 2009
Verpa conica	Fruit body	Methanolic extracts	Elmastas *et al.*, 2007

Mushrooms are currently available in Taiwan, including *Agaricus blazei, Agrocybe cylindracea* and *Boletus edulis*. Ethanolic extracts were more effective than hot water in antioxidant activity using the conjugated diene method and scavenging ability on 2,2-diphenyl-1-picrylhydrazyl (DPPH) radicals whereas hot water extracts were more effective in reducing power, scavenging ability on hydroxyl radicals, and chelating ability on ferrous ions [102]. According to Sarikurkcu *et al.* studies about the antioxidant activity of *Lacttarius deterrimus, Suillus collitinus, Boletus edulis, Xerocomus chrysenteron. L. deterrimus. B. edulis* showed activities as strong as the positive controls. The reducing power of the species was excellent. Chelating capacity of the extracts was proportional to the increasing concentration.

Northeast of Portugal is recognized as one of the richest regions of Europe in wild edible mushroom species, which have considerable gastronomic relevance. *Russula cyanoxantha, Amanita rubescens, Suillus granulates* and *Boletus edulis* are among more common and marketed species. Four species studied are rich in organic acids, but phenolics compounds are present in low amounts. Organic acids, phenolics compounds and alkaloids composition is insufficient to justify the antioxidant potential of analyzed species. Other compounds also participate in the observed activity. These species present antioxidant potential, especially high for *Boletus edulis*, which is also the richest specie in alkaloids [83]. *Boletus edulis* was a popular edible mushroom in Europe, North America and Asia. This mushroom presented interesting results for antioxidant effects cited by three authors from different regions of world.

Metabolisms, physiology of mushroom and environmental conditions are important and determinant to the production of antioxidant compounds. However, the extraction methodologies are primordial to remove compounds produced intracellular and/or others substances resulting of metabolism (Table 2).

The antioxidant activity of five mushroom species: *Agaricus bisporus, Agaricus arvensis* Schaeffer, *Agaricus romagnesi* Wasser, *Agaricus silvaticus* Schaeffer and *Agaricus silvicola* were analyzed. All the species proved to have antioxidant activity, especially *A. silvaticus* [95].

Soares *et al.* investigated the young and mature extracts of fruiting bodies of *Agaricus brasiliensis*. Both extracts showed antioxidants activities, except the chelating ability ferrous ions. Consumption of fruiting bodies of *A. brasiliensis* might be beneficial to the human antioxidant protection system against oxidative damage.

Hypsizigus marmoreus (peck) Bigelow, also known bunashimeji and hon-shimeji is a mushroom cultivated and commercially available in Taiwan. Naturally occurring antioxidants components, including tocopherols and total phenols, were found in extracts from *H. mamoreus*. The major antioxidant components found in hot water extracts were total phenols and in ethanolic extracts were total tocopherols [47].

Extraction use polar and non-polar solvents and different temperatures, considering solubility and thermostability of each substance, respectively (Table 2).

Shiitake (*Lentinus edodes*) have several compounds including bioactive polysaccharides (lentinan), dietary fiber, ergosterol, vitamin B_1, B_2 and C and minerals have been isolated from de fruiting body, mycelia and culture medium of this mushroom. Heat treatment of Shiitake sample increased the overall content of free polyphenolic and flavonoid compounds. The heat treatment can produce changes in their extractability due to disruption of the cell wall thus bound polyphenolic and flavonoid compounds may be released more easily relative to those of raw materials [14].

Clitocybe maxima, Pleurotus ferulae, and *Pleurotus ostreatus* were used to study antioxidant properties. Ethanolic extracts and hot water extracts from *P. ferulae* and *P. ostreatus* were more effective than *C. maxima* cap and stipe antioxidants activities. Total phenols were major naturally occurring antioxidant components found in the range of 5.51-9.66 gallic acid equivalents/g and 5.10-11.1mg gallic acid equivalents/g for ethanolic and hot water extracts, respectively. These mushrooms could be used in grams levels as food or food ingredient. Therefore, these three mushrooms might serve as possible protective agents in human diets to help human reduce oxidative damage [102].

The ethanolic extract of the *Pleurotus ostreatus* and various known antioxidants showed concentration-dependent anti-oxidant activity by inhibiting lipid peroxidation, scavenging hydroxyl and superoxide radicals, reducing power and chelating ferrous ions when compared to different standards such as ascorbic acid, BHT and EDTA [35].

TABLE 2   Contents of total phenols in extracts from mushrooms

Mushroom	Extraction (mg/g)			References
	Hot Water	Ethanolic	Methanolic	
Agaricus arvensis	-	-	2,72	Barros *et al.*, 2008b
Agaricus bisporus	-	-	4,49	Barros *et al.*, 2008b
Agaricus romagnesii	-	-	6,18	Barros *et al.*, 2008b
Agaricus silvaticus	-	-	8,95	Barros *et al.*, 2008b
Agaricus silvicola	-	-	6,45	Barros *et al.*, 2008b
Agaricus blazei	5,67	5,80	-	Tsai *et al.*, 2007
Agrocybe cylindracea	5,6	5,7	-	Tsai *et al.*, 2007
Boletus edulis	5,81	5,73	-	Tsai *et al.*, 2007
Hypsizigus marmoreus	10,01	6,89	-	Lee *et al.*, 2008
Clitocybe maxima (cap)	9,71	9,66	-	Tsai *et al.*, 2008
Clitocybe maxima (stipe)	5,1	5,51	-	Tsai *et al.*, 2008
Pleurotus ferulae	7,73	6,71	-	Tsai *et al.*, 2008
Pleurotus ostreatus	11,1	7,11	-	Tsai *et al.*, 2008
Pleurotus ostreatus	-	5,49		Jayakumar *et al.*, 2009

7.2.2   ANTI-INFLAMMATORY ACTIVITY

Inflammatory response is succession of cellular reactions involving the generation and release of cellular mediators such as cytokines. The excessive amount or duration of the production of cytokines, especially TNF-α, can cause serious harm to the body [19].

Excessive or unregulated production of these mediators has been implicated in mediating a number of diseases including rheumatoid, arthritis, osteoarthritis, sepsis, chronic pulmonary inflammatory disease, Crohn's disease, ulcerative colitis, and also carcinogenesis. Inflammation is inherent to pathogenesis of a variety of diseases. Inhibition of activation and the proliferation of these inflammatory cells appears to be an important therapeutic target for small molecular drugs in the treatment of inflammatory diseases and cancer [19,105].

TNF-α is a major pro-inflammatory cytokines with diverse biological activities. Large quantities of TNF-α may induce intravascular thrombosis, shock and cachexia.

Macrophages play important role in host-defense mechanism, and inflammation. The overproduction of inflammatory mediators by macrophages has been implicated in several inflammatory diseases and cancer. The activation of macrophages is important in the instigation of defensive response such as the production of interleukins IL-1ß, IL-6, TNF-α, reactive oxygen species, prostaglandin and nitric oxide. IL-1 ß are also a multifunctional cytokine which has been implicated in pain, fever, inflammation and autoimmune conditions. It stimulates acute phase protein synthesis in the liver and may cause rise in body temperature. It is also up-regulated in many inflammatory diseases. IL-6 is a multifunctional cytokine with pro-/anti-inflammatory properties.

Nitric oxide is an important messenger in diverse pathological functions, including neuronal transmission, vascular relaxation, immune modulation and cytotoxicity against tumor cells. Nitric oxide, secretor product of mammalian cells, produced by inducible nitric oxide synthase, endothelial nitric oxide synthase, and neuronal nitric oxide synthase is considered an important signaling molecule in inflammation [105]. Mushrooms have been applied in the treatment of infections in popular culture or medicine in many countries, such as China, Japan, Russia and Brazil. The anti-inflammatory properties of mushrooms have interested researchers and motivated the investigation of some species (Table 3).

TABLE 3   Potential Anti-inflammatory of mushrooms

Mushroom	Anti-inflammatory Activity		
	Source	Substance	Reference
Agrocybe cylindracea	Fruit body	Agrocybin	Ngai, Zhao and Ng, 2005
Amanita muscaria	Fruit body	Hot water, metha-nolic and ethanolic extracts	Michelot and Melendez-Howell, 2003
Fomitopsis pinicola	Submerged fermentation	Polysaccharides	Cheng *et al.*, 2008
Ganoderma lucidum	Fruit body	Triterpene	Dudhgaonkar *et al.*, 2009
Geastrum saccatum	Fruit body	Glucans	Guerra Dore *et al.*, 2007
Inonotus obliquus	Fruit body	Hot water	Van *et al.*, 2009
Phellinus linteus	Fruit body	Butanol fraction	Kim *et al.*, 2004
Poria cocos	Submerged fermentation	Polysaccharides	Lu *et al.*, 2009
Pleurotus nebrodensis	Fruit body	Nebrodolysin	Lv *et al.*, 2009
Pleurotus pulmonarius	Fruit body	Polysaccharides	Smirdele *et al.*, 2008

The *Geastrum saccatum,* a mushroom native from Brazil, is produced under natural conditions in the unexplored reserve of "Mata da Estrela-Rio Grande do Norte". This basiodiomycete is a saprobic fungus and it is well adapted to tropical regions. The mushroom, known as "Star of the Land", is used in popular medicine by obstetri-

TABLE 3   *(Continued)*

cians and healers, and has curative properties for eye infections and diseases, such as asthma. The anti-inflammatory effects of glucans *G. saccatum* extract on carragennan-induced pleurisy and observed that the glucans extract decreased the number of cells from pleural fluid rats. There is evidence that inhibition of nitric oxide synthase reduces the production of prostaglandins by **ciclooxigenase** through reduced synthesis of oxide nitric, these decrease several inflammatory symptoms such as vessel dilation. Thus, it could be related to inhibition of diapedesis of cells as mononuclear leukocytes in the inflammation site when used the glucans extract. The animals treated with glucans extract decreased oxide nitric. These effects suggest an anti-inflammatory effect of glucans *G. saccatum* extract [17].

Lu *et al.*, demonstrated that *Poria cocos*, called Fu Ling in China, can participate in the regulation of the anti-inflammatory process [64]. *Poria cocos* is commercially available and is popularly used in the formulation of nutraceuticals, tea supplements, cosmetics, and functional foods in Asia. Chemical compounds found in *Poria cocos* include triterpenes and ß-pachyman, a polysaccharide composed of ß-pachimarose, pachymic acid, and poricoic acid. IFN-γ is one of the major mediators which predispose endothelial cells toward inflammatory/immunological responses. The pretreatment with the polysaccharide extracted of Poria *cocos* was dose-dependently and inhibited IFN-γ-induced inflammatory gene IP-10 protein release. It suggests that the effect of polysaccharide on IP-10 expression was regulated at the translational level and thus it may participate in regulating inflammatory-related diseases. This polysaccharide showed no toxicity to endothelials cells, indicating the safety of its use.

Inonotus obliquus also known as Chaga, is a black mushroom that grows on birch trees in northern climates such as in Russia. These mushrooms act as traditional medicine to treat gastrointestinal cancer, cardiovascular disease and diabetes. Polyphenolic compounds produced by Chaga can protect cells against oxidative stress. It also showed to inhibit platelet adhesion and aggregation, which plays an important role in thrombosis. Those platelets are important in hemostasis and modulation of the inflammatory response, including the released of cytokines, Chaga may be involved in various aspects in the inflammatory. Levels of nitrite, which is an indicator of oxide nitric concentration, displayed a significant decline when treated with Chaga. The inflammatory effect caused by Chaga may be a cascade effect with inhibition of oxide nitric production.

Dudhgaonkar and researchers showed that triterpene extract from medicinal mushroom *Ganoderma lucidum* markedly suppressed in the inflammatory response in LPS-active murine macrophages. Specifically, triterpene by *Ganoderma lucidum* suppressed LPS-dependent secretion of TNF-α, IL-6, oxide nitric and prostaglandin E2 from murine macrophages cells. The inhibition of production of oxide nitric and prostaglandin E2 by *Ganoderma lucidum* triterpene was mediated through the down-regulation of expression of Inductible Nitric Oxide Synthase and **ciclooxigenase**-2, respectively. Moreover, this triterpene inhibited LPS-dependent induction of NF-κB as well as expression, phosphorylation and nuclear translocation of p65 NF-κB subunit. Also, triterpene of *Ganoderma lucidum* seem to be potent in suppressing the key molecules responsible in the inflammatory response. Extract of *Ganoderma lucidum* containing triterpenes or isolated triterpenes (ganoderic acid A, F, DM, T-Q, lucidenic acid A, D_2, E_2, P, methyllcidenate A, D_2, E_2 Q and 20-hydroxylucidenic acid N) suppressed ear-edema inflammation in laboratories animals.

Phellinus linteus, traditional mushroom medicine in oriental countries, showed topical anti-inflammatory activity. Extract ethanolic was evaluated using croton oil-induced ear edema test and showed an inhibitory effect on inflammation. Among the subfractions, the butanol fraction appeared to be most effective in anti-inflammation, supposing that *Phellinus linteus* have hydrophilic compounds [42].

Curiously, *Amanita muscaria* is not considerable edible mushroom, but has been used for various purposes, mostly as a psycostimulant, by different ethnic groups from Mexico to Siberia to Eastern Asia. Slavic nations have their own traditions of *Amanita muscaria* use. Ethnic people are especially fond of the beneficial effects they achieve from topical application of ethanolic extract (or strong vodka). Hot water, methanolic and ethanolic extracts of *Amanita muscaria* have description of their use for reduction of the consequences of inflammatory processes in cases of rheumatic diseases, body injuries, insect bites, others [69].

7.2.3 ANTIMICROBIAL ACTIVITY

Antibiotic agents have been effective therapeutic use since their discovery in the 20th century. However, it has paradoxically resulted in the emergence and dissemination of multi-drug and resistant pathogens. Antibiotic resistance represents a prospect of therapeutic failure for life-saving treatments.

The search for new drugs that is able to inhibit the antibiotic resistance of bacteria. Mushrooms interestingly showed antimicrobial activities, some examples are table 4. Biologist and others researches related that some mushrooms need special attention in natural environmental because of the relation with other species, growing local, conditions of temperature, substrates and others factors environmental.

TABLE 4   Potential antimicrobial of fruit body of mushrooms

Biological Activity: Antimicrobial		
Mushroom	Substance	References
Ganoderma japonicum	Oil essential	Liu et al., 2009
Ganoderma lucidum	Water extracts	Wu et al., 2006b
Ganoderma lucidum	Ganodermin	Wang and Ng, 2006
Laetiporus sulphureus	Ethanol extracts	Turkoglu et al., 2007
Lentinula edodes	Water extracts	Hearst et al., 2009
Lentinula edodes	Chloroform extract	Hirasawa et al., 1999
Leucopaxillus giganteus	Methanol	Barros et al., 2008b
Pleurotus sajor-caju	Ribonuclease	Ngai and Ng, 2004
Russula delica	Ethanol extracts	Yaltirak et al., 2009

TABLE 4 *(Continued)*

Russula paludosa	Lacase	Wang, Wang and Ng, 2007
Tricholoma giganteum	Trichogin	Guo, Wang and Ng, 2005

Laetiporus sulphureus (Bull.) Murrill is a wood-roting basidiomycete, growing on several tree species and producing shelf-shaped fruit-bodies of pink-orange colour, except for the fleshy margin which is bright yellow. *Laetiporus* species contain N-methylated tyramine derivatives, polysaccharides, a number of lanostane, triterpenoids, laertiporic acids and other metabolites have reported that laetiporic acids might have potential as food colorants. The antimicrobial effect of ethanol extracts of *L. sulphureus* was tested against six species of Gram-positive bacteria, seven species of Gram-negative bacteria and one species of yeast. The most susceptible bacterium was *Micrococcus flavus* (23 ± 1 mm diameter). The ethanol extract of *L. sulphureus* showed no antibacterial activity against *Klebsiella pneumonie* at the concentration used. The ethanol extract exhibited high anticandidal activity on *C. albicans* [104].

Russula delica Fr. is used as food in Turkey and growth under coniferous and deciduous trees. The antimicrobial effect of ethanolic extract of *R. delica* was tested against three species of Gram positive bacteria and six species of Gram negative bacteria. Results showed inhibitory activity against *Shigella sonnei* and *Yersinia enterocolitica*. Natural Antimicrobials agents are more safety to the people and low risk for resistance development by pathogenic microorganisms [108].

Hirasawa *et al.* studied the antibacterial activity of shiitake extracts (*Lentinula edodes*) as a preventive agent against dental caries and adult periodontitis. Shiitake extracts were antibacterial effective against *Streptococcus* spp., *Actinomyces* spp., *Lactobacillus* spp., *Prevotella* spp., and *Porphymonas* spp., of oral origin. This extract of mushroom can be used to prevent dental caries and periodontitis because also supports the idea that inhibit the formation of water-insoluble glucans from sucrose by glucosyltransferase.

Rao, Millar and Moore studied the activities of shiitake freeze-dried powder. Bioassay of the extracts showed that all the fractions exhibited qualitative inhibitory activity against bacteria and fungi. Thirty-four compounds from extracts of the shiitake was identified, for example: Cycloheximide (antibiotic that acts as a plant growth regulator, but causes human liver toxicity and reduction in protein synthesis); Bostrycoidin (bioactive *in vitro* against *Mycobacterium tuberculosis*); Anticarcinogênica alkaloids (muscarine, choline); Tanins (epiafzelechin); Terpenoids (adiantone); Cyclopiazonic acid (a natural food contaminant); Aspergillomarasmine; Disulphides, lenthionine compounds in the organic extracts.

Shiitake mushroom extract had extensive antimicrobial activity against 85% of the organisms it was tested on, including 50% of the yeast and mould species in the trial. This compared with the results from both the positive control [31] and Oyster mushroom, in terms of the number species inhibited by the activity of the metabolites inherent to the shiitake mushroom [120].

Mushroom proteins have important play antimicrobial activity. Mushroom compositions have contained 2–40% protein according with species. Each protein of mushroom has specific sequence of amino acids and weight molecular. Antifungal proteins have function of protecting organisms from the deleterious consequences of fungal assault; they display a spectacular diversity of structures. An antifungal peptide with a molecular of 9 KDa was isolated from fresh fruiting bodies of the mushroom *Agrocybe cylindracea*. The antifungal peptide, designated as agrocybin, exhibited remarkable homology to RPI 3, a cysteine-rich-protein that is expressed during fruiting initiation in *Agrocybe chaxingu*. Agrocybin is also similar to grape (*Vitis vinifera*) antifungal peptide in N-terminal sequence. The data suggest that antifungal function of agrocybin is important during fruiting initiation for protecting the fruiting bodies. Agrocybin inhibits mycelial growth in *Mycosphaerella arachidicola*, in line with the majority of fungal proteins and peptides that exert their antifungal action against a number fungal species and is not effective against a variety of bacteria [71].

Pleurotus nebrodensis produce hemolysin that can be implicated as a virulence factor. This hemolysin was named nebrodolysin and showed antiviral activity, inclusive exhibits a suppressive action on HIV-1 and reproducible antiviral effect. The mechanism of the antiviral suggested that nebrodeolysin might act in a different way by interaction the infection of the virus [65].

Antifungal protein from *Ganoderma lucidum*, ganodermin, inhibits mycelial growth in the phytopatogenic fungi *Botrytis cinerea*, *Fusarium oxysporum* and *Physalospora piricola* [117].

Niohshimeji (*Tricholoma giganteum*), produced trichogin, antifungal proteins monomeric and have N-terminal sequence. This protein showed antifungal activity against *Fusarium oxysporum*, *Mycosphoerella arachidicola* and *Physolospora piricola*. The antifungal activity of protein is high compared with others antifungal proteins. Trichogin inhibits HIV-1 reverse transcriptase [24].

Ngai and Ng demonstrated antimicrobial activity of ribonuclease from the extract of *Pleurotus sajor-caju*. The ribonuclease inhibited mycelial growth in the fungi *Fusarium oxysporum* and *Mycosphaerella arachidicola* and bacteria as *Pseudomonas aeruginosa* and *Staphylococcus aureus*. The molecular mass of *P. sajor-caju* RNase is 12 KDa and poly U-specific and high activity, compared with others mushrooms. The N-terminal sequence of *P. sajor-caju* RNase bears resemblance to the terminal sequence of a bacteriocin peptide and to a portion of the sequence in two enzymes involved in RNA-specific editase. This structural feature of *P. sajor-caju* RNase may be related to its antibacterial and RNase activities.

Tricholoma giganteum produce lacase and characterized with N-terminal sequence that dissimilar from reported N-terminal sequences of mushroom lacases. Its molecular mass (43 kDa) is smaller than most of the reported mushroom lacases which are around 60kDa [24].

Russula paludosa is a wild edible mushroom collected from Chine. Its fruiting bodies are abundant in the summer. Extracts from fruiting bodies of *R. paludosa* exhibited an inhibitory effect on HIV-1. The peptide was devoid of hemagglutinating, ribonuclease, antifungal, protease, protease inhibitor and lacase activities [72].

Ganoderma japonicum is found in China and has been used for the treatment of various diseases. The essential oil of *G. japonicum* has pharmacologicals effects and contains bactericidal components, such as (E)-nerolidol, linalool and (2E, 4E)-decadienal. The antimicrobial results indicated that oil inhibited mainly Methicillin-resistant

Staphylococcus aureus (MRSA). This component has been confirmed to have bacteriostatic and bactericidal activity, causing changes in cell membrane permeability and bacterial death [61].

7.2.4 ANTITUMORAL ACTIVITY

The National Cancer Institute (US National Institutes of Health) define Cancer is a term used for diseases in which abnormal cells divide without control and are able to invade other tissues. Cancer cells can spread to other parts of the body through the blood and lymph systems. There are more than 100 different types of cancer.

Cancer types can be grouped in main categories (Table 5):

TABLE 5   Classification of cancer types

Categories	Definition
Carcinoma	Begins in the skin or in tissues that line or cover internal organs
Sarcoma	Begins in bone, cartilage, fat, muscle, blood vessels, or other connective or supportive tissue
Leukemia	Starts in blood-forming tissue such as the bone marrow and causes large numbers of abnormal blood cells to be produced and enter the blood
Lymphoma and myeloma	Begin in the cells of the immune system
Central nervous system cancers	Begin in the tissues of the brain and spinal cord

Source: http://www.cancer.gov - September, 2009

According to World Health Organization related lung, stomach, liver, colon and breast cancer cause the most cancer deaths each year. About 30% of cancer deaths can be prevented. It is estimated that in 2030 there will be 26 million cases of cancer worldwide. Deaths from cancer worldwide are projected to continue rising, with an estimated 12 million deaths in 2030 (http://www.who.int/en/).

International Union against Cancer reported that Africa is less than 0.1% and Asia is only 8.5% of the population is covered by cancer registration. The Chernobyl disaster was a nuclear reactor accident that occurred on 26 April 1986 at the Chernobyl Nuclear Power Plant in Ukraine and it is now estimated that by 2065 there will be 16 000 cases of thyroid cancer and 28 000 cases of other cancers in Europe as a result of this accident (http://www.uicc.org/). Numerous mushroom species are studied and purified substances such as polysaccharides and proteo-polysaccharides are recognized to be the potent immunomodulatory and antitumor (Table 6).

TABLE 6 Potential antitumoral of mushrooms

Mushroom	Biological Activity: Antitumoral		
	Source	Substance	**References**
Agaricus brasiliensis	Submerged fermentation	Polysaccharides	Fan *et al.*, 2007
Agrocybe aegerita	Fruit body	Methanolic extracts	Diyabalanage *et al.*, 2008
Albatrellus confluens	-	Grifolin	Ye *et al.*, 2007
Cordyceps sinensis	Submerged fermentation	Polysaccharide	Yang *et al.*, 2005
Coriolus versicolor	Fruit body	Terpenoids and polyphenols	Harhaji *et al.*, 2008
Fomes fomentarius	Fruit body	Polysaccharides	Chen *et al.*, 2008
Ganoderma capense	Fruit body	Lectin	Ngai and Ng, 2004
Ganoderma lucidum	Solid fermentation	Polysaccharides	Rubel *et al.*, 2008
Ganoderma tsugae	Submerged fermentation	Polysaccharide	Peng *et al.*, 2005
Grifola frondosa	Submerged fermentation	Polysaccharides	Cui *et al.*, 2007
Inonotus obliquus	Submerged fermentation	Polysaccharide	Kim *et al.*, 2006
Lentinula edodes	Fruit body	Fiber	Choi *et al.*, 2006
Lentinula edodes	Fruit body	Polysaccharides	Frank *et al.*, 2006
Lentinula edodes	Fruit body	Ethanolic	Hatvani, 2001
Poria cocos	Submerged fermentation	Polysaccharides	Huang *et al.*, 2007
Pleurotus citrinopileatus	Fruit body	Lectin	Li *et al.*, 2008
Pleurotus ostreatus	Fruit body	Methanolic	Tsai *et al.*, 2008

Agaricus blazei Murril also called *Agaricus brasiliensis* is native to southern Brazil, popularly known as "Himematsuke" in Japan, or "Cogumelo do Sol" in Brazil. Consumption has increased in Brazil, Japan, Korea, Canada and United States because of its medicinal properties. Mechanism studies have demonstrated that antitumor activities of *A. blazei* extracts can be related to induction of apoptosis, cell-cycle arrest and inhibition of tumor-induced, neovascularization, immunopotentiation and restoration of tumor-suppressed host immune system.

The exopolysaccharide produced by *A. brasiliensis* showed strong inhibition against Sarcoma 180. The complete regression ratio was 50% and the suppression ratio percentage was 72.19%. Exopolysaccharide was characterized a mannan-protein complex, with its molecular weight being 10^5–10^7 by gel filtration and contained small amounts of glucose, galactose and ribulose [21].

Kim *et al.*, optimized the extraction of *Agaricus blazei* for isolation of bioactives components with antitumor effects. Extracts of mushroom was obtained with different polarities and solubilities. One fraction, extracted at 80°C using 70% water-ethanol (v/v) shoed tumor inhibitory activity against the human promyelotic leukemia cells *in vitro.*

However, has been reported fresh mushroom fed at high amounts can be carcinogenic in mice. Lee *et al.* showed dietary intake of *A. blazei* Murril fed at 6250, 12500 and 25000 ppm for two years of dry powder appears to enhance survival in males and not appear to be carcinogenic in rats.

Grifolin is a natural, biologically active substance isolated from fresh fruiting bodies of the mushroom *Albatrellus confluens.* Studies showed that grifolin is able to inhibit the growth of some cancer cell lines *in vitro* by induction of apoptosis. Ye *et al.* showed that grifolin inhibits the proliferation of nasopharyngeal carcinoma cell line through G1 phase arrest, which mediated by regulation of G1-related protein.

Cordyceps militaris is the best-known and most frequently collected bug-killing *Cordyceps*, but there are dozens of "entomogenous" species in North America [45].

Park *et al.* demonstrated that water extracts of *C. militaris* may increase mitochondrial dysfunction, and results in the activation of caspase-9, leading to the activation of caspase-3 target proteins. The caspase family proteins are known to be one of the key executioners of apoptosis. Water extracts of *C. militaris* induces apoptosis in human lung carcinoma cells.

Coriolus versicolor, also known as Yun-Zhi produce bioactive compounds. Hot water and ethanol extracts from *C. versicolor* demonstrated activities antitumoral to treatment of melanoma cells. According to analysis, the predominant compounds are terpenoids and polyphenols. The prevention of tumor growth was exerted through diverse mechanism including cell cycle suspend, induction of tumor cell death by apoptosis and secondary necrosis, together with stimulation of the anti-tumor activity of macrophages [28].

Peng *et al.* also studied *Coriolus versicolor* and showed inhibitory effect on the growth of Sarcoma 180 solid tumors. Anti-tumor activity of polysaccharopeptide resides in its anti-angiogenic properties, via a suppression of vascular endothelial growth factor gene expression, resulting in a deprivation of angiogenic stimulation to the tumor growth.

Ganoderma lucidum is known as "mushroom of immortality" because enhancing longevity. Researchers have also demonstrated that *G. lucidum* inhibits the migration of breast cancer cells and prostate cancer cells, suggesting is potency to reduce tumor invasiveness. Since integrins are the major cell surface adhesion molecules expressed by all cell types. Tests showed that incubation with *G. lucidum* polysaccharides reduced integrin expression. Integrins are composed of α and ß transmembrane subunits. Each α and ß combinations has its own binding specificity and signaling properties [76].

G. lucidum polysaccharides used dose-dependently treatment enhance catalase activity in the polysaccharides-treatment groups when compared to the ovarian cancer model [90].

Fruiting body of *Ganoderma tsugae* is used to promote health and longevity in Orientals countries. Peng *et al.*, related that results indicate that ß-D-galacto-α-D-mannan isolated from culture filtrate of *G. tsugae* mycelium also exhibited significant antitumor activities. In 2005, the same researchers demonstrated anti-tumor activities were observed in three polysaccharides fractions with inhibition ratio 50%.

Mushrooms have antitumoral properties is controversy, sometimes researchers describe collateral effects. Sadava *et al.* analyzed cytotoxicity of twelve species of *Ganoderma* and just four species showed non-cytotoxic effects. However, active *Ganoderma* extracts induced apoptosis.

Grifola frondosa, has been reported to posses many biologically active compounds. Especially, the antitumor and immune-stimulating activities of polysaccharide D-fraction, a branched ß-(1-6)-D-glucan isolated from the fruiting body. Most reports conformed that mushroom polysaccharides exerted their antitumor action via activation on the immune response of the host organism, and mushroom polysaccharides were regarded as biological response modifiers. Polysaccharides from different strains have different antitumor activities *in vitro*. The data suggest that the polysaccharides fractions from *G. frondosa* had selective antitumor activities on the different tumor cell lines [86].

Surenjav *et al.* (2006), studied lentinan, (1-3)-ß-D-glucan, a antitumor polysaccharide, has been isolated from the fruiting body of *Lentinus edodes*. The triple helical (1-3)-ß-D-glucan antitumor containing protein showed activities against the growth of Sarcoma 180. The triple helical conformation plays an important role in the enhancement of the antitumor activities. Data suggesting that the antitumor activity of polysaccharide is also related to their molecular weight and content of the bound protein.

Li *et al.* describe antitumoral activity of *Hedysarum polybotrys* Hand.-Maz (HP). In China, is used in the treatment of diseases, such as cancer, glycemy and immunomodulatory, anti-aging, anti-oxidation activities. The α-(1→4) - D-glucan showed that inhibit proliferation of human hepatocellular carcinoma and human gastric cancer.

Inonotus obliquus is a white rot fungus, called Chaga, is a medicinal mushroom that has been used in Siberian and Russia folk medicine to treat stomach discomforts. Extracts of *I. obliquus* are known to inhibit the growth and protein synthesis of tumor cells. Alpha-linked-fucoglucomannan isolated from cultivated mycelia of *I. obliquus* can inhibit tumor growth *in vivo*. The endopolysaccharide-mediated inhibition of tumor growth is apparently caused by an induced humoral immunity of the host defense system rather than by a direct cytotoxic effect against tumor cells (Kim *et al.*, 2006).

Huang *et al.*, studied *Poria cocos* called Fu Ling, it is collected between July and September in China. Polysaccharides fractions was tested and showed strong inhibition against leukemia cell proliferation at all concentrations. The three water-soluble fractions presented significantly high inhibition ratio of more 80% at concentration of 200μg/mL.

Phellinus ignarius, an orange color mushroom is used to improve health and remedy various diseases, such as gastroenteric disorders, lymphatic diseases and cancer. Extracts from fruiting body of *Phellinus igniarus* inhibited the proliferation of SK-Hep 1 cells and RHE cells in a concentration-dependent manner, with IC_{50} values of 72 and 103 μg/mL [110].

Pleurotus is important mushroom because have importance gastronomic, nutritional, commercially and medicinal properties. *Pleurotus citrinopileatus* is a widely used edible mushroom, delicious taste and rich in nutrients. Antitumor activities of lecitin from *P. citrinopileatus* are similar to those *Pleurotus ostreatus* lecitin. *P. citrinopileatus* lecitin is dimeric, like lecitins from mushrooms [41]. Sarangi *et al.* (2006) demonstrated that two fractions of *Pleurotus ostreatus* can directly kill Sarcoma 180 cells *in vitro*. Cell-cell adhesion determines the polarity of cells and participates in the maintenance of the cell societies called tissues. Adhesion is generally reduced in human cancer cells. Reduce intercellular adhesion allows cancer cells to disobey the social order, resulting in the destruction of histological structure, which is the morphological hallmark of malignant tumors. Reduced intercellular adhesiveness is also indispensable for cancer invasion and metastasis. Tong *et al.* (2009) observed also antitumor activity against HeLa tumor cell *in vitro*, in a dose-dependent manner and exhibited lower cytotoxicity to human embryo kidney cells.

Leucopaxillus giganteus is enormous mushroom is often found growing in large fairy rings or arcs in woodland clearings. It is apparently widely distributed, but most common in the Pacific Northwest and Rocky Mountains [35].

Ren *et al.* demonstrated that clitocine isolated from *L. giganteus* have proliferation inhibitory activity against HeLa cells in a dose-dependent manner by mechanism involved the induction of apoptosis.

7.2.5 IMMUNOMODULADOR ACTIVITY

Molecules of macrofungi and secondary metabolites are known as bioactive compounds that belong to polysaccharides, glycoproteins, involved in the innate and adaptive immunity, resulting in the production of cytokines. The therapeutic effects of these compounds such as antitumor and anti-infective activity and suppression of autoimmune diseases have been associated in many cases with their immunomodulating effects (Table 7).

Compounds that are capable of interacting with the immune system to up regulate or down regulate specific aspects of the host response can be classified as immunomodulators or biologic response modifiers. These agents can be applied in treating and preventing diseases and illnesses and with regard to the increase in diseases involving immune dysfunction, such as compounds with remedy potential without side effects, pathogenic resistance or affecting normal cell division.

Innate immunity serves as an essential first line of defense against microbial pathogens and may also influence the nature of the subsequent adaptive immune response. Phagocytic cells, such as macrophages and neutrophils, play a key role in innate immunity because of their ability to recognize, ingest, and destroy many pathogens by oxidative and non-oxidative mechanisms. Bioactive polysaccharides and polysaccharide-protein complexes have been isolated from mushrooms (Table 7), yeast, algae, lichens and plants, and these compounds have attracted significant attention because of their immunomodulatory and antitumor effects (Xie, Schepetkin and Quinn, 2007).

Mushroom polymers have immunotherapeutic properties by facilitating growth inhibition and destruction of tumors cells. Fungal ß-glucans-induced immune responses are different in their actions in immune therapies based on supplementation of elements of the immune system and stimulating the immune system, can be option in treatment of diseases. This compounds are not synthesized by humans and inducing both innate and adaptive immune response [27].

Proteoglycans and polysaccharide have high molecular mass, constituent of β-glucans, cannot penetrate cells, so the first step in the modulation of cellular activity is binding to immune cell receptors. The mechanism by which the innate immune system recognizes and responds to compounds of mushroom is complex and multifactorial process. After this activation and signaling is humoral- and cell-mediate immunity induction.

TABLE 7   Potential immunomodulatory activity

	Immunomodulator Activity		
	Source	Extraction/compound	References
Agaricus blazei	Fruit body	Water soluble compounds	Kasai et al., 2004
Agaricus blazei	Fruit body/mycelium	Biocompounds	Shimizu et al., 2002
Coriolus versicolor	Submerged fermentation	Polysaccharide	Lee et al., 2006
Ganoderma lucidum	-	Fractions	Ji et al., 2007
Ganoderma lucidum	-	Polysaccharide	Zhu et al., 2007
Grifola frondosa	Submerged fermentation	Polysaccharide	Yang et al., 2007
Grifola frondosa	Submerged fermentation	Polysaccharide	Wu et al., 2006
Inonotus obliquus	Submerged fermentation	Polysaccharide	Kim et al., 2006
Lentinula edodes	Fruit body	Polysaccharide	Gu and Belury, 2005
Lentinula edodes	Fruit body	Polysaccharide	Zheng et al., 2005

Agaricus blazei is a medicinal mushroom originating from Brazilian subtropical regions and is produced on an industrial scale in some countries such as China, Japan and Brazil [59].

Extracts of *Agaricus blazei* from the fruiting body and the mycelium were effective in activation of the human complement pathway. Both bioactive compounds have been demonstrated to be potent activators of the complement system in human serum in a dose and time dependent manner [95].

Kasai *et al.* analyzed *Agaricus blazei* fraction induced expression of IL-12, a cytokine known to be a critical regulator of cellular immune responses. According to Kimura *et al.* (2006), supplementation of *A. blazei* was effective in the activation of enzymes related to energetic metabolism in leukocytes of calves. *A. blazei* extract was water-soluble and easy to deal with as a food additive.

Lentinan, schizophyllan and krestin have been accepted as immunoceuticals in several oriental countries. Increase of natural killer cells, cytotoxic T lymphocytes and delayed type hypersensitivity responses against tumor antigen were observed after administration of lentinan. *Lentinula edodes* was claimed to have a range of health benefits. The concentration of TNF-α, IFN-$\square$ in serum increased significantly in the polysaccharide groups compared with the model control group. Also, polysaccharide of *L. edodes* increased oxide nitric production in peritoneal macrophages and catalase activity of macrophage (Zheng *et al.*, 2005).

Coriolus versicolor, known as Yun Zhi, has been used in China for treatment of cancer and immunodeficient. Polysaccharopeptides produced for *C. versicolor* can stimulate cytokines production but also demonstrated that a critical time for culture harvesting is essential for obtaining optimal bioactivities of the fungi (Lee *et al.*, 2006).

Ganoderma lucidum is a Chinese medicinal fungus, which has been clinically used in East Asia and is given considerable attention for treatment for various diseases. Medicinal functions have been assigned to crude extracts and isolated components of *G. lucidum*. The potential immunomodulating activity is the capacity of a particular substance to influence specific immune functions such as activating individual components of the immune system and promoting cytokine synthesis. Ji *et al.* (2007) showed that extracts of *G. lucidum* activated mouse macrophages in a dose dependent manner, increased the levels of IL-1, IL-12p35 and IL-12p40 gene expression, and significantly enhanced oxide nitric production. These immunomodulatory functions suggest that *G. lucidum* may also interfere with the growth of certain tumors.

Bao *et al.* (2001) demonstrated that (1-3)-D-Glucans of *Ganoderma lucidum* have immunomodulating and antitumor activities. The structural and physicochemical properties and lymphocyte proliferation activity of all samples varied with the functionalized groups and the degree of substitution. The results of immunological assays indicated that some modified derivatives had stimulating effects on lymphocyte proliferation and antibody production and the introduction of carboxymethyl group with low degree of substitution was the best choice on the improvement of the immunostimulating activity.

Ganoderma lucidum mycelia stimulated moderate levels of TNF, IL-6 and IFN-release in human whole blood and moderately stimulate cytokine production without potentiating oxide nitric release. The ineffectiveness in inducing oxide nitric release by *G. lucidum* mycelia indicates that the compositions and structures of glucan in mycelia and fruiting body may be different, and this might result in enhancing innate immune response through different receptors or pathways [19].

Ganoderma lucidum polysaccharides enhanced the activity of immunological effectors cells in immunosuppressed mice and promoted phagocytosis and cytotoxicity of macrophages. The above beneficial effects induced by the low-dose of polysaccharide treatment did not result in any side effects [119].

Polysaccharides obtained from fermented and fruiting body of *Grifola frondosa* have demonstrated many interesting biological activities. Exopolymers fractions of *G. frondosa* can be enhancers of innate response and considered as potent materials for immune system (Shi *et. al.*, 2008; Yang *et al.*, 2007).

Grifolan, polysaccharide of *Grifola frondosa,* showed that hot water-soluble fractions (polysaccharides) of mycelia from submerged fermentation can effectively induce innate immunity and therefore enhance pro-inflammatory cytokine release, phagocytosis, and Natural Killer cytotoxicity activity *in vitro* (Wu *et al.*, 2006).

Inonotus obliquus is a white rot fungus widely distributed over Europe, Asia, and North America. The polysaccharide yield of species of *Inonotus* increased proportionally with an increasing cell mass of the fermentation broth. However, the immunostimulating activities were not proportional to the corresponding polysaccharide yield. High specific activities of endopolysaccharide were obtained during both the late lag and the late stationary phases, but not during the active cell growth phase, indicating that the polysaccharide activity is probably closely related to cell age. During the late lag phase, low total activities were obtained due to low cell masses in spite of high specific activities. The specific activity of endopolysaccharide at late lag phase appeared to be highly similar to the activity at late stationary phase. The endopolysaccharide of *I. obliquus* showed much higher splenic cell activities than the corresponding exopolysaccharide.

7.2.6 ANTI-ANGIOGENIC ACTIVITY

Angiogenesis can be characterized as an integrate set of cellular, biochemical and molecular processes in which new blood vessels are formed from pre-existing vessels. This occurs physiologically during reproductive and developmental processes as well as during the late phases of wound healing following tissue damage [15]. Blood vessels run through every organ in the body (except cornea and cartilage), assuring metabolic homeostasis by supplying oxygen and nutrients and removing waste products. Therefore, angiogenesis is known to be essential in several physiologic processes, such as organ growth and development, wound healing and post-ischemic tissue repair. However, inappropriate or aberrant angiogenesis contributes to the development and progression of various pathological conditions including tumor growth and metastasis, diabetic retinopathy, cardiovascular diseases, inflammatory disease and psoriasis. Angiogenesis can be separated into several main steps, such as degradation of the basement membrane of exiting blood vessels, migration, proliferation and rearrangement of endothelial cells, and formation of new blood vessels (Makrilia *et al.*, 2009; Ramjaum and Hodivala-Dilke, 2009; Ribatti, 2009).

This switch clearly involves more than simple up-regulation of angiogenic activity and is known to be the result of net balance between positive and negative regulators. There are three particularly important stimulators of angiogenesis: i) vascular endothelial growth factor ii) fibroblast growth factor; iii) angiopoetins; between many others, like platelet derived growth factor, epidermal growth factor, ephrins; transforming growth factors alpha and beta, interleukins, chemokines, and small molecules, such as sphingosine 1-phosphate, that are known to promote cell proliferation, survival and differentiation of endothelial cell (Duarte, Longatto Filho and Schmitt, 2007; Jung *et al.*, 2008; Stupack, Storgard and Cheresh, 1999).

Anti-angiogenesis strategies are based on inhibition of endothelial cell proliferation, interference with endothelial cell adhesion and migration, and interference with metalloproteases. Down-regulation of angiogenesis has been considered to be advantageous for the prevention of tumors [6,37,97].

TABLE 8   Anti-angiogenesis Activity of mushrooms

Mushroom	Anti-angiogenesis Activity		
	Source	Extraction/compound	References
Antrodia cinnamomea	Mycelia	Hot water	Yang et al., 2009
Ganoderma tsugae	Fruit body	Methanol extracts	Hsu et al., 2009
Grifola frondosa	Fruit body	Water extracts	Lee et al, 2008b
Fomitopsis pinicola	Fruit body	Polysaccharides	Cheng et al., 2008
Phellinus linteus	Fruit body	Ethanolic extract	Song et al., 2003

In Taiwan, the mushroom *Antrodia cinnamomea* is known as "niu-cha-ku" or "chang-chih" and produced triterpene acids, steroid acids and polysaccharides with biological activities. The fraction >100kDa of polysaccharide from *A. cinnamomea* showed potential anti-angiogenic *in vivo* and *ex vivo* indirectly by immunomodulation (Table 8) [109].

Ganoderma sp contains numerous bioactive natural components and the two categories of those are polysaccharides and tripernoids, both of them are potent inhibitors of *in vitro* and *in vivo* tumor growth. The epidermal growth factor receptor activation is often linked with angiogenesis. Methanol extracts of *Ganoderma tsugae* showed antiangiogenic effects on the cancer cells by the downregulation of vascular endothelial growth factor [32].

Fomitopsis pinicola is marketed as a tea and food supplement. Chemical compounds found in *F. pinicola* include steroids, sesquiterpenes, lanostane tripertenoids and triterpene glycosides. Ethanolic extracts and polysaccharides of *F. pinicola* were effective for the anti-angiogenesis at 10 µg/mL concentration and showed no toxicity up to concentration of 1 mg/mL [11].

Grifola frondosa demonstrated anti-angiogenic activity by inhibit vascular endothelial growth factor induced angiogenesis *in vivo* and *in vitro*. Vascular endothelial growth factor is the most angiogenic factor associated with inflammatory diseases and cancer [48]. Ethanolic extracts from *Phellinus linteus* presented anti-angiogenic activity and can be used as adjuvant chemotherapy for the treatment of cancer [97].

7.2.7   HYPOGLYCEMIC ACTIVITY

The World Health Organization describes diabetes as a chronic condition that occurs when the pancreas does not produce enough insulin or when the body cannot effectively use the insulin it produced. Hyperglycemia and other related disturbances in the body's metabolism can lead to serious damage to many body's systems, especially to the nerves and blood vessels. Diabetes is a life threatening condition. World Health Organization indicates that worldwide almost 3 million deaths per year are attributable to diabetes.

Roglic *et al.* (2005) describe that excess mortality attributable to diabetes accounted for 2–3% of deaths in poorest countries and over 8% in the U.S., Canada, and the Middle East. In people 35–64 years old, 6–27% of deaths were attributable to diabetes. Functional foods and naturals compounds have become a popular approach to prevent occurrence of diabetes mellitus. Several mushroom species have been described to have anti-diabetic properties, because were found compounds as fibers source, polysaccharides and other biological activities (Table 9).

Lima *et al.* (2008) the diet supplemented with exopolysaccharides of *Agaricus brasiliensis* provided during 8 weeks to the mice produced a reduction in glucose plasma concentration around 22% [59]. It has been also demonstrated in another study that the supplementation with β-glucans and oligosaccharides obtained from the fruiting body of *A. brasiliensis* caused reduction in the glucose serum concentration in rats. Mushroom could have an anti-diabetic activity by promoting the insulin release by the Langerhans cells in the pancreas. The total cholesterol ratio in the mice was reduced around 27% with *A. brasiliensis* exopolysaccharide supplementation. Fruiting body biomass and the mushroom polysaccharides have significant anti-hyperglycemic activity and the abilities to increase glucose metabolism and insulin secretion in type 2 diabetes mellitus. However, the mechanisms of action for the exopolysaccharide of *A. brasiliensis* on cholesterol and glucose metabolism are still unknown. The effect of *A. brasiliensis* on the enzyme activities was considered to be sustained for at least 2 to 3 months after the supplementation in the calves, although the absorption of *A. brasiliensis* was considered satisfactory.

Vanadium compounds have the ability to imitate action of insulin. Oral administration of inorganic vanadium salts, have shown anti-diabetic activity *in vitro, in vivo* and even in patients. Vanadium at lower doses (0.18 mg/kg/d) was absorbed by fermented mushroom of *Coprinus comatus*, which is one rare edible fungus that is able to absorb and accumulate trace elements. *C. comatus* is a mushroom claimed to benefit glycemic control in diabetes and others properties. *C. comatus*, on a dry weight basis, contains, on the average, 58.8% carbohydrate, 25.4% protein and 3.3% fats, with the rest constituted of minerals. It indicates that *C. comatus* could supplement nutrients to the mice as well as lower blood glucose of hyperglycemic mice. *C. comatus* has the ability to take up and accumulate trace metals. However, the toxicity associated with vanadium limits is therapeutic efficacy [27].

Tremella mesenterica contains up 20% of polysaccharide glucuronoxylomannan in the fruiting bodies, is a popular, edible and medicinal mushroom in orient. Fruiting bodies, submerged fermentation and the acid polysaccharide of *T. mesenterica* have significant anti-hyperglycemic activity. Consumption of *T. mesenterica* by rats had significantly improved short-and long-term glycemic responses, as evidence by significantly decreased blood glucose concentrations in oral glucose tolerance test and serum fructosamine concentrations, respectively. This mushroom has potential anti-hyperglicemic functional food or nutraceuticals for diabetes with daily ingestion of 1 g/Kg of fruiting bodies, biomass and glucuronoxylomannan. Also, *T. mesenterica* has the ability to increase the insulin sensitivity, instead of increasing the insulin secretion in normal rats. Consumption of *T. mesenterica* may act functional food in improving the short and long-term glycemic control in persons with high risk of diabetes or type 2 diabetes mellitus, not in persons with type 1 diabete mellitus [63].

Phellinus baumii is a mushroom used as a folk medicine for a variety of human diseases in several Asian countries. The plasma glucose level in the exopolisaccharide-fed rats was substantially reduced by 52.3% as compared to the diabetic rats, which is the highest hypoglycemic effect among mushroom-derived. The activities of alanine aminotransferase and asparate aminotransferase were significantly decreased by administration of *P. baumii* exopolysaccharide, thereby exhibiting a remedial role in liver function. *P. baumii* of exopolysaccharide administration led to the diabetogenic effect and significantly reduced the degree of diabetes. Oral administration of *P. baumii* of exopolysaccharide may have a potential benefit in preventing diabetes, since pancreatic damage induced by environmental chemicals and other factors is a cause of diabetes [34].

Exopolysaccharides of *Ganoderma lucidum* (Lingzhi) have hypoglycemic effects. Studies showed that treatment with water extract of *Ganoderma lucidum* for 4-week oral gavages, 0.3 g/kg for consumption, lowered the plasma glucose level in mice. Phosphoenolpyruvate carboxykinase is a hepatic enzyme which is important in the regulation of gluconeogenesis. Inhibition of the hepatic phosphoenolpyruvate carboxykinase reduced blood glucose levels and improved glucose tolerance together with a decreased circulating free fatty acid and triacylglycerol levels in the diabetic mice. *G. lucidum* consumption caused a marked suppression of the hepatic phosphoenolpyruvate carboxykinase gene expression with a concomitant reduction of the serum glucose levels in mice [89].

Cordyceps, one of the most valued traditional Chinese medicines, consists of the dried fungus *Cordyceps sinensis* growing on caterpillar. Polysaccharide from *Cordyceps* protects against free radical induced neuronal cell toxicity. Polysaccharides of *Cordyceps* produced a drop in blood glucose level in both normal and diabetic animals, at doses of higher than 200 mg/kg body wt. Hypoglycemic effect is possibly because of the increase in blood insulin level, which may be due to the induced insulin release from the residual pancreatic cells and/or reduced insulin metabolism in body by polysaccharide [54,57].

Mechanisms which contribute to the formation of free radicals in diabetes include non-enzymatic and auto-oxidative glycosylation, metabolic stress resulting from changes in energy metabolism, levels of inflammatory mediators, and the status of antioxidant defense. Selective damage of islet cells in the pancreas may be one of the pathological mechanisms for Type I diabetes. Antioxidants could prevent the development of diabetes. Polysaccharide also has a strong antioxidant property which can protect cultured rat pheochromocytoma PC12 cells from being damaged by hydrogen peroxide. This antioxidant activity of polysaccharide may also play a protective role in the development of diabetes [57].

Kwon *et al.* (2009), described consequences caused for diabetes, such as difficult proliferation of cells, decreased collagen production, decreased chemotaxis and phagocytosis, reduction in the levels of growth factors, and the inhibition of fibroblast proliferation. They tested *Sparassis crispa*, an edible medicinal mushroom consumed in China and Japan. In experiments, the diabetes was induced and was accompanied by diabetic symptoms such as weight loss, polyuria, hyperglycemia, and neuroendocrine dysfunction. The oral administration of *S. crispa* increased migration of macrophages and fibroblasts, collagen regeneration, and epithelialization under hyperglycemic conditions.

7.2.8 ANTI-HYPERTENSIVE ACTIVITY

According to World Health Organization (WHO), cardiovascular diseases include coronary heart disease (heart attacks), cerebrovascular disease, raised blood pressure (hypertension), peripheral artery disease, rheumatic heart disease, congenital heart disease and heart failure. The major causes of cardiovascular disease are tobacco use, physical inactivity, and an unhealthy diet. Hypertension management and risk prediction based on diastolic blood pressure may be reasonably valuable for younger people and people with essential hypertension. The use of diastolic blood pressure as main treatment has been supported by the discovery that essential hypertension is characterized by increased peripheral vascular resistance and raised mean arterial pressure, which more closely correlates with diastolic blood pressure than with systolic blood pressure [74].

Angiotensin, an oligopeptide in the blood, causes vasoconstriction and increased blood pressure. It is part of the renin-angiotensin system, which is a major target for drugs that lower blood pressure. Angiotensin II causes the muscles surrounding blood vessels to contract, thereby narrowing the vessels. The narrowing of the vessels increases the pressure within the vessels causing high blood pressure (hypertension). Angiotensin II is formed from angiotensin I in the blood by the enzyme angiotensin converting enzyme. Angiotensin I converting enzyme is potentially of great importance for controlling blood pressure in the rennin-angiotensin system. The rennin-angiotensin system plays an important role in interrelated set of mechanism for the control of the volume, pressure and electrolyte composition of blood. Angiotensin I converting enzyme converts the inactive decapeptide angiotensin I to potent vessel pressure octopepetide angiotensin II, the main active component [26,60].

Hatakeshimeji mushroom (*Lyophyllum decastes* Sing.), which have a delicious taste, forms a family of Honshimeji (*Lyophyllum shimeji*) and belongs to a highly related genus of *Lyophyllum*. Hot water-extracts of Hatakeshimeji inhibited angiotensin converting enzyme activity and that increase of systolic blood pressure was inhibited by feeding dry powdered hot-water extracts of Hatakeshimeji fruit body to spontaneously hypertensive rats (Kokean *et al.*, 2005). On other hand, Kokean *et al.* (2005) studied the influence of cooking and processing. Suitable cooking method, without causing the destruction of both physiological effects and the taste, is important. It deep-frying did not adversely affect the taste, the texture and the antihypertensive property of Hatakeshimeji (Table 9).

Tamogi-take (*Pleurotus cornupiae*) is an edible mushroom that belongs to Hiratake family and grows on standing and fallen elm trees on the Siberian peninsula of Russia and in the eastern and northern parts of Hokkaido, Japan. This mushroom produced D-mannitol, which inhibits angiotensin converting enzyme activities and lowers the blood pressure of spontaneously hypertensive rats. The compound was identified as D-mannitol by direct comparison of spectral data with authentic compound. The inhibition of Angiotensin I converting enzyme is dose-dependent by various sugars. D-mannitol, D-sorbitol and D-dulcitol are classified as sugar alcohols (IC_{50}:16.4 mM) and were the most effective inhibitor of Angiotensin I converting enzyme. Although sugar alcohol might prevent an increased in blood pressure by mechanism such as osmotic diuretic effect, more studies are necessary to explain the mechanism *in vivo* [44].

Pleurotus nebrodensis is native from Southern Europe, Central Asia and China, and have been shown to prevent hypertension. Compounds of *P. nebrodensis* were administered orally, and antihypertensive actions were measuring blood pressure. Two hours after administration, the blood pressure decreased around 85% and increased gradually after 6 until 48 hours pos-administration [68].

Marasmius androsaceus, a traditional chinese mushroom, is usually used in tendon relaxation, pain alleviation and anti-hypertension. The 3,3,5,5 Tetramethyl-4-piperidone is an active compound prepared from *Marasmius androsaceus* but, the action unclear. Study showed that 3,3,5,5 Tetramethyl-4-piperidone have effects reducing blood pressure in anesthetic tested in rats and cats. It can inhibit the automatic rhythmic contraction of ileum section in guinea pig. It also can inhibit the concentration of rabbit aorta smooth muscle caused by adrenalin. 3,3,5,5 Tetramethyl-4-piperidone has a simple structure with low molecular weight, which is suitable to serve as leading compound. This is an antihypertensive compound, and the effect is partially related to ganglionic blocking [117].

TABLE 9   Potential biological of Mushrooms

	Activity Hypocholesterolemic		
Auricularia auricular	Fruit body	Powder	Cheung, 1996
Pleurotus ostreatus	Mycelia	Proteo-glucan	Sarangi et al., 2006
Pleurotus ostreatus	Fruit body	Dried Fruit body	Bobek et al., 1998
Tremella fuciformes	Fruit body	Powder	Cheung, 1996
	Activity Hypoglycemic		
Cordyceps sinensis	Fruit body	Polysaccharide	Li et al., 2006
Coprinus comatus	Submerged fermentation	Vanadium	Han et al., 2006
Marasmius androsaceus	Mycelia	Ethanol extracts	Zhang et al., 2009
Phellinus baumii	Submerged fermentation	Polysaccharide	Hwang et al., 2005
Sparassis crispa	Fruit body	Powder	Kwon et al., 2009

Tremella mesenterica	Submerged fermen-tation and fruit body	Fruit body Water extracts	Lo et al, 2006
	Activity Anti-hypertensive		
Lyophyllum decastes (Sing.)	Fruit body	Powder and water extracts	Korean et al., 2005
Pleurotus nebrodensis	Fruit body	Polysaccharide and water extracts	Miyazawa, Okazaki and Ohga, 2008
	Biosynthesis of collagen		
Grifola frondosa	Submerged Fermen-tation	Polysaccharides	Lee et al., 2003
	Activity Mitogenic		
Pleurotus citrinopileatus	Fruit body	Lectin	Li et al., 2008
	Activity Prebiotic		
Pleurotus ostreatus	Fruit body	Glucans	Synytsya et al., 2009

7.2.9   PREBIOTICS

Prebiotic is as "selectively fermented ingredients that allow specific changes, both in the composition and/or activity in the gastrointestinal microbiota that confers benefits upon host well-being and health". The effect of a prebiotic is indirect because it selectively feeds one or a limited number of microorganisms thus causing a selective modification of the host's intestinal microflora. Intestinal bacteria leads towards a consideration of factors that may influence the flora composition in a manner than can impact upon health. The criteria for consider prebiotic are: resistance to the upper gut tract, fermentation by intestinal microbiota, beneficial to the host health, selective stimulation of probiotics, stability to food processing treatments [106].

Ingestion of prebiotic was believed to enhance immune function, improve colonic integrity, decrease incidence and duration of intestinal infections, down-regulated allergic response as well as improve digestion and elimination of faeces [2].

Mushrooms are consumed as a delicacy, and particularly for their specific aroma and texture. Digestibility and bioavailability of mushroom constituents have been missing from the knowledge of mushroom nutritional value. The dry matter of mushroom fruit bodies is about 5–15%, they have a low fat content and contain 19-35% proteins. The content of carbohydrates, which are mainly present as polysaccharides or glycoproteins, ranges 50–90%. Most abundant mushroom polysaccharides are chitin, hemicelluloses, ß- and α-glucans, mannans, xylans and galactans. Mushroom polysaccharides are present as linear and branched glucans with different types of glycosidic linkages, such as (1-3), (1-6) - ß-glucans and (1-3) α-glucans, but some are heteroglicans containing glucorinic acid, xylose, galactose, mannose, arabinose or ribose. Like polysaccharides originated from other food products, they contribute to the digestion process as soluble or insoluble dietary fibers depending on their molecular structure and conformation. ß-glucans are recognized as immunological activators and are effective in treating diseases like cancer, diabetes, hypercholesterolemia and others.

Mushroom seem to be a potential in probiotics candidate for prebiotics as it contains carbohydrates like chitin, hemicelulose, ß- and α-glucans, mannans, xylans and galactans (Table 9). Chitin is a water-insoluble structural polysaccharide, accounting for up to 80–90% of dry matter in mushroom cell walls. A high proportion of indigestible chitin apparently limits availability of other components [2,38,100].

Synytsya *et al.* studied extract of *Pleurotus ostreatus* and *Pleurotus eryngii* with potential prebiotic activity. Specific soluble glucans were isolated from mushrooms by boiling water and alkali extraction. Potential prebiotic activity of extracts aqueous and alkali extracts was testing using nine probiotic strains of *Lactobacillus*, *Bifidobacterium* and *Enterococcus*. The utilization of both extracts by different manner affirms different chemical structure of polysaccharides, such as prebiotics. Extracts of *Pleurotus* can be used for symbiotic construction with select probiotic strains.

7.3   CONCLUSION

Mushrooms are traditionally used in oriental countries like food and several diseases treatment. The science widens opportunity of progress in the development of new diseases treatments, by discovering the natural compounds.

KEYWORDS

- **Edible Mushrooms**
- ***Ganoderma Lucidum***
- **Bioactive Substances**
- **Antibiotic Resistance**
- **Nutraceutical**

REFERENCES

1. Agrahar-Murugkar, D. and Subbulakshmi, G. Nutritional value of edible wild mushrooms collected from the Khasi hills of Meghalaya. *Food Chemistry*, **89**, 599–603 (2005).
2. Aida, F. M. N. A., Shuhaimi, M., Yazid, M., and Maaruf, A. G. Mushroom as a potential source of prebiotics: A review. *Trends in Food Science & Technology*, (2009), doi: 10,1016/j.tifs.2009.07.007.
3. Bao, X., Duan, J., Fang, X., and Fang, J. Chemical modifications of the (1–3)-α-D-glucan from sporos of *Ganoderma lucidum* and investigation of their physicochemical properties and immunological activity. *Carbohydrate Research*, **336**, 127–140 (2001).
4. Barros, L., Falcão, S., Baptista, P., Freire, C. Vilas-Boas, M., and Ferreira, I. C. F. R. Antioxidant activity of *Agaricus* sp Mushroom by chemical, biochemical and eletrochemical assays. *Food Chemistry*, **111**, 61–66 (2008).

5. Barros, L., Cruz, T., Baptista, P., Estevinho, L. M., and Ferreira, I. C. F. R. Wild and commercial mushrooms as source of nutrients and nutraceuticals. *Food and Chemical Toxicology*, **46**, 2742–2747 (2008).

6. Bhat, T. A. and Singh, R. P. Tumor angiogenesis – A potential target in cancer chemoprevention. *Food and Chemical Toxicology*, **46**, 1334–1345 (2008).

7. Barros , L., Baptista, P., Estevinho, L. M.,and Ferreira, I. C. F. R. Bioactive properties of the medicinal mushroom *Leucopaxillus giganteus* mycelium obtained in the presence of different nitrogen sources. *Food Chemistry*, **105**, 179–186 (2007).

8. Bobek, P., Lubomir, O., and Galbavy, S. Dose and Time dependent hypocholesterolemic effect of Oyster mushroom (*Pleurotus ostreatus*) in rats. *Basic Nutritional Investigation*, **14**(3), 282–286 (1998).

9. Carbonero, E. R., Gracher, A. H. P., Smiderle, F. R.; Rosado, F. R.; Sassaki, G. L.; Gorin, P. A. J.; and Iacomini, M. A b-glucan from the fruit bodies of edible mushrooms *Pleurotus eryngii* and *Pleurotus ostreatoroseus*. *Carbohydrate Polymers*, **66**, 252–257 (2006).

10. Chattopadhyay, N., Ghosh, T., Sinka, S., Chattopadhyay, K., Karmakar, P., and Ray,B. Polysaccharides from *Turbinaria conoides*: Structural features and antioxidant capacity. *Food Chemistry*, doi: 10.1016/jfoodchem.200905.069 (2009).

11. Chen, J. and Seviour, R. Medicinal importance of fungal ß-(1→3), (1→6)-glucans. *Mycological Research*, **111**, 635–652 (2007).

12. Chen, W., Zhao, Z., Chen, S-F., and LI, Y. Q. Optimization for the productionof exopolysaccharide from *Fomes fomentarius* in submerged culture and its antitumor effect *in vitro*. *Biosource Technology*, **99**, 3187–3194 (2008).

13. Cheng, J. J., Lin, C. Y., LUR, H. S., Chen, H. P.,and LU, M. K. Properties and biological functions of polysaccharides and ethanolic extracts isolated from medicinal fungus, *Fomitopsis pinicola*. *Process Biochemistry*, **43**, 829–834 (2008).

14. Cheung, P. C. K. The hypocholesterolemic effect of two edible mushrooms: *Auricularia auricula* (Tree-ear) and *Tremella fuciformes* (White Jelly-leaf) in hypercholesterolemic rats. *National Research*, **6**(10), 1721–1725 (1996).

15. Choi, Y.; Lee, S.M.; Chun, J.; Lee, H.B., and Lee, J. Influence of heat treatment on the antioxidant activities and polyphenolic compounds of Shiitake (*Lentinus edodes*) mushroom. *Food Chemistry*, **99**, 381–387 (2006),

16. Contois, L., Akalu A., Brooks, P. C. Integrins as "functional hubs" in the regulation of pathological angiogenesis. *Seminars in Cancer Biology*, (2009), doi:10.1016/j.semcancer.2009.05.002.

17. Cui, F. J., Tao, W. Y., Xu, Z. H., Guo, W. J., XU, H. Y., Ao, Z. H., Jin, J., and Wei, Y. Q. Structural analysis of anti-tumor heteropolysaccharide GFPS1b from the cultured mycelia of *Grifola frondosa* GF9801. *Bioresource Technology*, (2007).

18. Dore, C. M. P.G., Azevedo, T. C. G., Souza, M. C. R., Rego, L. A., Dantas, J. C. M., Silva, F. R.F., Rocha, H. A. O., Baseia, I. G., and Leite, E. L. Anti-inflammatory, antioxidant and cytotoxic actions of ß-glucan-rich extract from *Geastrum saccatum* mushroom. *International Immunopharmacology*, **7**, 1160–1169 (2007).

19. Duarte, M., LONGATTO FILHO, A.; SCHMITT, F. C. Angiogenesis, haemostasis and cancer: new paradigms old concerns *J. Bras. Patol. Med. Lab.*, 2007, **43**(6) 441–449.

20. Dudhgaonkar, S., Thyagarajan, A.,and Sliva, D. Suppression of the inflammatory response by triterpenes isolated from the mushroom *Ganoderma lucidum*. *International Immunopharmacology*, (2009), doi: 10.1016/j.intimp.2009.07.011.

21. Dyiabalanage, T., Mulabagal, V., Mills, G., Dewitt, D. L., M., and NAIR, M. G. Health-beneficial qualities of the edible mushroom, *Agrocybe aegerita*. *Food Chemistry*, **108**, 97–102 (2008).

22. Elmastas, M., Isildak, O., Turkekul, I., and Temur, N. Determination of antioxidant activity and antioxidant compounds in wild edible mushrooms. *Journal of Food Composition and Analysis*, 2007, **20**, 337–345.

23. Fan, L., Soccol, A. T., Pandey, A., and Soccol, C. R. Effect of nutrional and environmental conditions on the production of exo-polysaccharide of *Agaricus brasiliensis* by submerged fermentation and its antitumor activity. *LTW*, **40**, 30–35 (2007).

24. Frank, J. A., Xiao, R., Yu, S., Fergunson, M., Hennings, L. J., Simpson, P. M., Ronis, M.J. J., Fang, N. Badger, T. M., and Simmen, F. A. Effect of Shiitake mushroom dose on colon tumorigenisis in azoxymethane-treatd male Sprague-Dawley rats. *Nutrition Research*, **26**, 138–145 (2006).

25. Ghorai, S., Banik, S. P., Verma, D., Chowdhury, S., Mukherjee, S.,and Khowala, S. Fungal biotechnology in food and feed processing. *Food Research International*, **42**, 577–587, (2009).

26. Gu, Y. H. and Belury, M. A. Selective induction of apoptosis in murine skin carcinoma cells (CH72) by an ethanol extract of *Lentinula edodes*. *Cancer Letters*, **220**, 21–28 (2005).

27. Guo, Y., Wang, H., and NG, T. B. Isolation of trichogin, an antifungal protein from fresh fruiting bodies of the edible mushroom *Tricholoma giganteum*. *Peptides*, **26**, 575–580 (2005).

28. Hagiwara, S., Takahashi, M., Shen, Y., Kaihou, S., Tomiyama, T., Yazawa, M., Tamai,Y.; Sin, Y., Kazusaka, A., and Terezawa, M. A phytochemical in the edible Tamogi-take mushroom (*Pleurotus cornucopiae*), D-manitol, inhibits ACE activity and lowers the blood pressure of spontaneously hypertensive rats. *Biosc. Biotech. Bioch.* **69**(8), 1603–1605 (2005).

29. Han, C., Yuan, J., Wang, Y.,and Li, L. Hypoglycemic activity of fermented mushroom of *Coprinus comatus* rich in vanadium. *Journal of Trace Elements in Medicine and Biology*, **20**, 191–196 (2006).

30. Harhaji, Lj., Mijatovic, S., Maksimovic-Ivanic, D., Stojanovic, I., Momcilovic, M., Maksimovic, V., Tufegdzie, S., Marjanovic, Z., Mostarica-Stojkovic, M., VUCINIC, Z., and Stosic-Grujicic, S. Antitumor effect of *Coriolus versicolor* methanol extract against mouse B16 melanoma cells: *in vitro* and *in vivo* study. *Food and Chemical Toxicology*, **46**, 1825–1833 (2008).

31. Hatvani, N. Antibacterial effect of the culture fluid of *Lentinus edodes* mycelium grown in submerged liquid culture. *International Journal of antimicrobial Agents*, **17**, 71–74 (2001).

32. Hearst, R., Nelson, D., MCCOLLUM, G., Millar, B. C., MAEDA, Y., Goldsmith, C. E., Rooney, P. J., Loughrey, A., Rao, J. R., and Moore, J. E. An examination of antibacterial and antifungical properties of constituents of Shiitake (*Lentinula edodes*) and Oyster (*Pleurotus ostreatus*) mushrooms. *Complementary Therapies in Clinical Practice*, **15**, 5–7 (2009).

33. Hirasawa, M., Shouji, N., Neta, T., Fukushima, K., and Takada, K. Three kinds of antibacterial substances from *Lentinus edodes* (Berk.) Sing. (Shiitake, an edible mushroom). *International Journal of antimicrobial Agents*, **11**, 151–157 (1999).

34. Hsu, S. C., Ou, C. C., Chuang, T. C., Li, J. W., Lee, Y. J., Wang, V., Liu, J. Y., Chen, C. S., Lin, S.C.,and Kao, M.C. *Ganoderma tsugae* extract inhibits expression of epidermal growth factor receptor and angiogenesis in human epidermoid carcinoma cells: *in vitro* and *in vivo*. *Cancer Letters*, **28**, 108–116 (2009).

35. Huang, Q., Jin, Y., Zhang, L., Cheung, P. C. K., and Kennedy, J. F. Structure, molecular size and antitumor activities of polysaccharides from *Poria cocos* mycelia produced in fermented. *Carbohydrate Polymers*, **70**, 324–333 (2007).

36. Hwang, H. J., Kim, S. W., LIM, J. M., Joo, J. H., Kim, H. O., Kim, H. M., Yun, J. W. Hypoglicemic effect of crude exopolysaccharides produced by a medicinal mushroom *Phellinus baumii* in streptozotocin-induced diabetic rats. *Life Science*, **76**, 3069–3080 (2005).

37. Jayakumar, T., Thomas, P. A., and Geraldine, P. *In vitro* antioxidant activities of an ethanolic extract of the oyster mushroom *Pleurotus ostreatus. Innovative Food Science and Emerging Technologies*, **10**, 228–234 (2009).

38. Ji, Z., Tang, Q., Zhang, J., Yang, Y., Jia, W., and Pan, Y. Immunomodulation of RAW264.7 macrophages by GLIS, a proteopolysaccharide from *Ganoderma lucidum. Journal of Ethnopharmacology*, **112**, 445–450 (2007).

39. Jung, H. J., Kang, H. J. Song. Y. S., Park, E. H., Kim, Y. M., and Lim, C. J. Anti-inflammatory, anti-angiogenic and anti-nociceptive activities of *Sedum sarmentosum* extract. *Journal of Ethnopharmacology*, **116**, 138–143 (2008).

40. Kalac, P., Chemical composition and nutritional value of European species of wild growing mushrooms: *A review. Food Chemistry*, 113, 9–16 (2009).

41. Kasai, H., He, L. M., Kawamura, M., Yang, P.T., Deng, X. W.; Munkanta, M., Yamashita, A., Terunuma, H., Hirama, M., Horiuchi, I., Natori, T., Koga, T., Amano, Y.; Yamaguchi, N., and ITO, M. IL-12 Production Induced by *Agaricus blazei* Fraction H (ABH) Involves Toll-like Receptor (TLR). *eCAM*, **1**(3), 259–267 (2004).

42. Kim, C. F., Jiang, J. J., Leung, K. L., FUNG, K.P., and Lau, C. B. S. Inhibitory effects of *Agaricus blazei* extracts on human myeloid leukemia cells. *Journal of Ethnopharmacology*, **122**, 320–326 (2009).

43. Kim, Y. O., Park, H. W., Kim, J. H., Lee, J. Y., Moon, S. H., and Shin, C. S. Anti-cancer effect and structural characterization of endo-polysaccharide from cultivated mycelia of *Inonotus obliquus. Life Sciences*, **79**, 72–80 (2006).

44. Kim, S. H., Song, Y. S., Kim, S. K., Kim, B. C., Lim, C. J., and Park, E. H. Anti-inflammatory and related pharmacological activities of the n-BuOH subfraction of mushroom *Phellinus linteus. Journal of Ethnopharmacology*, **93**, 141–146 (2004).

45. Kimura, N., Fujino, E., Urabe, S., Mizutani, H., Sako, T., Imai, S., Toyoda, Y., and Arai, T. Effect of supplementation of *Agaricus* mushroom meal extracts on enzyme activities in peripheral leukocytes of calves. *Research in Veterinary Science*, (2006), doi:10.1016/j.rvsc.2006.02.003.

46. Kokean, Y., Nishi, T., Sakakura, H., and Furuichi, Y. Effect of frying with edible oil on antihypertensive properties of Hatakeshimeji (*Lyophyllum decastes* Sing.) Mushroom. *Food Science Technology Research*, **11**(3), 339–343 (2005).

47. Kuo, M. C., Weng, C. Y., Hab, C. L., and Wu, M. J. *Ganoderma lucidum* mycelia enhance innate immunity by activating NF-B. *Journal of Ethnopharmacology*, **103**, 217–222 (2006)

48. Kwon, A. H., Qiu, Z., Hashimoto, M., Yamamoto, K., and Kimura, T. Effects of medicinal mushroom (*Sparassis crispa*) on wound healing in streptozotocin-induced diabetics rats. *The American Journal of Surgery*, **197**(4), 503–509 (2009).

49. Lee, Y. L., Jian, S. Y., Lian, P. Y.,and MAU, J. L. Antioxidant properties of extracts from a white mutant of the mushroom *Hypsizigus marmoreus*. *Journal of Food Composition and Analysis*, **21**, 116-124a (2008).

50. Lee, J. S., Park, B. C., Ko, Y. J., Choi, M. K., Choi, H. G., Yong, C. S., Lee, J. S., and Kim, J. A. *Grifola frondosa* (Maitake Mushroom) water extract inhibits vascular endothelial growth factor-induced angiogenesis through inhibition of reactive oxygen species and extracellular signal-regulated kinase phosphorylation. *Journal of Medicinal Food*, **11**(4), 643-651b (2008).

51. Lee, I. P.; Kang, B. H., ROH, J. K., and KIM, J.R. Lack of carcinogenicity of lyophilized *Agaricus blazei* Murill in a F344 rat two year bioassay. *Food and Chemical Toxicology*, **46**, 87–95c (2008).

52. Lee, I. K., Kim, Y. S., Jang, Y. W., Jung, J. Y., and Yun, B. S. New antioxidant polyphenols from medicinal mushroom *Inonotus obliquus*. *Bioorganic & Medicinal Chemistry Letters*, **17**, 6678–6681b (2007).

53. Lee, Y. L., Yen, M. T., and Mau, J. L. Antioxidant properties of various extracts from *Hypsizigus marmoreus*. *Food Chemistry*, **104**, 1–9a (2007).

54. Lee, C. L., Yang, X., and Wan, J. M. F. The culture duration affects the immunomodulatory and anticancer effect of polysaccharopeptide derived from *Coriolus versicolor*. *Enzyme and Microbial Technology*, **38**, 14–21(2006).

55. Lee, B. C., Bae, J. T., Pyo, H. B., Choe, T. B., Kim, S. W., Hwang, H. J., and Yun, J. W. Biological activities of the polysaccharides produced from submerged culture of the edible Basidiomycete *Grifola frondosa*. *Enzyme and Microbial Technology*, **32**, 574–581 (2003).

56. Leung, P. H., Zhao, S., Ho, K. P., and Wu, J. Y. Chemical properties and antioxidant activity of polysaccharides from mycelial culture of *Cordyceps sinensis* fungus Cs-HK1. *Food Chemistry*, **114**, 1251–1256 (2009).

57. Li, S. G., Wang, D. G., Tian, W., Wang, X. X., Zhao, J-X., Liu, Z, and Chen, R. Characterization and anti-tumor activity of a polysaccharide from *Hedysarum polybotrys* Hand.-Mazz. *Carbohydrate Polymers*, **73**, 344–350a (2008).

57. Li, Y. R., Liu, Q. H., Wang, H. X., and Ng, T. B. A novel lectin with potent antitumor, mitogenic and HIV-1 reverse transcriptase inhibitory activities from edible mushroom *Pleurotus citrinopileatus*. *Biochimica et Biophysica Acta*, **1780**, 51–57b (2008).

58. Li, S. P., Zhang, G. H., Zeng, Q., Huang, Z. G., Wang. Y. T., Dong, T. T. X., and Tsim, K. W.K. Hypoglycemic activity of polysaccharide, with antioxidation, isolated from cultured *Cordyceps* mycelia. *Phytomedicine*, **13**, 428–433a (2006).

59. Li, S. P.; yang, F. Q., and Tsim, K. W. K. Quality control of *Cordyceps sinensis*, a valued traditional Chinese medicine. *Journal of Pharmaceutical and Biomedical Analysis*, 2006b.

60. Lima, L. F. O., Habu, S., Gern, J. C., Nascimento, B. M., Parada, J. L., Noseda, M. D., Gonçalvez, A. G., Nisha, V. R., Pandey, A., Soccol, V. T., and Soccol, C. R. Production and Characterization of the Exopolysaccharides Produced by *Agaricus brasiliensis* in Submerged Fermentation. *Appl Biochem Biotechnol*, (2008), DOI 10.1007/s12010-008-8187-2.

61. Lima, D. P. Synthesis of angiotensin-converting enzyme (ACE) inhibitors: an important class of antihypertensive drugs. *Química Nova*, **22**(3), 375–381 (1998).

62. Liu, D.,Zheng, H., Liu, Z., Yang, B., Tu, W., and Li, L. Chemical composition and anti-microbial activity of essential oil isolated from the culture mycelia of *Ganoderma japonicum*. *Journal of Nanjing Medical University*, **23**(3), 168–172 (2009).

63. Liu, F., Ooi, E. C., and Chang, S. T. Free radical scavening activities of mushroom polysaccharide extracts. *Life Sciences*, **60**, 763–771 (1997).

64. Lo, H. C., Tsai, F. A., Wasser, S. P., Yang, J. G., Huang, B. M. Effects of ingested fruiting bodies, submerged culture biomass, and acidic polysaccharide glucuronoxylomannan of *Tremella mesenterica* Retz.:Fr. On glycemic responses in normal and diabetics rats. *Life Sciences*, **78**, 1957–1966 (2006).

65. Lu, M. K., Cheng, J. J., Lin, C. Y., Chang, C. C. Purification structural elucidation, and anti-inflammatory effect of a water-soluble 1, 6-branched 1, 3-α-d-galactan from cultured mycelia of *Poria cocos*. *Food Chemistry*, (2009), doi: 10.1016/j.foodchem.2009.04.126.

66. Lv, H., Kong, Y., Yao, Q., Zhang, B., Leng, F. W., Bian, H. J., Balzarini, J., DAMME, E. V., and Bao, J. K. Nebrodeolysin, a novel hemolytic protein from mushroom *Pleurotus nebrodensis* with apoptosis-inducing and anti-HIV-1 effects. *Phytomedicine*, **16**, 198–205 (2009).

67. Maiti, S., Bhutia, S. K., Mallick, S. K., Niyati Khadgi, A. K., and Maiti, T. K. Antiproliferative and immunostimulatory protein fraction from edible mushrooms. *Environmental Toxicology and Pharmacology*, **26**, 187–191 (2008).

68. Makrilia, N., Lappa, T., Xyla, V., Nikolaidis, I., and Syrigos, K. The role of angiogenesis in solid tumours. *Europe Journal International Medicine*, (2009), doi:10.1016/j.ejim.2009.07.009.

69. Miyazawa, N., Okazaki, M., and Ohga, S. Antihypertensive effect of *Pleurotus nebrodensis* in spontaneously hypertensive rats. *Journal of Oleo Science*, 2008, **57**(12), 675–681.

70. Michelot, D. and Melendez-Howell, L. M. *Amanita muscaria*: chemistry, biology, toxicology, and ethnomycology. *Mycol. Res.*, **107**(2), 131–146 (2003).

71. Moradali, M. F., Mostafavi, H., Ghods, S., and Hedjaroude, G. A. Immunomodulating and anticancer agents in the realm of macromycetes fungi (macrofungi). *International Immunopharmacology*, **7**, 701–724 (2007).

72. Ngai, P. H. K., Zhao, Z., and Ng, T. B. Agrocybin, an antifungal peptide from the edible mushroom *Agrocybe cylindracea*. *Peptides*, **26**, 191–196 (2005).

73. Ngai, P. H. K., Ng, T. B. A ribonuclease with antimicrobial, antimitogenic and antiptoliferative activities from edible mushroom *Pleurotus sajor-caju Peptides*, **25**, 11–17 (2004).

74. Ngai, P. H. K. and Ng, T. B. A mushroom (*Ganoderma capense*) lectin with spectacular thermostability, potent mitogenic activity on splenocytes and antiproliferative activity toward tumor cells. *Biochemical and Biophysical Research Communications*, **314**, 988–993 (2004).

75. Ono, A. E., Oyekigho, E. W., Adeleke, O. A. Isolated systolic hypertension: primary care practice patterns in a Nigerian high-risk subpopulation. *Sao Paulo Med J.*, **124**(2), 105–9 (2006).

76. Park, S. E., Yoo, H. S., Jin, C. Y., HONG, S. H., Lee, Y. W., KIM, B. W., Lee, S. H., Kim, W. J., Cho, C. K., and Choi, Y. H. Induction of apoptosis and inhibition of telomerase activity in human lung carcinoma cells by the water extract of *Cordyceps militaris*. *Food and Chemical Toxicology*, **47**, 1667–1675 (2009).

77. Peng, Y., Zhang, L., Zeng, F., and Kennedy, J. F. Structure and antitumor activities of the water-soluble polysaccharides from *Ganoderma tsugae* mycelium. *Carbohydrate polymers*, **59**, 385–392 (2005).

78. Peng, Y., Zhang, L., Zeng, F., and Xu, Y. Structure and antitumor activity of extracellular polysaccharides from mycelium. *Carbohydrate Polymers*, 2003, **54**, 297–303.

79. Ramjaum, A. R. and Hodivala-Dilke, K. The role of cell adhesion pathways in angiogenesis. *The International Journal of Biochemistry & Cell Biology*, **41**, 521–530 (2009).

80. Rao, J. R., Millar, B. C., and Moore, J. E. Antimicrobial properties of shiitake mushroom (*Lentinula edodes*). *International Journal of Antimicrobial Agents*, **33**, 591–592 (2009).

81. Reid, G. Probiotics and prebiotics, progress and challenges. *International Dairy Journal*, doi: 10.1016/j.idairyj.2007.11.025.

82. Ren, G. Zhao, Y. P, Yang, L., and Fu, C. X. Anti-proliferative effect of clitocine from the mushroom *Leucopaxillus giganteus* on human cervical cancer HeLa cells by inducing apoptosis. *Cancer Letters*, **262**, 190–200 (2008).

83. Ribatti, D. Endogenous inhibitors of angiogenesis A historical. *Leukemia Research*, **33**, 638–644 (2009).

84. Ribeiro, B., Lopes, R., andrade, P. B., seabra, R. M., Gonçalves, R. F., Baptista, P., Quelhas, I., and Valentão, P. Comparative study of phytochemicals and antioxidant potential of wild edible mushroom caps and stipes. *Food Chemistry*, **110**, 47–56 (2008).

85. Roglic, G., Unwin, N., Bennett, P. H., Mathers, C., Tuomilehto, J. Nag, S. Connolly, V. King, H. The Burden of Mortality Attributable to Diabetes. *Diabetes Care*, September, **28**(9), 2130–2135 (2005).

86. Rubel, R., Dalla Santa, H. S., Fernandes, L. C., Lima Filho, J. H. C., Figueiredo, B. C., Di Bernardi, R., Moreno, A. N., LEIFA, F., and Soccol, C. R. High Immunomodulatory and Preventive Effects against Sarcoma 180 in Mice Feed with Ling Zhi or Reishi Mushroom *Ganoderma lucidum* (W. Curt.:Fr.) P. Karst. (Aphyllophoromycetideae) Mycelium. *International Journal of Medicinal Mushrooms*, **10**, 37–48 (2008).

87. Sadava, D. Still, D. W., Mudry, R. R., and Kane, S. E. Effect of *Ganoderma* on drug-sensitive and multidrug-resistant small-cell lung carcinoma cells. *Cancer Letters*, **277**, 182–189 (2009).

88. Sarangi, I., Ghosh, D., Bhutia, S. K., Mallick S. K., and Maiti, T. K. Anti-tumor and immunomodulating effects of *Pleurotus ostreatus* mycelia-derived proteoglycans. *International Immunopharmacology*, **6**, 1287–1297 (2006).

89. Sarikurkcu, C., Tepe, B., and Yamac, M. Evaluation of the antioxidant activity of four edible mushrooms from the Central Anatolia, Eikisehir – Turkey: *Lactarius deterrimus, Suillus collitinus, Boletus edulis, Xerocomus chrysenteron. Biosource Technology*, **99**, 6651–6655 (2008).

90. Seto, S. W., Lam, T. Y., Tam, H. L., Au, A. L. S., Chan, S. W., Wu, J. H., Yu, P. H. F., Leung, G. P. H., Ngai, S. M., Yeung, J. H. K., Leung, P. S., Lee, S. M. Y., and Kwan, Y. W. Novel hypoglycemic effects of *Ganoderma lucidum* water-extract in obese/diabetic (+*db*/+*db*) mice. *Phytomedicine*, **16**, 426–436 (2009).

91. Shih, I. L., Chou, B. W., Chen, C. C., Wu, J. W., and Hsieh, C. Study of mycelial growth and bioactive polysaccharide production in batch and fed-batch culture of *Grifola frondosa. Bioresource Technology*, **99**, 785–793 (2008).

92. Shimizu, S., Kitada, H.; Yokota, H., Yamakawa, J., MURAYAMA, T., Sugiyama, K., Izumi, H., and Yamaguchi, N. Activation of the alternative complement pathway by *Agaricus blazei* Murill. *Phytomedicine*, 2002, **9**, 536–545.

93. Shu, Ch., and Lung, M. Y. Effect of culture pH on the antioxidant properties of *Antrodia camphorata* id submerged culture. *Journal of the Chinese Institute of Chemical Engineers*, **39**, 1–8 (2008).

94. Sicoli,G., Rana, G. L., Marino, R., Sisto, D., Lerario, P., and Luisi, N. Forest Fungi as Bioindicators of a Healthful Environment and as Producers of Bioactive Metabolites Useful For Therapeutic Purposes. 1st European Cost E39 Working Group 2 Workshop: "*Forest Products, Forest Environment, and Human Health: Tradition, Reality, and Perspectives*", Christos Gallis (Ed.), Firenze, Italy (April 20–22, 2005).

95. Smirdele, F. R., Olsen, L. M., Carbonero, E. R., Baggio, C. H., Freitas, C. S., Marcon, R., Santos, A. R. S., Gorin, P. A. J., and Iacomini, M. Anti-inflammatory and analgesic properties in a rodent modelo f a (1→3),(1→6)-liked ß-glucan isolated from *Pleurotus pulmonarius*. *European Journal of Pharmacology*, **597**, 86–91 (2008).

96. Soares, A. A., Souza, C. G. M., Daniel, F. M., Ferrari, G. P., Costa, S. M. G., and Peralta, R. M. Antioxidant activity and total phenolic content of *Agaricus brasiliensis* (*Agaricus blazei* Murril) in two stages of maturity. *Food Chemistry*, **112**, 775–781 (2009).

97. Song, T. Y. Lin, H. C. Yang, N. C., and Hu, M. L. Antiproliferative and antimetastatic effects of the ethanolic extract of *Phellinus igniarius* (Linnearus: Fries) Quelet. *Journal of Ethnopharmacology*, **115**, 50–56 (2008).

98. Song, Y. S., Kim, S. H., Sa, J. H., Jin, C., Lim, C. J., and Park, E. H. Anti-angiogenic, antioxidant and xantine oxidase inhibition activities of the mushroom *Phellinus linteus*. *Journal of Ethnopharmacology*, **88**, 113–116 (2003).

99. Stupack, D. G., Storgard C. M. and Cheresh, D.A. A role for angiogenesis in rheumatoid arthritis. *Brazilian Journal of Medical and Biological Research*, **32**.573–581 (1999).

100. Surenjav, U., Zhang, L., Xu, X., Zhang, X., and Zeng, F. Effects of molecular structure on antitumor activities of (1-3)-b-D-glucans from different *Lentinus edodes*. *Carbohydrate Polymers*, **63**, 97–104 (2006).

101. Synytsya, A., Mickvá, K., Synytsya, A., Jablonsky, I., Spevacek, J., Erban, V., Kovarikova, E., and Copikova, J. Glucans from fruit bodies of cultivated mushrooms *Pleurotus ostreatus* and *Pleurotus eryngii*: Structure and potential prebiotic activity. *Carbohydrate Polymers*, **76**, 548–556 (2009).

102. Tong, H., Xia, F., Feng, K., Sun, G., Gao, X., Sun, L., Jiang, R., Tian, D., and Sun, X. Structural characterization and in vitro antitumor activity of a novel polysaccharide isolated from the fruiting bodies of *Pleurotus ostreatus*. Bioresource *Technology*, **100**, 1682–1686 (2009).

103. Tsai, S. Y., Huang, S. J., Lo, S. H., Wu, T. P., Lian, P. L., Mau, J. L. Flavor components and antioxidant properties of several cultivated mushrooms. *Food Chemistry*, (2008), doi: 10.1016/j.foodchem.2008.08.034.

104. Tsai, S. Y., Tsai, H. L., and Mau, J. L. Antioxidant properties of *Agaricus blazei, Agrocybe cylindracea*, and *Boletus edulis*. *LWT*, **40**, 1392–1402 (2007).

105. Turkoglu, A., Duru, M. E., Mercan, N., Kivrak, I., and Gezer, K. Antioxidant and antimicrobial of *Laetiporus sulphureus* (Bull.) Murril. *Food Chemistry*, **101**, 267–273 (2007).

106. Van, Q., Nayak, B. N., Reimer, M., Jones, P. J. H., Fulcher, R. G. and Rempel, C. B. Anti-inflammatory effect of *Inonotus obliquus, Polygala senega* L., and *Viburnum tribolum* in a cell screening assay. *Journal of Ethnopharmacology*, 2009, doi: 101016/j.jep.2009.06.026.

107. Vasiljevic, T. and Shah, N. P. Probiotics - From Metchnikoff to bioactives. International Dairy Journal, **18**, 714– 728 (2008).

108. Xie, G., Schepetkin, I. A., and Quinn, M. T. Immunomodulatory activity of acidic polysaccharides isolated from *Tanacetum vulgare* L. *International Immunopharmacology*, (2007), doi: 10.1016/j.intimp.2007.08.013.

109. Yaltirak, T., Aslim, B., Ozturk, S., and Alii, H. Antimicrobial and antioxidant activities of *Russula delica* Fr. *Food and Chemical Toxicology*, **47**, 2052–2056 (2009).

110. Yang, C. M., Zhou, Y. J., Wang, R. J., Hu, M. L. Anti-angiogenic effects and mechanisms of polysaccharides from *Antrodia cinnamomea* with different molecular weights. *Journal Ethnopharmacology*, **123**, 407–412 (2009).

111. Yang, B. K., Gu, Y. A., Jeong, Y. T., Jeong, H., and Song, C. H. Chemical characterics and immune-modulating activities of exo-biopolymers produced by *Grifola frondosa* during submerged fermentation process. *International Journal of Biological Macromoleculaes*, **41**, 227–233 (2007).

112. Yang, J.; Zhan, W.; Shi, P.; Chen, J.; Han, X.; Wang, Y. Animal and *in Vitro* Models in Human Diseases Effects of exopolysaccharide fraction (EPSF) from a cultivated *Cordyceps sinensis* fungus on c-Myc, c-Fos, and VEGF expression in B16 melanoma-bearing mice. *Pathology – Research and Practice*, **201**, 745–750 (2005).

113. Ye, M., Luo, X., Li, L., shi, Y., Tan, M., Weng, X., Li, W., Liu, J., Cao, Y. Grifolin, a potential antitumor natural product from the mushroom *Albatrellus confluens*, induces cell-cycle arrest in G1 phase via the ERK1/2 pathway. *Cancer Letters*, **258**, 199–207 (2007).

114. Wang, J., Wang, T. B., Ng, T. B. A peptide with HIV-1 reverse transcriptase inhibitory activity from the medicinal mushroom *Russula paludosa. Peptides*, **28**, 560–565 (2007).

115. Wang, H. and Ng, T. B. Ganodermin, an antifungal protein from fruiting bodies of medicinal mushroom *Ganoderma lucidum. Peptides*, **27**, 27–30 (2006).

116. Youguo, C., Zongji, S., and Xiaoping, C. Modulatory effect of *Ganoderma lucidum* polysaccharides on serum antioxidant enzymes activities in ovarian cancer rats. *Carbohydrate Polymers*, (2009), doi:10.1016/j.carbpol.2009.03.030.

117. Wang, H. X. and Ng, T. B. Purification of a novel low-molecular-mass laccase with HIV-1 reverse transcriptase inhibitory activity from the mushroom *Tricholoma giganteum. Biochemical and Biophysical Research Communications*, **315**, 450–454 (2004).

118. Wang, J., Wang, H. X., and Ng, T. B. A peptide with HIV-1 reverse transcriptase inhibitory activity from the medicinal mushroom *Russula paludosa. Peptides*, **28**, 560–565 (2007).

119. Wang, Y. Prebiotics: Present and future in food science and technology. *Food Research International*, **42**, 8–12 (2009).

120. Wu, J. Y., Zhang, Q. X., and Leung, P. H. Inhibitory effects of ethyl acetate extract of *Cordyceps sinensis* mycelium on various cancer cells in culture and B16 melanoma in C57BL/6 mice. *Phytomedicine* (2006).

121. Wu, M. J., Cheng, T. L., Cheng, S. Y., Lian, T. W., Wang, L., Chiou, S. Y. Immunomodulatory Properties of *Grifola frondosa* in Submerged Culture. *J. Agric. Food Chem.*, **54**, 2906–2914 (2006).

122. Wu, P. Q., Xie, Y. Z., Li, S. Z., La Pierre, D. P., Deng, Z., Chen, Q., Li, C., Zhang, Z., Guo, J., Wong, C. K. A., Lee, D. Y., Yee, A. , and Yang, B. B. Tumor cell adhesion and

integrin expression affected by *Ganoderma lucidum*. *Enzyme and Microbial Technology*, **40**, 32–41 (2006).

123. Zhang, L., Yang, M., Song, Y., Sun, Z., Peng, Y., Qu, K., and Zhu, H. Antihypertensive effect of 3,3,5,5-tetramethyl-4-piperidone, a new compound extracted from *Marasmius androceus*. *Journal of Ethnopharmacology*, **123**, 34–39 (2009).

124. Zheng, R., Jie, S., Hanchuan, D., and Moucheng, W. Characterization and immunomodulating activities of polysaccharide from *Lentinus edodes*. *International Immunopharmacology*, **5**, 811– 820 (2005).

125. Zhu, X. L., Chen, A. F., and Lin, Z. B. *Ganoderma lucidum* polysaccharides enhance the function of immunological effector cells in immunosuppressed mice. *Journal of Ethnopharmacology*, **111**, 219–226 (2007).

MOLECULAR AND IMMUNOLOGICAL APPROACHES FOR THE DETECTION OF IMPORTANT PATHOGENS IN FOODS OF ANIMAL ORIGIN

PORTEEN KANNAN and NITHYA QUINTOIL

CONTENTS

8.1   INTRODUCTION

The foodborne diseases are a widespread and growing public health problem, both in developed and in developing countries. While most foodborne diseases are sporadic and often not reported, foodborne disease outbreaks may take on massive proportions. For example, in 1994, an outbreak of salmonellosis due to contaminated ice cream occurred in the USA, affecting an estimated 2,24,000 persons. In 1988, an outbreak of hepatitis A, resulting from the consumption of contaminated clams, affected some 3,00,000 individuals in China [1]. Zoonotic pathogens originate in live animals and can, without farm to fork care, spread throughout the food chain causing infection in humans [2]. In addition, a number of well-known and preventable zoonoses are transmissible to humans through food such as salmonellosis, campylobacteriosis, and brucellosis. All major zoonotic diseases, emerging, re-emerging or endemic, in addition to being direct public health problems by affecting the health and well-being of millions of people, also prevent the efficient production of food, particularly of much-needed proteins, and create obstacles to international trade in animals and animal products. The rejection of the many food consignments destined for export on account of contamination by foodborne pathogens is the major cause of huge economic losses to the developed as well as developing countries.

Detection and isolation of these bacteria from food are often difficult due to the high number of contaminating and indigenous bacteria and a low number of the pathogenic bacteria of concern. Traditional methods of identification of food-borne pathogens, which cause disease in humans, are time-consuming and laborious, so there is a need for the development of innovative methods for the rapid identification of food-borne pathogens. Current trends in the nutrition and food technology are increasing the demand on food microbiologist to ensure a safe food supply. Today, there are more than 250 known diseases caused by different foodborne pathogenic microorganisms, including pathogenic viruses, bacteria, fungi, mycoplasma, parasites, marine phytoplankton, and cyanobacteria, etc. Among these microorganisms, bacteria are the most common foodborne pathogens, accounting for more than 90% of the total outbreaks of foodborne illness in the world [3,4].

Bacterial pathogens encountered to human illness in the last decades are through consumption of undercooked or minimally processed ready-to-eat meats (hotdogs, sliced luncheon meats and salami), dairy products(softcheeses made with unpasteurized milk, ice cream, butter, etc.) [5] or fruits (apple cider, strawberries, cantaloupe, etc.) and vegetables [6-10]. Though the global incidence of food borne disease is difficult to estimate, it hasbeen reported that in 2005 alone 1.8 million people died from diarrhoeal diseases and a great proportion of these cases can be attributed to contamination of food and drinking water [11]. Foodborne pathogens cause an estimated 76 million illnesses, accounting for325,000 hospitalizations and more than 5000 deaths in the United States each year [12]. Some foodborne diseases are well recognized, but are considered emerging because they have recently become more common. Though here are various food borne pathogens that have been identified for food borne illness, *Campylobacter, Salmonella, Listeria monocytogenes,* and *Escherichia coli* O_{157}:H_7 have been generally found to beresponsible for majority of food-borne outbreaks

[13,14]. Also foodborne diseases and associated complications resulted extreme economic losses and abnormal health implications. The medical costs and productivity losses associated with five major pathogens such as *E. coli* O_{157}:H_7, non-O_{157} *STEC (Shiga Toxin-Producing Escherichia coli), Salmonella (non-typhoidal serotypes only), Listeria monocytogenes and Campylobacter*, is atleast \$6.9 billion annually [15]. Data published in 2006 by the CDC suggested that infections due to *Yersinia, Shigella, Listeria, Campylobacter, Escherichia coli* O_{157}:H_7 and *Salmonella* have decreased dramatically ,while infections due to *Vibrio* have increased [16].

The contamination of ready to eat food products with pathogenic as well as spoilage bacteria is a serious concern nowadays because these products are generally not receive any further treatment before consumption. Food animals and poultry are most important reservoirs for many of the food borne pathogens [17], while animal byproducts such asfeed supplements, may also transmit pathogens to other animals.Seafoods are another potential source of pathogens, such as *Vibrio, Listeria, Yersinia, Salmonella, Shigella, Clostridium, Campylobacter* and Hepatitis A [18,19]. The infectious doses of many of these pathogens are very low (10-1000 bacterial cells). Food safety practiceshave vastly improved nowadays and consumers have become much more aware of food safety issues in the food processing environment [20]. There are many methodical programs like good agricultural practices [21,22], good manufacturing practices [22,23], hazard analysis and critical control point (HACCP) [23,24] and the food code indicating approaches [64], which can significantly reduce thepathogenic microorganisms in food. But still, the role of pathogendetection technology is vital, which is the key to the prevention and identification of problems related to health and safety.

The last 20–30 years have seen many developments intechniques and also the dawning of technologies, which were predicted to change our ways of detecting pathogenic bacteria in food.This is an ideal situation wherein rapid methods such as on-line monitoring system can be useful to quickly screen large number of samples and thereby enhancing the processing efficiency. The analysis of food for the presence of both pathogenic and spoilage bacteria is a standard practice for ensuring food safety and quality [25]. However, the advent of biotechnology has greatly altered food testing methods and there are numerous companies that are actively developing assays that are specific, faster and often more sensitive than conventional methods in testing for microbial contaminants in food [26]. These rapid method not only deals with the early detection and enumeration of microorganisms, but also with the characterization of isolates by use of microbiological, chemical, biochemical, biophysical, molecular biological, immunological and serological methods [17,27-29]. The degree to which rapid method and automation are accepted and used for microbiological analysis is determined by the range and type of testing required, volume throughput of samples to be tested, availability of trained laboratory staff and the nature of manufacturing practices [30].

THE PATHOGENS TRANSMITTED THROUGH FOODS OF ANIMAL ORIGIN

The presence of microorganisms in food is a natural and unavoidable occurrence. Cooking generally destroys most of the harmful bacteria, but undercooked foods, processed ready-to-eat foods, and minimally processed foods can contain some harmful bacteria that are serious health threats. The foodborne pathogens could be bacterial, viral or parasitic. The established pathogens include *Salmonella, Staphylococcus aureus, Clostridium perfringens* and *Cl. botulinum* etc. Additional challenge lies in confronting the emerging foodborne pathogens like the Norwalk virus, *Listeria, Campylobacter, Escherichia coli* O157 H7 etc. Meat, dairy, and poultry products are important reservoirs for many of the food-borne pathogens, including *Salmonella, Campylobacter, Listeria,* and *E. coli* O157:H7. Animal by-products, such as feed supplements, may also transmit pathogens to food animals (for example, *Salmonella* and bovine spongiform encephalopathy). Seafood is another potential source of food-borne pathogens, such as *Vibrio, Listeria,* and Hepatitis A. Infectious doses of many of these pathogens are very low (10 bacterial cells), increasing the vulnerability of the elderly, infants, and people with immunological deficiencies or organ transplants.

The incidence of foodborne disease during recent years has increased significantly, mainly due to epidemics caused by *Salmonella* spp. and *E. coli* [31]. Many industrialized countries are also experiencing outbreaks of diseases due to relatively newly recognized types of foodborne pathogens such as *Campylobacter jejuni, Listeria monocytogenes* and *E. coli* 0157:H7. Infection due to *E. coli* 0157:H7 causing permanent kidney damage and death is emerging as an important public health problem both because of the gravity of the disease as well as its increasing trend. *Campylobacter jejuni* infection has increased to such an extent that it is now the leading foodborne disease in several industrialized countries.

COMMON SOURCES OF IMPORTANT FOODBORNE PATHOGENS

Food-producing animals (e.g., cattle, chickens, pigs, and turkeys) and their produce are the major sources for many of zoonotic organisms, which include *Salmonella* spp., *Campylobacter* species, non-Typhi serotypes of *Salmonella enterica*, Shiga toxin–producing strains of *E. coli,* and *Yersinia enterocolitica* [32]. The other important zoonotic pathogens transmitted through foods of animal origin include *Shigella* spp., *Mycobacterium* spp., *Coxiella burnetii, Brucella* spp., *Aeromonas* spp., *Plesiomonas* spp., *Cl. Perfringens* Type A and *Cl. botulinum* type E. These organisms can contaminate animal/poultry carcasses at slaughter or cross-contaminate other food items, leading to human illness.

Human salmonellosis cases have been attributed to consumption of meat or poultry products contaminated during slaughter [33,34]. The poultry, eggs and food containing eggs have been identified as predominant sources of *Salmonella* Enteritidis. The pathogen infects eggs in hens' ovaries, and its symptoms include fever, abdominal cramps, and diarrhea. In some countries, as much as 60–80% of poultry meat is reported to be contaminated with *Salmonella* Enteritidis. Animal feed is frequently contaminated with non-Typhi serotypes of *Salmonella enterica* and may lead to infection or colonization of food animals. Cattle, poultry, pigs, and other food animals are colonized with non-Typhi serotypes of *S. enterica*, which have multiple routes into the food supply.

Another emerging foodborne pathogen- *E. coli* 0157:H7 is found in the intestines of cattle. It is shed in the manure of cattle and contaminates beef during slaughter, and normally transmitted in ground beef [. *Campylobacter jejuni* can cause severe gastroenteritis, arthritis, and a rare disease of the peripheral nervous system- Gullien barr syndrome. It is the most common pathogen in uncooked chicken; and as with salmonellosis, the main vehicles for transmission are poultry meat and raw milk. Listeriosis, caused by *L. monocytogenes* is particularly dangerous for the elderly, pregnant women/foetus, neonates, and immunoimpaired persons. The soft cheese and raw meat products such as pate contaminated with faeces are the main sources of the pathogen. The overall fatality rate is 30%, but may be as high as 70% with 1-10% of humans are carriers.

Among the other important food borne zoonotic pathogens, *Shigella* spp. spreads through faecal contamination of food while *Mycobacterium* spp., *Coxiella burnetii* and *Brucella spp.* are primarily transmitted through ingestion of contaminated raw milk and milk products from cattle, goat and sheep. *Cl. botulinum* causes the severe intoxication, botulism, which may lead to respiratory paralysis, with a fatality rate of 35-65%. The zoonotic disease usually associated with consumption of fish include the infections caused by *Aeromonas* spp., *Plesiomonas* spp., *Cl. botulinum* type E.

DETECTION OF FOODBORNE PATHOGENS

Assessment of the quality and safety of foods is important in human health [31]. Rapid and easy detection of pathogenic organisms will facilitate precautionary measures to maintain healthy food [37]. An important step in the control of foodborne diseases is timely detection of the pathogen involved, and thereby, adopting suitable control strategy. The rapid detection of pathogens in food is critical for ensuring the safety of consumers. Recent advances in technology make detection and identification faster, more convenient, sensitive, and specific than conventional assays.

Rapid methods, a subjective term used loosely to describe a vast array of tests that includes miniaturized biochemical kits, antibody- and DNA-based tests, and assays that are modifications of conventional tests to speed up analysis. Currently available rapid methods are often unsuitable for use in industrial laboratories. They lack sensitivity, are expensive and complex to perform, often requiring specialized personnel and significant capital expenditure. The absence of rapid cost effective methods for bacterial detection poses particular difficulties for food items with a short shelf-life and for the implementation of effective HACCP management systems. Automated systems are currently being used for rapid detection of pathogens, and they are of immense use especially in food processing plants where rapid detection is of great help in quality assurance of the product. Two major categories of rapid methods include immunologic or antibody-based assays and genetic-based assays such as the polymerase chain reaction. Next generation assays under development include biosensors and DNA chips that potentially have the capability for near real-time and on-line monitoring for multiple pathogens in foods. In recent years, there has been much research activity in the area of sensor development for detecting pathogenic microorganisms.

8.2   CONVENTIONAL MICROBIOLOGICAL METHODS

Conventional culture methods remain the most reliable and accurate techniques for food-borne pathogen detection. However, these traditional methods are laborious and time consuming, requiring a few days to a week or longer to complete. These methods include blending of the food product with a selective enrichment medium to increase the population of the target organism; plating onto selective or differential agar plates to isolate pure cultures; and examining the cultures by phenotypic analysis or metabolic fingerprinting (monitoring of carbon or nitrogen utilization). A major drawback is that these methods are labor-intensive and take 2–3 days for results, and up to 7–10 days for confirmation. There is a need for the development of innovative methods for the rapid identification of food-borne pathogens. To avoid delays, many of the modern detection tools use a conventional method along with an automated or semi-automated DNA, antibody, or biochemical-based method. These methods allow detection in 3–4 days [38]. Recent advances in molecular cloning and recombinant DNA techniques have revolutionized the detection of pathogens in foods [29].

8.3   IMMUNOLOGICAL METHODS

The basic principle of antibody-based detection (immunoassay) is the binding of antibodies to a target antigen, followed by the detection of the antigen-antibody complex. Antibodies are produced by the body in response to a specific invading pathogen. Experimentally, these molecules are produced in laboratory animals against a specific antigenic component of the pathogen or toxin. The most important characteristic of an antibody is its ability to recognize only the target antigen in the presence of other organisms and interfering food components. In addition, the successful use of antibodies to detect pathogens depends on the stable expression of target antigens in a pathogen, which are often influenced by temperature, preservatives, acids, salts, or other chemicals found in foods. The highly specific binding of antibody to antigen, especially monoclonal antibody, plus the simplicity and versatility of this reaction, has facilitated the design of a variety of antibody assays and formats, and they comprise the largest group of rapid methods being used in food testing. In general, immunological tests are simpler and faster to perform, but are less specific than DNA tests like PCR. Among these, the enzyme-linked immunosorbent assay (ELISA) and the latex agglutination (LA) are most suited for detection of microorganisms based on their production of specific antigens and for quantitative detection of bacterial toxins.

The various formats of antibody assays are described as:

i.   ***Immunodiffusion***: In the ID test format, an enrichment sample is placed in a gel matrix with the antibody; and if the specific antigen is present, a visible line of precipitation is formed.

ii. ***Agglutination***: In this format, the colored insoluble (particulate) antigen binds with the corresponding antibodies to develop flakes that are visible by naked eye. Zoonotic pathogens like *Coxiella burnetii* and *Brucella* spp. are excreted in the milk of infected animal(s), therefore, the detection of antibodies against them in the milk by capillary agglutination test (CAT) and Rose Bengal Plate Test (RBPT), respectively, can serve as an indirect measure to detect their presence in the milk.

iii. ***Latex Agglutination***: Employs antibody-coated colored latex beads or colloidal gold particles for quick serological identification or typing of pure culture isolates of bacteria from foods. A modification of LA, known as reverse passive latex agglutination (RPLA), is used for detecting   soluble antigens, mostly the toxins produced by pure cultures or toxins in food extracts.

iv. ***Immunomagnetic Separation (IMS)***: It is the use of antibody-coated magnetic beads, which are able to capture and concentrate bacterial cells from complex food matrices. These captured bacterial cells are then detected by various methods, including biosensors. These antibodies are also used to capture pathogens from pre-enrichment media. IMS is analogous to selective enrichment, but instead of using antibiotics or harsh reagents that can cause stress-injury, an antibody is used to capture the antigen, which is a much milder alternative.

v.   ***Direct Immunoassay***: In this method, antibodies are immobilized onto a solid support and then test samples are added. Once specific antigen-antibody binding takes place, the complex is detected. The most common detection method is to use another antibody (sandwich format of antigen between two antibodies) conjugated to an enzyme or a fluorescent dye specific for the antigen. When enzyme activity is quantified by using a substrate that produces a colored product, the assay is called an enzyme-linked immunosorbent assay, or ELISA. Instead of an enzyme, a fluorescent-labeled antibody is used in immunofluorescence assay (IFA). The sensitivities of these methods are in the range of 10^4–10^7 bacterial cells, and the assays take about 3–4 h to complete [38].

*   ***Enzyme-linked Immunosorbent Assay***: ELISA is the most prevalent antibody assay format used for pathogen detection in foods. Usually designed as a "sandwich" assay, an antibody bound to a solid matrix is used to capture the antigen from enrichment cultures and a second antibody conjugated to an enzyme is used for detection. The walls of wells in microtiter plates are the most commonly used solid support; but ELISAs have also been designed using dipsticks, paddles, membranes, pipette tips or other solid matrices.

The detection of *Listeria* spp. by ELISA has been achieved using either polyclonal antibodies, or monoclonal antibodies directed against different listerial antigens or, specific cell surface proteins of the organism. Monoclonal antibody specific for the detection of *L. monocytogenes* serotype 4b has been developed, but its efficacy has not been evaluated for detection of organism from the food. Some of the bacterial cell surface proteins have been reported to be species-specific and have been reported to be involved in the virulence of the organism. However, use of polyclonal antibodies [39] and monoclonal antibodies [40,41] and against cell surface antigens in ELISA have been found to show cross-reactions with *L. innocua* and some strains and *L. welshimeri*. The commercially available ELISA kits using MAbs [62] as well as polyclonal antibodies [42] are useful for detection of *Listeria* spp.

Rapid detection of *Salmonella* is also carried out by using commercial ELISA kits e.g., BioEnzabead™ Screen kit, *Salmonella*-Tek™ ELISA test system, BacTrace™ Microwell ELISA and Q-TROL™ *Salmonella* Test (Blackburn, 1993). The immunoassay screening procedure (ELISA) method commercialized by Organon Teknika (Durham, North Carolina) uses two monoclonal antibodies specific for *Salmonella* detection. This test is used for the detection of *Listeria* spp. and *E. coli* also.

Staphylococcus aureus enterotoxins (SEs) namely A, B, C_1, C_2, C_3, D, E can be detected in the foods with the help of commercial kits. Some assays like Bommeli SET-EIA, a solid phase ELISA that uses polystyrene beads coated with antibodies against SEs A-D, require as long as 24 h due to the need for overnight incubation with the antibody-coated beads, while kits like -TECRA staphylococcal enterotoxin visual immunoassay (Bio-enterprises Pty. Ltd., Roseville, Australia), a rapid microtitre plate-based screening test employing polyvalent antisera can detect SEs A-E at concentrations down to 1 ng/ml in foods and culture filtrates, and has the added advantage of reading the results visually with a colour card, besides measuring the absorbance at 414 nm in a plate reader. The Ridascreen™ SET (R-Biopharm GmbH, Darmstabt, Germany) is a sandwich ELISA for the qualitative analysis of SEs A-E that uses monovalent antibodies. The TRANSIA immunoassay for SEs A-E is a tube-based sandwich assay, which uses a mixture of monoclonal and polyclonal antibodies. Results are determined visually by comparison with a positive control or photo metrically at 450 nm (Lapeyre *et al.,* 1996). The VIDAS staphylococcal enterotoxin (SET) assay (bio merieux, Marcy-I Etoile, France) is an automated enzyme-linked fluorescent immunoassay (ELFA) for the simultaneous detection of all the SEs in foods with the sensitivity ranging from <1 ng/ml to 1.5 ng/ml, depending on the toxin and food.

ELISAs have been developed for the detection of *A. hydrophila* haemolysin and, O11 serotypes of *Aeromonas* (using antibody against surface array protein specific to O11 serotypes). However, reports on using ELISA for the detection of aeromonads are quite limited.

Rapid sandwich ELISA, indirect ELISA, Immunomagnetic separation ELISA IMS-ELISA, has been reported for identification and quantification of the enterotoxin for *Cl. perfringens*. Rapid sandwich ELISA and indirect ELISA can easily detect 1 ng/ml and 25 ng/ml of enterotoxin, respectively, whereas, IMS-ELISA can detect 2.5 ng/ml of enterotoxin.

Detection of parasite in food sample remains a tedious task. Various assays for detection of *Taenia solium* in pork, including antigen (AG)-ELISA, antibody-ELISA, enzyme linked immunoelectro transfer blot (EITB) and tongue inspection were found to show low sensitivity in rural pigs infected naturally with low levels of the parasite.

This observation was also found true for *T. saginata* infections in cattle as only a small percentage (13–22%) of cattle carrying fewer than 30–50 viable cysticerci are detected by AG-ELISA (OIE, 2004). Conversely, antibody has proven most useful for detecting cysts that are no longer viable. Nonetheless, AG-ELISAs do have a use in field-based epidemiological studies for indicating transmission. Detection of *Trichinella spiralis* antibodies by ELISA has been reported for the surveillance of selected pig breeding farms in the Republic of Korea [46].

- *•Immunofluorescence Assay*: In case of IFA, a fluorescent-labeled antibody is used to detect the pathogen in food and a spectrofluorometer or an epifluorescence microscope is used to measure emission of fluorescence.
- *•Immunoprecipitation or Immunochromatography or Lateral Flow Device*: This is based on the technology developed for home pregnancy tests. It is also a "sandwich" procedure but, instead of enzyme conjugates, the detection antibody is coupled to colored latex beads or to colloidal gold. Using only a 0.1 ml aliquot, the enrichment sample is wicked across a series of chambers to obtain results. These assays are extremely simple, require no washing or manipulation and are completed within 10 minutes after cultural enrichment. When the assay is applied to a food sample, the bacterial pathogens or antigens are detected by a double-antibody sandwich format on a membrane. When a sample is introduced, it binds to antibody-gold conjugates. This antigen-antibody complex migrates on the absorbing membrane and binds to another antibody (antigen-specific) resulting in a visible band. This type of assay is very fast and simple, and results can be obtained in 10–15 min. The assay, however, requires a large number of cells (10^7–10^9 cells) and may lack specificity.

vi. *Electro immunoassay Technology*: It is composed of a circuit with a capture antibody attached to the solid surface in the area of the electrode. Upon addition of a sample, the target antigen binds to the capture antibody. In the next step, a colloidal gold–labeled detection antibody is bound, creating a capture-target-detector sandwich. The final step is the deposition of silver ions onto the colloidal gold, which produces a conductive silver bridge, closing the circuit and resulting in a measured drop in resistance. This detection technology can also be applied to nucleic acid hybridization assays [47].

8.4   MOLECULAR DIAGNOSTIC METHODS

Nucleic acid-based methods, which incorporate an amplification step for the target DNA are now widely used. The most popular method of nucleic acid amplification is the polymerase chain reaction (PCR) [48]. The PCR techniques are highly specific and currently being validated against traditional cultural methods using a wide range of food sample types. The low number of target bacteria present in foods necessitates an enrichment period of up to 48 h prior to detection, which limits the rapidity of the test. Interference from competing bacteria and the food sample matrix can also be problematic. Continued development in this field will fully automate PCR methods and reduce costs encouraging the uptake of rapid tests for routine analysis of food samples.

1. **DNA-based Assays:** There are many DNA-based assay formats, but only probes, PCR, real-time PCR and bacteriophages have been developed commercially for detecting foodborne pathogens.

   - . *Probe Assays*: Probe-based methods have been developed for the detection and enumeration of food borne pathogens like *Salmonella, Staphylococcus* species, *Listeria* spp. and hepatitis A virus [49]. These method s employ nucleic acid (DNA or RNA) probes that are labeled either by radio isotopes, digoxigenin or biotin; and all the three types have been found to be equally sensitive.

   - . *Target DNA Probe*: Colony hybridization using a radiolabeled DNA probe, consisting of some 500 bp of the presumptive haemolysin gene [50] and, *hly* and *inlA* gene probes has been reported for the detection of *L. monocytogenes*. The disadvantage with the hybridization technique is that it needs a large number of target cells for the direct detection of pathogenic bacteria (10^{-4}–10^5). However, this technique is useful for culture confirmation and for typing of isolates.

A DNA-DNA hybridization (DNAH) method, using probes consisting of radiola-belled DNA fragments of *Salmonella* Typhimurium has been in use for detecting 10^8 salmonellae/ml of foods. A biotin-labeled probe of 1.8 kilo base pair (kb) has been found to detect all the 40 serovars of *Salmonella* in broth cultures without showing any cross reactions with 15 non-*Salmonella* strains. Similarly, another 2.3 kb *Salmonella* probe has been identified, which hybridizes to all of 396 strains of *Salmonella* belonging to 214 serovars of all subspecies, but not to 178 non-*Salmonella* strains of 52 species belonging to 23 genera of the family Enterobacteriaceae. The rRNA directed oligonucleotide probe also detected 355 out of 362 strains of *Salmonella*, representing 213 serovars, however, false-positive reactions were observed in 7% of more than 100 non-*Salmonella* strains tested [51]. A second generation biometric hybridization (color DNAH) assay developed commercially based on the rRNA probes of Wilson *et al.* gave positive results within 48 h with only 1.5% false-positive reactions and 0.8% false-negative reactions on inoculated samples.

The gene probes are cumbersome to perform and their sensitivity is low for detection of enterotoxigenic strains of *Clostridium perfringens,* however, nucleic acid probes directed against enterotoxin gene (*cpe*) of *Cl. perfringens* are most commonly used. The sensitivity of chromogenic, non-radioactive colony hybridization assay is reported to be ≤ 10 cfu/g of raw beef for *Cl. perfringens*. *L. monocytogenes* has also been successfully detected in artificially inoculated soft cheese and ground chicken using a hydrophobic grid membrane filter DNA probe. The dot-blot meat speciation test based on a DNA probe can also identify the origin of meat.

   - *Target RNA Probe:* Another major change in this area is the development of probes to detect target RNA. In a cell there is only one copy of DNA, however, there may be 1000 to 10000 copies of ribosomal RNA (rRNA). Thus, the new generation of probes is designed to probe target RNA. Currently, kits are available for *Salmonella, Listeria, Campylobacter, Yersinia,* etc [52]. *L. monocytogenes*-specific nucleic acid probes that are not virulence gene-based have been developed from rRNA sequences. The probes have also been developed for the genes thought to be involved in pathogenicity.

2. *Polymerase chain reaction (PCR)*: The basic principle of DNA hybridization is also being utilized in PCR assay for rapid and reliable detection of pathogens in foods. In this method, short fragments of DNA (probes) or primers are hybridized to a specific sequence or template, which is then enzymatically amplified by *Taq* polymerase using a thermocycler. Theoretically, PCR can amplify a single copy of DNA by a million fold in less than 2 h; hence it's potential to eliminate, or greatly reduce the need for cultural enrichment. However, the presence of inhibitors in foods and in many culture media can prevent primer binding and diminish amplification efficiency, so that the extreme sensitivity achievable by PCR with pure cultures is often reduced when testing foods. Therefore, some cultural enrichment is still required prior to analysis. In the first reported PCR for the detection of *L. monocytogenes*, the haemolysin (*hlyA*) sequence was used. Since then, genes of different virulence markers of *L. monocytogenes* have been used for the species-specific identification of *L. monocytogenes*. The PCR based on the virulence genes of *L. monocytogenes* (*hlyA, plcA, iapA, actA* and *prfA*) has been used for detection of pathogen in the milk of mastitic cattle [53].

Several PCR methods for detecting *Salmonella* have also been developed utilizing primers based on specific gene sequences namely, replicons, gene coding for DNA binding protein, gene coding for abequose and paratose synthase, gene coding for their aggregative fimbriae, *agfA*, rRNA operon, random genomic fragment, genes from *spvR* of the virulence plasmid, *H-li* and *H-in* flagellin genes, 16S DNA and *ompC* gene. The *inv*ABC operon has been reported to be highly conserved in all pathogenic salmonellae and this gene was found to code for the invasive property of *Salmonella*. Therefore, specific primers derived from a conserved sequence of the *invA* gene, which amplified a 389 bp fragment from 33 *Salmonella* serovars but not from 16 non-*Salmonella* bacteria, was used for the PCR-based detection of *Salmonella* from food [54]. The PCR targeting repetitive sequences of the transposons (trans-PCR) in the genome of *Coxiella burnetii* has been employed for detection of pathogen in the milk samples of dairy herds [55,56].

Some modifications of the PCR assay have also been developed like Amplisensor PCR assay, which is an automated fluorescence-based system for detection of PCR products targeting 284 bp sequence in the *invA* gene of *Salmonella*. Magnetic immuno-PCR assay [57] employs magnetic beads to concentrate bacteria and colorimetric detection procedures to enhance the sensitivity and specificity and also makes PCR suitable for automation. PCR-based commercial systems like Probelia™ *Salmonella* spp., Taqman™ *Salmonella* PCR amplification detection kit and BAX™ are also available, in which PCR may be performed on the pre-enrichment medium after 16-20 h of incubation, however, these are expensive and not easily available. RT-PCR and multiplex PCR developed for detection of *Salmonella* are also cumbersome, thereby, making these unsuitable for large scale screening of samples.

For detection of *Cl. perfringens* by PCR, two most commonly targeted genes are phospholipase (*plc*) and enterotoxins (*cpe*) genes. The duplex PCR developed for simultaneous amplification of both the genes had the detection levels of approximate 10^5 $Cl.$ $perfringens$ cells/g of stool or food sample. Another modification of simple PCR i.e. nested PCR, used for detection of low levels of $Cl.$ $perfringens$ in feces and meat has been reported to be 10^3 times more sensitive than single PCR. Several other molecular methods like plasmid isolation, and ribotyping have also been used successfully for $Cl.$ $perfringens$ strain differentiation [58].

Detection of $S.$ $aureus$ in foods by PCR targets amplification of the ent B, $entC$, and nuc genes. Since SEA is the most common toxin implicated in staphylococcal food poisoning, $entA$ gene has remained one of the most targeted genes for designing primers. PCR is also used to distinguish the staphylococcal enterotoxins (SEA to SEE). New types of staphylococcal enterotoxins such as SEG, SEH, SEI and SEJ have also been identified in the foods by using PCR.

Virulence factor-based PCR method for the detection of $Aeromonas$ spp. is routinely followed nowadays. A 316 bp fragment of 16SrDNA target PCR primer was used for rapid (within 8 h) detection of $A.$ $caviae$ and $A.$ $trota$ from sea food and water samples, and the detection level was found to be 50-100 /g crab meat.

Amplified fragment length polymorphism (AFLP) profiling is a technique, which has proven useful for speciation and outbreak investigation of the related species of $C.$ $jejuni$ and; Penner serotyping as an epidemiological marker for $C.$ $coli$ [59]. Molecular typing and virulence markers were used to evaluate the genetic profiles and virulence potential of 106 $Yersinia$ $enterocolitica$ strains. The strains were better discriminated by PFGE than by enterobacterial repetitive intergenic consensus PCR (ERIC-PCR) [60].

Several PCR-based assays have been developed for detection of DNA from $T.$ $gondii$. The main target regions are the B1 repetitive sequence, the P30 (SAG1) gene or ribosomal RNA (rRNA). The B1 and P30 PCR are both widely used techniques and good diagnostic aids, but are better used in conjunction with another test, as they are insufficiently robust when used alone. A nested PCR assay based on the detection of 18S rRNA, amplifying the B1 repetitive sequence of DNA has proved to be more sensitive for detection of pathogen's DNA in several tissues, including placenta, the central nervous system, heart and skeletal muscle.

- •$Real$-$time$ PCR: Real-time PCR allows fast detection of specific pathogens through greatly improved gene amplification efficiency. Therefore, it is rapidly becoming an accepted detection technique for monitoring viral and bacterial pathogens that may threaten public health. Initial efforts have also been undertaken to develop real-time PCR assays directed at human bacterial pathogens such as $E.$ $coli$ $O157$, $Campylobacter,$ $Listeria$ $monocytogenes$ and $Salmonella$ $Enteritidis$, all of which are serious food borne zoonotic pathogens. Standard laboratory protocols are being developed for an early detection of zoonotic agents such as $Anthrax,$ $Yersinia$ $pestis$ and Q-fever in animal origin food products within the United States and for the cross-border monitoring of livestock coming into the United States [63]. Diagnostic real-time PCR for the specific detection of $Salmonella$ in foods is increasingly being used as a rapid and reliable tool for the control of contaminated samples along the food production chain. A robust 5' nuclease

(TaqMan) real-time PCR has been developed and validated for the specific detection of *Salmonella* in food [61] and, detection of *C. burnetii* in the bulk tank milk samples from dairy herds as well as shedding of the pathogen in the milk of cattle [55,56].

3. ***Microarray Analysis***: This technique uses high-density microscopic array elements, planar glass substrates, low reaction volume, multicolour fluorescent labeling, high binding specificity and high-speed instrumentation for detection several pathogens at a time. The array elements react specifically with labeled mixtures, producing signals that reveal identity and concentration of each labeled species in solution. These attributes provide miniature biological assays that allow exploration of any organism on genomic scale. Microarray analysis, similar to recombinant DNA and PCR, is a foundation technology with broad application in safety assessment, proteomics and diagnostics.

4. ***Biosensors***: Biosensors are devices that detect biological or chemical complexes in the form of antigen-antibody, enzyme-substrate, or receptor-ligand. Most biosensors that have been developed for pathogens have been tested only with pure bacterial cultures. For such applications, the pathogens are first separated from the food and then applied to the sensor. Very few studies have actually attempted to detect bacteria directly from food. Food is extremely complex material consisting of fats, proteins, carbohydrates, chemicals, and preservatives, as well as different acidities, salt concentrations, and colors. The application of nanotechnology (interrogating nanosize particles on sensors) to detect pathogens from complex food systems is an incredible challenge. Moreover, the populations of target microorganisms are often extremely small compared with the indigenous ones. Consequently, clever strategies need to be used to detect such low numbers of pathogens directly from food.

A variety of biosensors is available in the market to monitor pathogens. A biosensor has been developed that can detect the potentially deadly bacteria *L. monocytogenes* in less than 24 h at concentrations as low as 1,000 cells per milliliter of fluid - an amount about the size of a pencil eraser. The sensor also is selective enough to recognize only the species *L. monocytogenes*. The sensor also is selective enough to recognize cells of *L. monocytogenes* when other types of foodborne contaminants, such as *Salmonella* or *E. coli*, are present. Known as an "optical biosensor," the device uses light to detect the presence of a target organism or molecule. The sensor is made of a small piece of optical fiber - a clear, solid, plastic material that transmits light through its core. The fiber is coated with a type of molecule called an antibody, which specifically recognizes *L. monocytogenes* and captures it, binding it to the fiber. When the fiber is placed in a liquid food solution, any *L. monocytogenes* in the sample will stick to the fiber. The presence of *L. monocytogenes* is verified by the addition of a second antibody, which not only recognizes *L. monocytogenes* but also carries a molecule that produces a fluorescent glow when exposed to laser light. This antibody attaches to the *L. monocytogenes* bound to the fiber and acts as a flag, signaling the pathogen's presence when laser light is passed through the liquid.

- *. Fiber-optic biosensor*: The use of optical fiber is being investigated for the real-time detection of biological agents (such as bacterial cells, toxins, or spores) in the air, soil, food or environment. The fiber-optic biosensor operates by covalently linking a specific antibody to the fiber-optic cable, binding a target antigen to the antibody, and then detecting the antigen-antibody complex by means of a secondary antibody conjugated to molecules that can be stimulated to emit fluorescent light, which is measured by a laser detector. Antibody-coupled fiber-optic biosensors are being developed for the detection of botulinum toxin, staphylococcal enterotoxin, *E. coli* O157:H7, *Listeria*, and *Salmonella*.

- . *Genetically engineered biosensor*: It is another type of biosensor e.g. BIO FLASH, and the technology behind it is the 'cellular analysis and notification of antigens risks and yields' (CANARY) technology. It is a complete biosensing system engineered to express membrane bound pathogen specific antibody and a calcium sensitive bioluminescent protein. Cross linking of antibody by even a small number of pathogens leads to increase of intracellular calcium and resulting bioluminescence is detected by luminometer. Biosensors will surely be in place for all HACCP programmes.

- . *Surface plasmon resonance sensor*: A surface plasmon resonance sensor is an optical sensor that is capable of characterizing the binding event of biomolecules in real time (few seconds to minutes) without the need for labeling molecules, by detecting differences in the intensity of light reflecting off a sensing surface. Antibodies or receptors for detection of food-borne pathogens or toxins are immobilized in a biolayer attached to a sensing surface. When binding takes place, it alters the angle of light reflected off the medium, resulting in a signal. If the compound cannot be accommodated within the boundaries of the biolayer, it will not provide a strong signal. Although this system has been used for detection of whole cells of *E. coli* O157:H7, *Salmonella*, and *Listeria* at variable concentrations, it showed strong signals with small toxin molecules, such as staphylococcal or botulinum.

- . *Electrochemical immunosensors*: Electrochemical immunosensors are based on conventional antibody-based enzyme immunoassays. In these applications, catalysis of substrates by an enzyme conjugated to an antibody produces products (ions, pH change, or oxygen consumption) capable of generating an electrical signal on a transducer. Potentiometric, capacitive, and amperometric transducers have been used for such applications. In amperometric detection, alkaline phosphatase conjugated to an antibody hydrolyzes nitrophenyl phosphate to produce phenol, which could be detected by voltammetry. In light-addressable potentiometric sensors (LAPS), urease attached to the antibody hydrolyzes urea, resulting in the production of carbon dioxide and ammonia, causing a change in the pH of the solution. A silicon chip coated with a pH-sensitive insulator and an electrochemical circuit detects the alternating photocurrent from a light-emitting photodiode on a silicon chip. These sensors are very sensitive and have been used to detect *Salmonella* and *E. coli* O157:H7 in 30–90 min.

- .*Microfluidic biochips*: Biochips are microelectronic or microfabricated electronic devices (such as semiconductor chips) that are used for monitoring the activities of biomolecules. Biochips have been used for monitoring eukaryotic or prokaryotic cell growth and DNA hybridization. Microbial growth on a biochip is measured by an impedance analyzer on a microscale level. The microbial metabolism of sugars or other substrates produces acids and ionic species. If the electrical property of the medium is monitored over time, the conductance and the capacitance increases while the bulk impedance decreases. A concentration level of 50–100 cells per nanoliter can be detected on a chip. In its current setup, the specificity of this type of sensor depends on the specific growth medium used. Alternatively, a specific antibody can be used to capture a target bacterium on the chip.

- .*Cell-based sensor*: One key feature of pathogenic bacteria is that they interact with eukaryotic cells as part of the infectious process. A gross change in the eukaryotic cell structure by a pathogen can be observed microscopically. However, if a sensor can be used to measure the change occurring at a single cell, the sensitivity of the detection is greatly enhanced. In principle, the compartments of a cell are separated from the surrounding medium by a cell membrane, which consists mainly of highly structured electrically insulating phospholipids. The electrical properties of the biological membrane can be modeled as a resistor and capacitor network. These membrane properties affect the conductivity of the cell system at alternating-current frequencies. Any external factors, such as live bacteria or active cytotoxins that affect the integrity of the membrane, alter the conductivity and provide a signal. The signal (impedance of the cells) can be measured with an interdigitated microsensor electrode. This type of sensor shows great promise for detecting potential pathogenic food-borne microorganisms and toxins.

- .*Piezoelectric (PZ) biosensors*: These sensors detect changes in the mass on the surface of a quartz crystal [38]. Antibodies are used for specific binding of the analytes, which increase the mass of the crystal while the resonance frequency of the oscillation decreases proportionately. *Salmonella* and *Listeria* have been detected with this system. A variation of the piezoelectric system called quartz crystal microbalance consists of a thin quartz disc with implanted electrodes.

5. **Biofinger/ nano detector**: This is one of the recent applications of nanotechnology in food safety and quality assurance. The machine detects and analyses molecules in fluids using nano and micro cantilevers. In the medical world, it is a way to rapidly and accurately diagnose disease. When coated with antibodies the cantilevers bend and resonate to changes in surface tension and mass when fluids containing disease-related protein molecules attach to them. By seeing whether or not the cantilevers react, it could be determined whether or not a pathogen is present. BioFinger incorporates the cantilevers on a disposable microchip, allowing it to be reconfigured with new on-chip canti-

levers to detect different substances. The analysis, which can be performed anywhere, anytime, takes between 15 and 20 minutes.

6. **Nano Bioluminescence Spray**: The spray will be a low cost system that will readily identify a broad range of food related pathogens, such as common *Salmonella* and *E. coli*. The spray, when applied to a food or beverage will react with the pathogen strain and produce a visual glow for detection. It uses nanotechnology to detect the harmful bacteria. The technique uses a small, luminescent protein molecule that has been modified so that it attaches itself to the surface of the target bacterium. The process would work in a similar fashion to an immune system antibody, designed to lock on to a particular feature on the "coat" of the microbe. In this case the higher the number of connections between bacteria and molecules, the more intense the glow produced.

7. **Electronic Nose**: This has been developed to detect volatile chemicals released by pathogens, some of which may be specific for the pathogen. More work on these lines is needed before it can be applied commercially.

8.5 CONCLUSION

It is concluded that the applications of immunochemical techniques, biosensors, DNA probes and the polymerase chain reaction are upcoming and reliable approaches for rapid and foolproof analysis of foods. Researchers are continuously searching for sensitive tools that are fast, accurate, and ultrasensitive. In addition to the identification and classification of microorganisms associated with foods or the food chain, global analysis methods are becoming increasingly available for analyzing these microorganisms and their environments in ways that will lead to a more complete understanding of how these organisms respond to their environments.

KEYWORDS

- **Animal Feed**
- **Foodborne Illness**
- **Monoclonal Antibody**
- **Polymerase**
- **Polymerase Chain Reaction**

REFERENCES

1. 1.Alice, A., Bungay, C., Methusyla, J., Estacio and Calvin, S., Reyes, de los. The "Emerging" Foodborne Bacterial Zoon. Philippine J Sci., 131(1), 29–35 (2002)
2. 2.Alocilja E.C. and Radke S.M. Market analysis of biosensors for food safety. Biosensors & Bioelectronics, 18, 841–846 (2003).

3.   3.Altekruse, S. F. N., Bauer, A. Chanlongbutra, R. DeSagun, A. Naugle, W. Schlosser, R. Umholtz and P. White. Salmonella Enteritidis in broiler chickens, United States, 2000-2005. Emerg. Infect. Dis., 12, 1848–1852 (2006).

4.   4.Beran, G. W., Shoeman, H. P., Anderson, K. F. Food safety: an overview of problems. Dairy Food Environ Sanit., 11, 189–94 (1991).

5.   5.Bhunia, A. K. and Johnson, M. G. Monoclonal antibody specific for Listeria monocyto-genes associated with a 66-kilodalton cell surface antigen. Appl. Environ. Microbiol., 58, 1924–1929 (1992).

6.   6.Bhunia, A. K. and Lathrop, A. McGraw-Hill Encyclopedia of Science & Technology. The McGraw-Hill Comp, Inc. Pp. 1-3 (2003).

7.   Bhunia, A. K., Ball, P. H., Fuad, A. T., Kurz, B. W., Emerson, J. W. and Johnson, M. G. Development and characterization of a monoclonal antibody specific for Listeria monocy-togenes and Listeria innocua. Infect. Immun., 59, 3176–3184 (1991).

8.   Bhunia, A.K. Biosensors and bio-based methods for the separation and detection of food-borne pathogens. Adv. Food Nutr. Res., 54, 1–44 (2007).

9.   Biswas, A.K., N. Kondaiah, K.N. Bheilegaonkar, A.S.R. Anjaneyulu and S.K. Mendiratta et al.,. Microbial profiles of frozen trimmings and silver sides prepared at Indian buffalo meatpacking plants. Meat Sci., 80, 418–422 (2008).

10.   Blackburn, C. de W. Rapid and alternative methods for the detection of salmonellas in foods. J. Appl. Bacteriol., 75, 199–214 (1993).

11.   Boening, D. W. and Tarr, P. I. Proposed method for isolation of Escherichia coli 0157: H7 from environmental samples. J. Environ. Health, 57, 9–21 (1995).

12.   Brunelle, S. Electroimmunoassay technology for food-borne-pathogen detection, IVD Technology, Medical Device Link (2001).

13.   Carter, M. J. Enterically infecting viruses: Pathogenicity, transmission and significance for food and waterborne infection. J. Applied Microbiol., 98, 1354–1380 (2005).

14.   CDC.,. Update on Multi-State Outbreak of E. coli 0157: H7 Infections from Fresh Spinach E. coli 0157: H7 Outbreak in Spinach. CDC., New York, New Jersey (2006).

15.   Chemburu, S., Wilkins, E., and Abdel-Hamid, I. Detection of pathogenic bacteria in food samples using highly-dispersed carbon particles. BiosensBioelectron, 21, 491–499 (2005).

16.   Chen, S. C. and Chang, T. C. Identification of Listeria monocytogenes based on the detec-tion of a 68-kilodalton cell surface antigen. J. Food. Prot., 59(11), 1176–1181 (1996).

17.   Cocolin, L., Manzano, M., Cantoni, C. and Comi, G. Use of polymerase chain reaction and restriction enzyme analysis to directly detect and identify Salmonella Typhimurium in food. J. Appl. Microbiol., 85, 673–677 (1998).

18.   Datta, A. R., Bary, A. W. and Hill, W. E. Detection of hemolytic Listeria monocytogenes by using DNA colony hybridization. Appl. Environ. Microbiol., 53(9), 2256–2259 (1987).

19.   De Boer, E. and Beumer, R. R. Methodology for detection and typing of foodborne micro-organisms. Int J Food Microbiol.,50, 119–30 (1999).

20.   Doyle, M. P. Food Microbiology: Fundamentals and Frontiers. 2nd Edn., ASM Press, Washillgton (2001).

21.   Doyle, M. P. and M. C. Erickson. Reducing the carriage of foodborne pathogens in live-stock and poultry. Poult. Sci., 85 960–973 (2006).

22.   Duffy, G., Kilbride, B., Fitzmaurice, J., and Sheridan, J. J. Routine diagnostic tests for Food-borne pathogens. The National Food Centre Teagasc, Dunsinea, Castleknock. pp.1–14 (2001).

23.   Falcao, J. P., Falcao, D. P., Pitondo-Silva, A., Malaspina, A. C. and Brocchi, M. Molecular typing and virulence markers of Yersinia enterocolitica strains from human, animal and

food origins isolated between 1968 and 2000 in Brazil. J Med Microbiol., 55, 1539–1548 (2006).

24. FDA (Food and Drug Administration). Fact sheet: Foodborne diseases, emerging FDA Bad Bug Book and Partnership for Food Safety Education No. 124. www.cfsan.fda.gov/~mow/intro.html (2002).

25. Feldhusen, F. The role of seafood in bacterial foodborne diseases. Microbes Infect., 2,1651–1660 (2000).

26. Feng, P. Commercial assay systems for detecting food-borne Salmonella: a review. J Food Prot, 55, 927–934 (1992).

27. Gangarosa, E. J., Barker, W. H. Jr., Baine, W. B., Morris, G. K. and Rice, P.A. Man v. animal feeds as the source of human salmonellosis. Lancet; 1, 878–879 (1973).

28. Groisman, E. A. and Ochman, H. The path to Salmonella. ASM News, 66, 21–26

29. Jin, S. S., Zhou, J., and Ye, J. Adoption of HACCP system in the Chinese food industry: a comparative analysis. Food Control, 19, 823–828 (2008).

30. Kay, D., Crowther, J., Fewtrell, L., Francis, C. A., Hopkins, M., Kay, C., et al. Quantification and control of microbial pollution from agriculture: a new policy challenge? Environ Sci Policy, 11:171–184 (2008).

31. Kerr, K. G., Rotowa, N. A., Hawkey, P. M. and Lacey, R. W. Evaluation of the Mast ID and API 50 CH systems for identification of Listeria spp. Appl. Environ. Microbiol., 56, 657–660 (1990).

32. Kim, S. G., Kim, E. H., Lafferty, C. J., and Dubovi, E. Coxiella burnetii in Bulk Tank Milk Samples, United States. Emerg. Infect. Dis. J., 11(4), (2005).

33. Lapeyre, C., de Solan, M. and Drouet, X. Immunoenzymatic detection of staphylococcal enterotoxins : International interlaboratory study. J. Assoc. off. Anal. Chem. Intl., 76, 1095–1101 (1996).

34. Li, X., Boudjellab, N. and Zhao, X. Combined PCR and slot blot assay for detection of Salmonella and Listeria monocytogenes. Int. J. Food. Microbiol., 56, 167–177 (2000).

35. Lynch, M. J., Painter, R. and Woodruff, C. Braden and Centers for Disease Control and Prevention,. Surveillance for foodborne disease outbreaks: United States, 1998–2002. MMWR. Surveill. Summ. Rep., 55, 1–42 (2006).

36. Malorny, B. Paccassoni, E. Fach, R. Bunge, C. Martin, A., and Helmuth, R. Diagnostic Real-Time PCR for Detection of Salmonella in Food. Appl. Environ. Microbiol., 70(12), 7046–7052 (2004).

37. Mandal, P. K., Biswas, A. K., Choi, K., and Pal U. K.,. Methods for Rapid Detection of Foodborne Pathogens: An Overview. American Journal of Food Technology, 6(2), 87–102 (2011).

38. Mattingly, J. A., Butman, B. T., Plank, M. C., Durham, R. J. and Robison, B. J. Rapid monoclonal antibody based enzyme linked immunosorbent assay for detection of Listeria in food products. J. AOAC, 71, 679–681 (1988).

39. Mead, P. S. and Griffin, P. M. Escherichia coli O157:H7. Lancet, 352, 1207–1212 (1998).

40. Mead, P. S., SlutskerL, DietzV, McCaig, L.F., Bresee, J.S., Shapiro,C., et al. (1999). Food-related illness and death in the United States. Emerg Infect Dis 5,607–25.

41. Mucchetti, G., Bonvini, B., Francolino, S., Neviani, E., Carminati, D. (2008). Effect of washing with a high pressure water spray on removal of Listeria innocua from Gorgonzola cheese rind. Food Control, 19:521–5.

42. Naravaneni, R. and Jamil, K. (2005). Rapid detection of food-borne pathogens by using molecular techniques. Med Microbiol., 54: 51-54.

43. OIE (2004). Manual of Diagnostic Tests and Vaccines for Terrestrial Animals. http://www.oie.int/fr/normes/mmanual/a_00138.htm.

44. Park, Y.H., (2001). Application of biotechnology in animal products safety. Proceeding of 3rd CJK Symposium on Biotechnology and Animal Production, (CSBAP'01), Jinju National University, Korea, pp: 33-69.

45. Pearson, A.M. and Dutson, T.R. (1994). Quality Attributes and their Measurement in Meat, Poultry and Fish Products. Advances in Meat Research Vol. 9, Blackie Academic and Professional, pp. 429-433.

46. Peterson, R. (2005). Biosensors using Real-time PCR Technology for Early Detection of Zoonotic Pathogens in Livestock and Animal Origin Food Products in the United States. In: Crossover. The Pennsylvania State University, University Park, PA.

47. Piatek, D.R. and Ramaen, D.L.J., (2001). Method for controlling the freshness of food products liable to pass an expiry date, uses a barcode reader device that reads in a conservation code when a product is opened and determines a new expiry date which is displayed [Patent Number: FR2809519-A1], 2001.

48. Potter ME, Gonzalez-Ayala S, Silarug N. (1997). Epidemiology of foodborne diseases. In: Doyle MP, Beuchat LR, Montville TJ, editors. Food Microbiology: Fundamentals and Frontiers. Washington, DC: ASM Press.

49. Rawool, D.B.; Malik, S.V.S.; Shakuntala, I.; Sahare, A.M. and Barbuddhe, S.B. (2006). Detection of multiple virulence associated genes in pathogenic Listeria monocytogenes from bovines with mastitis. Int. J. Food Microbiol. (Published online on Sciencedirect on 18 September, 2006).

50. Schalch, B., Bader, L., Schau, H., Bergmann, R., Rometsch, A. Maydl, G. and Keßler, S. (2003). Molecular Typing of Clostridium perfringens from a Food-Borne Disease Outbreak in a Nursing Home: Ribotyping versus Pulsed-Field Gel Electrophoresis. J Clin Microbiol., 41(2): 892–895.

51. Shiferaw, B., Yang, S., Cieslak, P., et al. (2000) Prevalence of high-risk food consumption and food-handling practices among adults: a multistate survey, 1996 to 1997. J. Food Prot., 63: 1538-43.

52. Siemer, B.L. Nielsen, E.M., and On, S.L.W. (2005). Identification and Molecular Epidemiology of Campylobacter coli Isolates from Human Gastroenteritis, Food, and Animal Sources by Amplified Fragment Length Polymorphism Analysis and Penner Serotyping. Appl Environ Microbiol., 71(4): 1953-1958.

53. Singhal, R.S.; Kulkarni, P.R. and Rege, D.V. (1997). New techniques for food analysis. In: Handbook of indices of food quality and authenticity. Woodhead publishing Limited. Cambridge England. Pp. 19-31.

54. Sockett, P.N. (1991). The economic implications of human Salmonella infection. J Appl. Bacteriol., 71: 289–295.

55. Swaminathan B, Gerner-Smidt P (2006). Foodborne Disease Trends and Reports .Foodborne Pathog Dis 3:220–221

56. Tauxe, R.V. and Pavia, A.T. (1998). Salmonellosis: nontyphoidal. In: Evans AS, Brachman PS, eds. Bacterial infections of humans: epidemiology and control. 3d ed. New York: Plenum, Pp. 613–30.

57. Taylor E. (2007). A new method of HACCP for the catering and food service industry. Food Control 19:126–34.

58. Umali-Deininger D, Sur M. (2007). Food safety in a globalizing world: opportunities and challenges for India. Agric Econ;37:135–47.

59. USDA/ERS (Economic Research Service). Economics of foodborne disease; 2002. http://www.ers.usda.gov/briefing/FoodborneDisease/.

60. Vaidya, V.M. (2006). Studies on prevalence of Q fever in man and domestic animals by molecular, serological and isolation methods. Thesis submitted to Indian Veterinary Reserach Institute, Deemed University (ICAR)., Izatnagar - 243 122.
61. Vasavada, P.C., (1993). Rapid methods and automation in dairy microbiology. J. Dairy Sci., 76: 3101-3113
62. Wee, S.; Lee, C.; Joo, H. and Kang, Y. (2001). Enzyme-linked immunosorbent assay for detection of Trichinella spiralis antibodies and the surveillance of selected pig breeding farms in the Republic of Korea. The Korean J Parasitolo. 39 (3): 261-264.
63. Wilson, S.G., Chan, S., Deroo, M., Veragarcia, M., Johnson, A., Lane, D. and Halbert, D.N. (1990). Development of a colorimetric, second generation nucleic acid method for detection of Salmonella in foods and a comparison with conventional culture procedure. J. Food Sci., 55: 1394-1398.
64. World Health Organization (2006). Zoonoses and Veterinary Public Health. http://www. veterinary-public-health.de/home_e/index_e.htm.
65. WHO. Food safety & food-borne illness. fact sheet no. 237 (reviewed March 2007). Geneva: World Health Organization; 2007.

CHAPTER 9

CROSS-LINKING OF FERULATED ARABINOXYLANS EXTRACTED FROM MEXICAN WHEAT FLOUR: RHEOLOGY AND MICROSTRUCTURE OF THE GEL

A. MORALES-ORTEGA, E. CARVAJAL-MILLAN, P. TORRES-CHAVEZ, A RASCÓN-CHU, J. LIZARDI-MENDOZA, and Y LÓPEZ-FRANCO

CONTENTS

9.1 INTRODUCTION

In the present chapter, water extractable arabinoxylans (WEAX) from Mexican spring wheat flour (cv. Tacupeto F2001) were enzymatically cross-linked. WEAX solution at 3% (w/v) formed gels induced by a laccase as cross-linking agent. Cured WEAX gels registered storage (G') and loss (G") modulus values of 71 and 16 Pa. The mechanical spectra of WEAX after 2 hr gelation was typical of solid-like materials with a linear G' independent of frequency and G" much smaller than G' and dependent of frequency.

Scanning electron microscopy (SEM) analysis of the lyophilized WEAX gels showed that this material resembles that of an imperfect honey comb.

Arabinoxylans (AX) are important cereal non-starch polysaccharide constituted of a linear backbone of a$(1\rightarrow4)$-linked D-xylopyranosyl units to which α-L-arabinofuranosyl substituents are attached through O-2 and/or O-3. Some of the arabinose residues are ester linked on (O)-5 by ferulic acid (FA) (3-methoxy, 4 hydroxy cinnamic acid) [1]. These polysaccharides have been classified as water extractable (WEAX) or water-unextractable (WUAX). The WEAX form highly viscous solutions and gel through ferulic acid covalent cross-linking upon oxidation by some chemical or enzymatic free-radicals generating agents [2] (Figure 1). Diferulic acids (di–FA) and triferulic acid (tri–FA) have been identified as covalently cross-linked structures in laccase-gelled AX. Both covalent bridges (di–FA, tri–FA) and physical interactions between AX chains have been reported to be involved in the gelation process and the final gel properties [3-5]. Most of polysaccharide gels are stabilized by physical interactions (hydrogen bonding and/or ionic interaction) but polysaccharide covalent networks such as WEAX gels are not very common. Covalently cross-linked gels are generally strong, form quickly, are not temperature dependent and exhibit no synersis after long time storage. Furthermore, WEAX gels have interesting functional properties, which have not been exploited even though the WEAX neutral taste and odor are desirable properties for industrial applications. The WEAX networks have a high water absorption capacity (up to 100 g of water per gram of polymer) and they are not sensible to electrolytes or pH. Tacupeto F2001 is a spring wheat variety developed by The International Maize and Wheat Improvement Center (CIMMYT) in Mexico, and provided to National Institute for Investigation in Forestry, Agriculture and Animal Production (INIFAP) for testing and release [1,2]. The INIFAP released this wheat variety for cultivation in Northwestern Mexico. Tacupeto F2001 presents resistance to leaf rust [6, 7] To improve the quality of wheat varieties (*Triticum aestivum* L.) have been developed improvement programs that require a comprehensive understanding of the constituents of grain as the biochemical constituents of wheat grain largely determine its end-use quality [8]. Nevertheless, research on WEAX from improved wheat varieties such as Tacupeto F2001 are not common even though WEAX are key constituents of wheat grain playing an important role on the grain functionality. Research on WEAX from Tacupeto F2001 can be useful to wheat-improvement programs where the aim is to develop wheat cultivars with superior and consistent endues quality. The objective of this study was to investigate the rheology and microstructure of covalent gels formed by using WEAX from Tacupeto F2001.

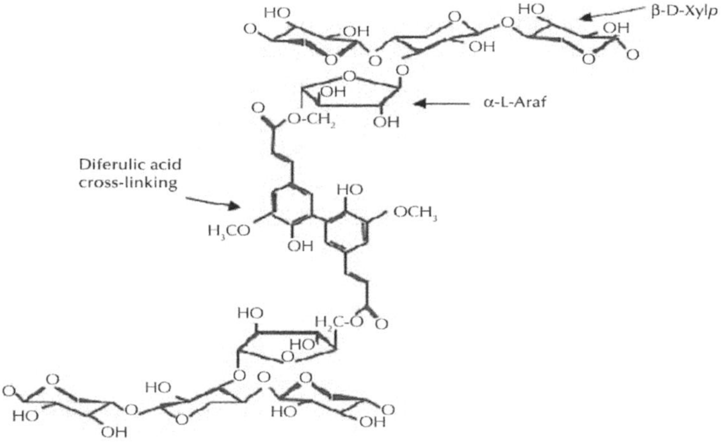

FIGURE 1   Chemical structure of covalent cross-linked WEAX.

9.2   MATERIALS AND METHODS

9.2.1   MATERIALS

The WEAX were extracted from the endosperm of a spring wheat variety (Tacupeto F2001). The WEAX presented an A/X ratio of 0.66, a Mw of 521 kDa, a [η] of 3.5 dL/g, and Mv of 504 kDa [7]. The FA content of WEAX was 0.526 μg/mg WEAX. Commercial laccase (benzenediol:oxygen oxidoreductase, E.C.1.10.3.2) was from *Trametes versicolor* was used as cross-linking agent. All chemical reagents were purchased from Sigma Chemical Co. (St Louis, MO, USA). The grain sample was milled by using Quadrumat Sr mill (Branbender, South Hackensack, NJ) according to the AACC method 26–10 [9].

9.2.2   METHODS

*WEAX Gelation*A WEAX solution (3 % w/v) was prepared in 0.05 M citrate phosphate buffer pH 5.5. Laccase (1.675 nkat per mg WEAX) was added to WEAX solution as cross-linking agent. Gels were allowed to develop for 2 hr at 25°C [10].

   *Laccase Activity*Laccase activity was measured at 25°C from a laccase solution at 0.125 mg/ml dissolved in 0.05 M citrate-phosphate buffer pH 5.5 as previously reported [4].

RHEOLOGY

*Small Deformation Measurements*Small amplitude oscillatory shear was used to follow the gelation process of WEAX solution. Cold (4°C) WEAX solution (3% w/v) in 0.05 M citrate phosphate buffer pH 5.5 was mixed with laccase and immediately poured on cone-plate geometry (5.0 cm in diameter and 0.04 rad in cone angle) of a strain controlled rheometer (Discovery Hybrid Rheometer, TA Instruments). Exposed edges were recovered with silicone to prevent evaporation. WEAX gelation was started by a sudden increase of temperature from 4 to 25°C and monitored at 25°C for 2 hr by recording the storage (G') and loss (G") moduli. Measurements were carried out at 1.0 Hz frequency and 10% strain. From strain sweep tests, WEAX gels showed a linear behavior from 0.02 to 100% strain. 10% strain was used in all the rheological measurements. The mechanical spectra of gels were obtained by frequency sweep from 0.01 to 10.0 Hz with a 10% strain at 25°C [3,10].

LARGE DEFORMATION MEASUREMENTS

The hardness of 3% (w/v) WEAX gels made in 6 mL glass flasks of 30 mm height and 25 mm internal diameter was analyzed with a TA.XT2 Texture Analyzer (Stable Micro Systems, Godalming, England). The gels were deformed by compression at a constant speed of 1.0 mm/s to a distance of 4 mm from the gel surface using a cylindrical plunger (diameter 15 mm). The maximum force obtained from the force versus distance curve was recorded as a measure of gel hardness [11].

STRUCTURE OF FREEZE DRIED WEAX GELS

The WEAX gels at 3% (w/v) were frozen at −20°C and lyophilized at −37°C/0.133 mbar overnight in a Freezone 6 freeze drier (Labconco, Kansas, MO). The external structure of the freeze-dried WEAX gel was analyzed with a stereo light microscope (Leica CLS 150 XE Leica Microsystems®, Switzerland) at a low magnification (10×). The internal structure of freeze-dried WEAX gel was studied by SEM (JEOL 5410LV) at low voltage (20 kV). The SEM image was obtained in secondary electrons image mode.

9.3   DISCUSSION AND RESULTS

9.3.1   WEAX GELATION

The cross-linking process of WEAX was rheologically investigated by small amplitude oscillatory shear. Figure 2 shows the development of storage (G') and loss (G") moduli versus time of 3% (w/v) WEAX solution undergoing oxidative gelation by laccase. G' and G" moduli rise to reach a pseudo plateau region. This behavior reflects an initial formation of covalent linkages between FA of adjacent WEAX molecules producing a three-dimensional network. Once sufficient cross-links have formed, movement of chains is impeded by the rigidity of the gel. The final G' and G" values of 3% (w/v) were 71 and 16 Pa, respectively. These results are in the range reported for WEAX gelled by laccase [3,4]. The gelation time (tg), calculated from the cross-over of the G' and G" curves (G' >G") was 3 min. The tg value indicates the sol/gel transition point and at this point G' = G" or tan δ = G"/G' = 1 [12]. The mechanical spectra of WEAX after 2 hr gelation (Figure 3) was typical of solid-like materials with a linear G' independent of frequency and G" much smaller than G' and dependent of frequency [13]. This behavior is similar to that previously reported for WEAX cross-linked by laccase or peroxidase/H_2O_2 system [14-16].

The WEAX solutions at 3% (w/v) produced firm and brittle gels in presence of laccase. The hardness of the WEAX gels is presented in Figure 4. Hardness is related to the strength of the gel structure under compression. The WEAX gel registered a hardness value of 3.5 N, which is higher than those previously reported for laccase induced wheat flour and maize bran arabinoxylan gels (0.5–1 N) prepared at different concentrations (1–8% w/v) [1,4]. Such behavior might have its origin in the structural and/or conformational characteristics of these macromolecules. Clearly, further studies on the distribution of arabinose and feruloyl groups along the polymer chain backbone are needed to establish relationships between the molecular structure and gelling ability of WEAX.

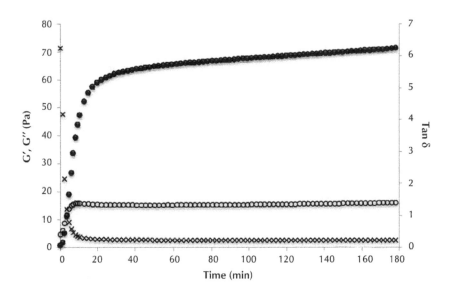

FIGURE 2   Rheological kinetics of 3% (w/v) WEAX solution gelation by laccase. G' (○), G" (●), and tan δ (×). Measurements at 25°C, 1 Hz and 10% strain.

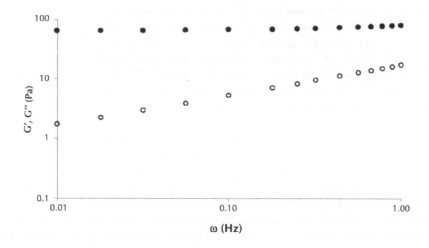

FIGURE 3   Mechanical spectrum of WEAX gel at 3% (w/v). G' (○), G" (●). Measurements at 25°C, 1 Hz and 10% strain.

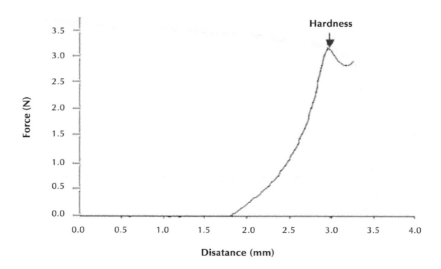

FIGURE 4   Experimental curve obtained from compression test of gels at 3% (w/v) in WEAX.

9.3.2   FREEZE DRIED WEAX GEL

In Figure 5 is shown a WEAX gel image after freeze drying (A). Figure 5 (B) shows a stereomicrograph of the freeze dried WEAX gel external structure. It is possible that frozen caused the crust formation of the sample. The internal structure of the lyophilized WEAX gel was observed by SEM (Figures 5 (C) and (D)). The WEAX gel network presents many connections and can be compared with an irregular honeycomb structure. In this study the average inner dimensions of the cell were approximately 100 x 200 µm (Figure 5 (D)). This morphological microstructure is similar to that reported before for lyophilized wheat and maize AX gels [7,17-20]. However, the SEM microstructure of freeze dried WEAX gels are different to that recently reported for supercritical CO_2-dried WEAX aerogels which present more spongy network structure [21].

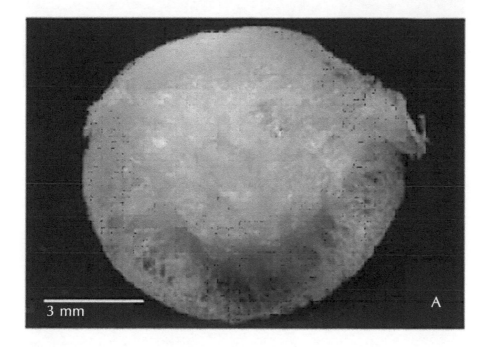

3 mm

A

FIGURE 5   *(Continued)*

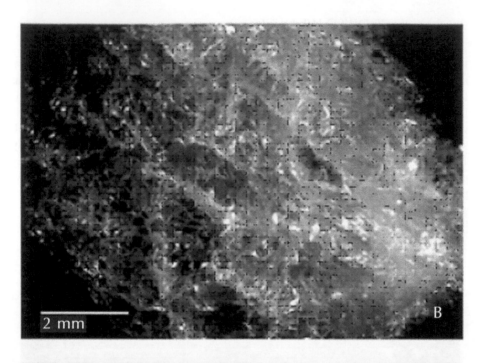

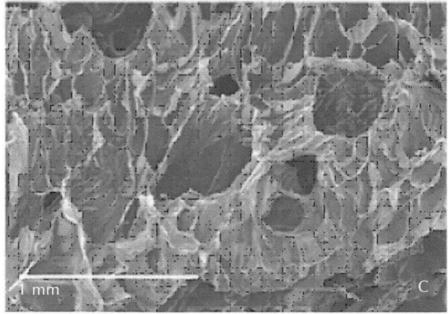

FIGURE 5   *(Continued)*

FIGURE 5   Lyophilized WEAX gel (A), stereomicrograph of lyophilized WEAX gel (B), SEM micrographs of lyophilized WEAX gel at 35× magnification (C), and 200× magnification (D).

9.4   CONCLUSION

The WEAX from the spring wheat variety Tacupeto F2001 are able to form gel in the presence of laccase as shown by dynamic rheometry. Freeze dried arabinoxylan gels present a porous network constituted by an irregular honeycomb structure. Understanding Tacupeto F2001 arabinoxylan gels characteristics can be useful to propose alternative uses of this improved wheat cultivar.

9.5   FUTURE CONSIDERATIONS

The WEAX gels continue to be investigated and functional properties are being proposed. Several questions remained to be elucidated, especially those concerning the effect of food process on health benefits of these compounds. In addition, arabinoxylan gels could be matrices suited for bioactive compounds or microorganisms.

KEYWORDS

- Ferulated arabinoxylans
- Gelation
- Macromolecule
- Scanning electron microscopy
- Trametes Versicolor

ACKNOWLEDGMENT

This research was supported by Fondo Proyectos y fortalecimiento de Redes Temáticas CONACYT de investigación formadas en 2009 (Grant 193949 to E. Carvajal-Millan).

REFERENCES

1. Izydorczyk, M. S. and Biliaderis, C. G. Cereal arabinoxylans: advances in structure and physicochemical properties. *Carbohydr Polym*, **28**, 33–48 (1995).
2. Niño-Medina, G., Carvajal-Millán, E., Rascon-Chu, A., Márquez-Escalante, J. A., Guerrero, V., and Salas-Muñoz, E. Feruloylated arabinoxylans and arabinoxylan gels: structure, sources, and applications. *Phytochem Rev*, **9**, 111–120 .
3. Vansteenkiste, E., Babot, C., Rouau, X., and Micard, V. Oxidative gelation of feruloylated arabinoxylan as affected by protein. Influence on protein enzymatic hydrolysis. *Food Hydrocolloids*, **18**, 557–564 (2004).
4. Carvajal-Millan, E., Guigliarelli, B., Belle, V., Rouau, X., and Micard, V. Storage stability of arabinoxylan gels. *Carbohydr Polym*, **59**, 181–188 .
5. Berlanga-Reyes, C. M., Carvajal-Millan, E., Lizardi-Mendoza, J., Islas-Rubio, A. R., and Rascón-Chu, A. Enzymatic cross-linking of alkali extracted arabinoxylans: gel rheological and structural characteristics. *Int J Mol Sci.*, **12**(9), 5853–61 (2011).
6. Camacho Casas, M. A., Singh, R. P., Figueroa López, P., Huerta Espino, J., Fuentes Dávila, G., and Ortiz-Monasterio Rosas, I. *Tacupeto F2001 nueva variedad de trigo harinero para el noroeste de México. Folleto Técnico No. 50*, Talleres Gráficos de CIRNO: Obregón, México, pp. 1–20 (2003).
7. Morales-Ortega, A., Carvajal-Millan, E., López-Franco, Y., Rascón-Chu, A., Lizardi-Mendoza, J., Torres-Chavez, P., and Campa-Mada, A. *Molecules. Submitted.*
8. Li, S., Morris, C. F., and Bettge, A. D. Genotype and environment variation for arabinoxylans in hard winter and spring wheats of the U.S. Pacific Northwest. *Cereal Chem*, **86**, 88–95 (2009).
9. AACC. *Approved Methods of the American Association of Cereal Chemists*. 10a ed., Am. Assoc. Cereal Chemists, St. Paul, MN, USA (2000).
10. Carvajal-Millan, E., Landillon, V., Morel, M. H., Rouau, X., Doublier, J. L., and Micard, V. Arabinoxylan gels: impact of the feruloylation degree on their structure and properties. *Biomacromolecules*, **6**, 309–317 (2005).
11. Urias-Orona, V., Huerta-Oros, J., Carvajal-Millán, E., Lizardi-Mendoza, J., Rascón-Chu, A., and Gardea, A. A. Component analysis and free radicals scavenging activity of *Cicer arietinum* L. husk pectin. *Molecules*, **15**, 6948–6955 (2010).

12. Doublier, J. L. and Cuvelier, G. Gums and hydrocolloids: functional aspects. In *Carbohydrates in Food*, A. C. Eliasson (Ed.), Marcel Dekker, New York, USA, pp. 283–318 (1996).
13. Ross-Murphy, S. B. Rheological methods. In *Biophysical methods in food research*. H. W. S. Chan (Ed.), Blackwell Scientific Publications, Oxford, United Kingdom, pp. 138–199 (1984).
14. Dervilly-Pinel, G., Rimsten, L., Saulnier, L., Andersson, R., and Aman, P. Water-extractable arabinoxylan from pearled flours of wheat, barley, rye, and triticale. Evidence for the presence of ferulic acid dimmers and their involvement in gel formation. *J Cereal Sci*, **34**, 207–214 (2001).
15. Izydorczyk, M. S., Biliaderis, C. G., and Bushuk, W. Oxidative gelation studies of water-soluble pentosans from wheat. *J Cereal Sci*, **11**, 153–169 (1990).
16. Dervilly, G., Saulnier, L., Roger, P., and Thibault, J. F. Isolation of homogeneous fractions of wheat water-soluble arabinoxylan. Influence of structure on their macromolecular characteristics. *J Agric Food Chem*, **48**, 270–278 (2000).
17. Pae☐s, G. and Chabbert, B. Characterization of Arabinoxylan/Cellulose Nanocrystals Gels to Investigate Fluorescent Probes Mobility in Bioinspired Models of Plant Secondary Cell Wall. *Biomacromolecules*, **13**, 206–214 (2012).
18. Iravani, S., Fitchett, C. S, and Georget, D. M. R. Physical characterization of arabinoxylan powder and its hydrogel containing a methyl xanthine. *Carbohydr Polym*, **85**, 201–207 (2011).
19. Martínez-López, A. L., Carvajal-Millan, E., Miki-Yoshida, M., Alvarez-Contreras, L., Rascón-Chu, A., Lizardi-Mendoza, J., and López-Franco, Y. Arabinoxylan Microspheres: Structural and Textural Characteristics. *Molecules*, **18**, 4640–4650 (2013).
20. Martínez-López, A. L., Carvajal-Millan, E., Rascón-Chu, A., Márquez-Escalante, J., and Martínez-Robinson, K. Gels of ferulated arabinoxylans extracted from nixtamalized and non-nixtamalized maize bran: rheological and structural characteristics. *CyTA-J Food*, DOI:10.1080/19476337.2013.781679 (2013).
21. Marquez-Escalante, J., Carvajal-Millan, E., Miki-Yoshida, M., Alvarez-Contreras, L., Toledo-Guillén, A. R., Lizardi-Mendoza, J., and Rascón-Chu, A. Water Extractable Arabinoxylan Aerogels Prepared by Supercritical CO_2 Drying. *Molecules*, **18**, 5531–5542 (2013).

CHAPTER 10

FREE AND ESTER-LINKED FERULIC ACID CONTENT IN A HARD-TO-COOK PINTO BEAN (*PHASEOLUS VULGARIS L.*) VARIETY

AGUSTHN RASCON-CHU, KARLA ESCARCEGA-LOYA, ELIZABETH CARVAJAL-MILLAN, and ALFONSO SÁNCHEZ

CONTENTS

10.1   INTRODUCTION

Consumption of beans is influenced by hard-to-cook (HTC) behavior and the existence of certain phytochemicals such as ferulic acid. In this study, free and ester-linked ferulic acid were quantified in a HTC and a soft (control) bean varieties. A higher content of free ferulic acid was present in the control sample (61 mg/100 g of dry matter) than in HTC beans (42 mg/100 g of dry matter). Ester-linked ferulic acid was higher in HTC than in control beans (88 mg/100 g of dry matter) than in control (29 mg/100 g of dry matter). The cooking process reduced free and ester-linked ferulic acid content by 60% and 49% and for

82% and 54% for HTC and control, respectively. The HTC beans presented a lower luminosity and a higher breaking hardness in comparison to the control beans.

Beans play an important role in human nutrition since they are rich in protein, fiber, certain minerals, and vitamins. Nevertheless, prolonged storage of common beans at high temperature and humidity causes the textural defect known as HTC [2]. The HTC beans are undesirable as they require longer soaking and cooking time compared to fresh beans and are less palatable. The HTC phenomenon is attributable to changes in the cotyledon tissue, which include formation of insoluble pectinates, lignification of the cell wall and middle lamella, interaction of condensed phenolic compounds with proteins and starch, and changes to the structure and functionality of the cellular proteins and starch [16]. Rozo described a decrease in the extraction of soluble tannins of the seed coat during hardening [11]. It was inferred that this could be related to an increase in condensed tannins in the seed coat and a possible migration of soluble tannins to the cotyledon.

Pinto Villa beans have traditionally been a staple food for the people in Northern Mexico. Nevertheless, Pinto Villa variety presents a high susceptibility to develop the HTC phenomenon. This defect results in economic losses either through the rejection of beans by consumers for its poor texture or due to the need for increased energy and time required for cooking. The relationship between HTC phenomenon and the changes in the phenolic compounds of the grain has not been yet assessed. In this paper we describe changes in the content of free and ester-linked ferulic acid in a HTC and a soft bean variety (*Phaseolus vulgaris*).

10.2   MATERIALS AND METHODS

10.2.1   MATERIALS

The HTC Pinto Villa and control bean (*Phaseolus vulgaris* L.) were supplied by a commercial proveyor in Northern Mexico. All chemicals were reagent pure grade.

10.2.2   COOKING PROCESS

The HTC Pinto beans and control beans were cooked in a pressure casserole during 1 hr. A bean:water ratio of 1:2 (w/v) was used.

10.2.3   FREE AND ESTER-LINKED FERULIC ACID CONTENT

Total ferulic acid content in the samples was quantified by HPLC after de-esterification step as described by Vansteenksite et al. [14]. 100 mg of sample was allowed to react with 1 mL of 2 N NaOH for 2 hr in the dark at 35°C under argon. After adding 3,4,5-trimethoxy-trans-cinnamic acid (TMCA, internal standard, 10 μg), the pH was adjusted to 2.0 ± 0.2 with 4N HCl. Phenolics were extracted twice with diethyl ether, and evaporated at 30°C under argon. The dried extracts were solubilized in 0.50 mL methanol/water/acetic acid (40/59/01), filtered (0.45 μm) and injected (20 μl) into HPLC using a Supelcosil LC-18-DB (250 ΄ 4.6 mm) (Supelco, Inc., Bellefont, PA) column. Detection was by UV absorbance at 280 nm. Isocratic elution was performed using methanol/water/acetic acid (40/59/01) at 0.6 mL/min at 35°C. A Varian 9012 photodiode array detector (Varian, St. Helens, Australia) was used to record the ferulic acid spectra. A Star Chromatography Workstation system control version 5.50 was used. Free ferulic acid content was quantified by using the same procedure excepting the alkali treatment. The ester-linked ferulic acid content was determined as the difference between saponified and non-saponified extracted phenolics.

10.2.4   COLOR

Cooked HTC and control beans were analyzed with a color difference meter (model CR-300, Minolta, Japan). Lightness (L), redness (a), and yellowness (b) values were recorded. Each value was an average of ten different independent measurements.

10.2.5   BREAKING STRENGTH

Breaking strength of the HTC and control cooked beans was determined with the TA-XT2 Texture Analyzer (Texture Technology Corp., Scarsdale, N.Y., U.S.A.) based on the method of Holguín-Acuña et al. [6]. Ten replications were conducted for samples from each treatment.

10.3   DISCUSSION AND RESULTS

In HTC Pinto Villa and control bean samples, the major monomeric phenolic acid was ferulic acid. Alkali treatment of beans released a large amount of ferulic acid. The concentration of ferulic acid present in HTC beans was 42 mg/100 g of dry matter (DM) without saponification and 88 mg/100 g DM after alkali treatment (total ferulic acid content). In the soft control bean lower total ferulic acid content was found with the free fraction being higher (61 mg/100 g DM). These values were noticeably higher than the quantity reported in studies carried on other beans varieties which ranged from 19.1 up to 48.3 mg/100 g DM [7]. After cooking, the content of free and ester-linked ferulic acid was 60% and 49% lower than in the raw sample, respectively for HTC bean. In control beans, free and ester-linked ferulic acid showed a reduction of 82% and 54% respectively. Ester-linked ferulic acid is apparently more stable to heat.

In this regard, ferulic acid is a ubiquitous grain component [5,15], which is bound by an ester linkage to some a-L-arabinose units of cell wall polysaccharides [12]. Two ester linked FA monomers may undergo an oxidative coupling to induce polysaccharides cross-linking. This is a known cell wall process in cereals as wheat [3], maize [4,8], and sugar beet [1]. Considerable interest has been shown in the involvement of ferulic acid in the mechanical properties of plant cell wall. Many studies have shown that light stimulates ferulate dimerisation in cell walls, and thereby increase the cell-wall rigidity resulting in limited cell extension [9,10,13]. Therefore, the high ester-linked ferulic acid content in Pinto Villa beans suggests that ferulate mediated cross-linking between cell wall components as pectin or hemicelluloses could be one of the contributing chemical changes related to grain hardening.

Color changes related to phenolics are also found darkening plant cell wall. Color L* value was lower in HTC cooked beans (Table 3) in comparison to that found in control cooked bean indicating a darker grain in comparison to the control. The latter could be related to the higher phenolics content in HTC beans. In a similar way, HTC cooked samples presented a higher breaking strength, which could be attributed to the little expansion capacity of HTC cell wall polysaccharides interwoven chains (Table 3).

TABLE 1 Free and ester-linked ferulic acid content in HTC and control beans

Ferulic acid (mg/100 g of dry matter)		
Sample	Free	Ester-linked
HTC	42 b	88 a
Control	61 a	29 b

Mean values in the same column with different letters are significantly different (P< 0.05).

TABLE 2 Free and ester-linked ferulic acid content in HTC and control beans

Ferulic acid (mg/100 g of dry matter)		
Sample	Free	Ester-linked
HTC	17 b	45 a
Control	11 b	13 a

Mean values in the same column with different letters are significantly different (P< 0.05).

TABLE 3 Cooked HTC and control beans varieties physical characteristics

Ferulic acid (mg/100 g of dry matter)		
Sample	Luminosity	Breaking Strength
HTC	68 b	75 a
Control	76 a	59 b

Mean values in the same column with different letters are significantly different (P< 0.05).

10.4 CONCLUSION

The HTC beans present a higher total ferulic acid in comparison to that reported in other beans varieties. In this bean variety, 68% of the ferulic acid content is esterified to other compounds and probably related to the cell wall components responsible for texture features. The implication of this esterified ferulic acid in the HTC phenomenon is suggested as complementary evaluations are in process in order to clarify this behavior. For the food industry, the knowledge on HTC beans cell wall changes could translate in new methods and techniques to manage bean storage and diminish this phenomenon. On the other hand, new biomaterials are always welcome in other industrial applications.

KEYWORDS

- **Beans**
- **Free ferulic acid**
- **Ester-linked ferulic acid**
- **Hard-to-cook**

ACKNOWLEDGMENT

This work has been carried out with grant from FOMIX CONACYT-Chihuahua (grant Chih-2006-C02-59228 to A. Rascon-Chu, PhD). Escarcega-Loya, K. thanks CONACYT, Mexico for providing an MSc scholarship. The authors wish to thank Nora Ponce de León, and Jorge A. Márquez-Escalante for technical assistance.

REFERENCES

1. Abdel-Massih, R. M., Baydoun, E. A. H., Waldron, K. W., and Brett, C. T. Effects of partial enzymic degradation of sugar beet pectin on oxidative coupling of pectin-linked ferulates in vitro. *Phytochemistry*, **68**, 1785–1790 .
2. Barron, J. M., Cota, A. G., Anduaga, R., and Rentería, T. R. Influence of the hard-to-cook defect in pinto beans on the germination capacity, cookability and hardness of newly harvested grains. *Trop. Sci.*, **36**, 1–5 .
3. Berlanga-Reyes, C. M., Carvajal-Millan, E., Lizardi-Mendoza, J., Islas-Rubio, A. R., and Rascon-Chu, A. Enzymatic cross-linking of alkali extracted arabinoxylans: Gel rheological and structural characteristics. *Int. J. Mol. Sci.*, **12**, 5853–5861 .
4. Carvajal-Millan, E., Rascon-Chu, A., Marquez-Escalante, J. A., Micard, V., Ponce de Leon, N., and Gardea, A. Maize bran gum: Extraction, characterization and functional properties. *Carbohydrates polymers*, **69**, 280–285 .
5. Fincher, G. B. Ferulic acid in barley cell walls: a fluorescence study. *J. Inst. Brew.* **82**, 347–349 .
6. Holguín-Acuña, A. L., Carvajal-Millán, E., Santana-Rodríguez, V., Rascón-Chu, A., Márquez-Escalante, , Ponce de León-Renova, ,and Gastelum-Franco, G. Maize bran/oat

flour extruded breakfast cereal: a novel source of complex polysaccharides and an antioxidant. *Food Chemistry*, **111**, 654–657 .

7. Luthria, D. L. and Pastor-Corrales, M. A. Phenolic acids content of fifteen dry edible bean (*Phaseolus vulgaris* L.) varieties. *Journal of food composition and analysis*, **19**, 205–211 Martínez-López, A. L., Carvajal-Millan, E., Rascón-Chu, A., Márquez-Escalante, J., and Martínez-Robinson, K. Gels of ferulated arabinoxylans extracted from nixtamalized and non-nixtamalized maize bran: rheological and structural characteristics, *CyTA Journal of Food*, DOI:10.1080/19476337.2013.781679

8. Miyamoto, K., Ueda, J., Takeda, S., Ida, K., Hoson, Y., Masuda, Y., and Kamisaka, S. Light-induced increase in the contents of ferulic and diferulic acids in cell walls of *Avena* coleoptiles: its relationship to growth inhibition by light. *Physiol. Plant.*, **92**, 350–355 .

9. Parvez, M. M., Wakabayashi, K., Hoson, T., and Kamisaka, S. Changes in cellular osmotic potential and mechanical properties of cell walls during light-induced inhibition of cell elongation in maize coleoptiles. *Plant Physiol.*, **96**,179–185 .

10. Rozo, C. Effect of extended storage on the degree of thermal softening during cooking cell wall components and polyphenolic compounds of red kidney beans (*Phaseolus vulgaris*). *Diss. Abstr. Int. B.*, **42**, 4732–4736 .

11. Smith, M. M. and Hartley, R. D. Occurrence and nature of ferulic acid substitution of cell-wall polysaccharides in graminaceous plants. *Carbohydr. Res.*, **118**, 65–80 .

12. Tan, K. S., Hoson, T., Masuda, Y., and Kamisaka, S. Involvement of cell wall bound diferulic acid in light-induced decrease in growth rate and cell wall extensibility of *Oryza* coleoptiles. *Plant Cell Physiol.*, **33**, 103–108 .

13. Vansteenkiste, E., Babot, C., Rouau, X., and Micard, V. Oxidative gelation of feruloylated arabinoxylan as affected by protein. Influence on protein enzymatic hydrolysis. *Food Hydrocolloids*, **18**, 557–564

14. Wetzel, D. L., Pussayanawin, V., and Fulcher, R. G.. Determination of ferulic acid in grain by HPLC and microspectrofluorometry. *Dev. Food Sci.*, **17**, 409–428 (1988).

15. Yousif, A. M., Kato, J., Deeth, H. C. Effect of storage on stored adzuki (*Vigna angularis* L.) beans coat colour changes. *Food Aust.*, **55**, 479–483 .

CHAPTER 11

POLYACRYLAMIDE-GRAFTED GELATIN: SWELLABLE HYDROGEL DELIVERY SYSTEM FOR AGRICULTURAL APPLICATIONS

M. S. MOHY ELDIN, A. M. OMER, E. A. SOLIMAN, and E. A.HASSAN

CONTENTS

11.1   INTRODUCTION

This study concerns the preparation, characterization and evaluation of water-swellable hydrogel *via* grafting cross-linked polyacrylamide chains onto gelatin backbone by free radical polymerization for agricultural applications. Characterizations by FT-IR, TGA, and SEM provide proofs of grafting process on the backbone of gelatin. The water holding capacity of grafted hydrogel was found depend on the concentrations of the feed compositions of the hydrogel. Moreover, the hydrogel particle size was found also of determining effect where maximum swelling was observed with particles size ranged from 500 μm to 1mm. The prepared cross-linked PAM-g-gelatin hydrogel shows good thermal stability and moderate sensitivity towards media pH changes in the range of 1–4. Finally the retention of water by the hydrogel graft copolymer in sandy soil has been used as monitor to evaluate its applicability as soil conditioners. The obtained results recommended our preparation for biotechnological and agricultural applications.

Highly swelling polymeric hydrogels are hydrophilic three-dimensional networks that can absorb water in the amount from 10% up to thousands of times their dry weight [1]. They are widely used in many technological and bio- technological fields, such as disposable diapers, feminine napkins, pharmaceuticals medical applications, agricultural and horticultural [2-5]. The productivity of sandy soils is mostly limited by their low water holding capacity and excessive deep percolation losses. Thus the management of these soils must aim at increasing their water holding capacity and reducing losses due to deep percolation. These polymers were developed to improve the physical properties of soil in view of: increasing their water-holding capacity, increasing water use efficiency, enhancing soil permeability, and infiltration rates. Hydrogels have been commonly utilized in agricultural field mainly as water storage granules [6]. The need for improving the physical properties of soil to increase productivity in the agricultural sector was visualized in 1950s [7]. This led to the development of water-soluble polymers such as CMC and PAM to function as soil conditioners [8]. The addition of hydrogels to a sandy soil changed the water holding capacity to be comparable to silty clay or loam [9]. Swellable hydrogel delivery systems are also commonly utilized for controlled release of agrochemicals and nutrients of importance in agricultural applications to enhance plant growth with reduced environmental pollution.

Polyacrylamide (PAM) is one of the most widely employed soil conditioner [10]. Linear PAM dissolves in water, cross-linked PAM is a granular crystal that absorbs hundreds of times its weight in water. Absorption of deionised water by polyacrylamides under laboratory conditions can vary between products in the range from 20 to1000 per g [11]. Several attempts have been made in the past to combine the best properties of both by grafting synthetic polymers onto natural polymers [12,13]. Recently, a new class of flocculating agents based on graft copolymers of natural polysaccharides and synthetic polymers has been reported [14-16]. Polyacrylamide has also been used in combination with natural polysaccharides for soil conditioning purposes. For example, Wallace et al. (1986) showed that a mixture of a galactomannan, extracted from guar bean, and polyacrylamide resulted in an additive response when applied to certain soils [17].

The present work reports a study on the preparation, characterization and evaluation of water-swellable hydrogel *via* grafting cross-linked polyacrylamide chains onto gelatin backbone by free radical polymerization. We aim to increase the water holding capacity of gelatin to wide its applications as soil conditioners.

11.2 EXPERIMENTAL DETAILS

11.2.1 MATERIALS

Gelatin (research grade), (purity 99%) obtained from El- Nasr Pharmaceutical Co. for Chemicals. (Egypt). Acrylamide (AM) (M wt. 71.08), purity 97%, was purchased from Sigma– Aldrich Chemie. N,N′-methylene bis-acrylamide(MBA) (M wt. 154.17, MP = 300) purchased from Sigma–Aldrich Chemie was used as cross-linking agent without any pre-treatment. Ammonium per sulfate(APS) (purity 99%, Mwt. 228.2) was purchased from Sigma–Aldrich Chemicals Ltd (Germany). Ammonium ferrous sulphate (extra pure AR), M.wt. 392.13, assay min. 99%, obtained from sisco research laboratories PVT.LTD, (India). Tetramethylethylenediamine (TEMED) (purity99%, M. wt.116.21) was obtained from merck-schuchardt (Germany). Sand Soil, was obtained from Alexandria desert (Egypt). Other chemicals were of analytical grade used throughout the experiments.

11.2.2 PREPARATION OF PAM-G -GELATIN HYDROGEL

Different concentrations of gelatin (2–10%),(w/v) were dissolved in a beaker 500ml using hot distilled water (pH 7.2). A known concentration of monomer AM (1–10%) (w/v) and crosslinker MBA (0.03–0.2%) (w/v) were added and finally a known concentration of initiator redox system

APS/FAS (0.025–0.125%)/(0.015:0.075%) (w/v) were added. Temed(25µM) was then injected in the mixture , the solution was stirred well quickly for 30 min. to avoid lumping and then left at (25–60°C) for 1 hr. The mixture was set aside undisturbed (overnight). The gels are then washed extensively with water to remove soluble moieties and unreacted monomers. Acetone was used as a squeezer of water. The gel so formed was dried overnight at 60°C. The dry gel was crushed and separated into different particle sizes (4mm–250µm) using a Sieve shakers. Scheme (1) describes the proposed mechanistic pathway for synthesis of gelatin-g-PAM hydrogel.

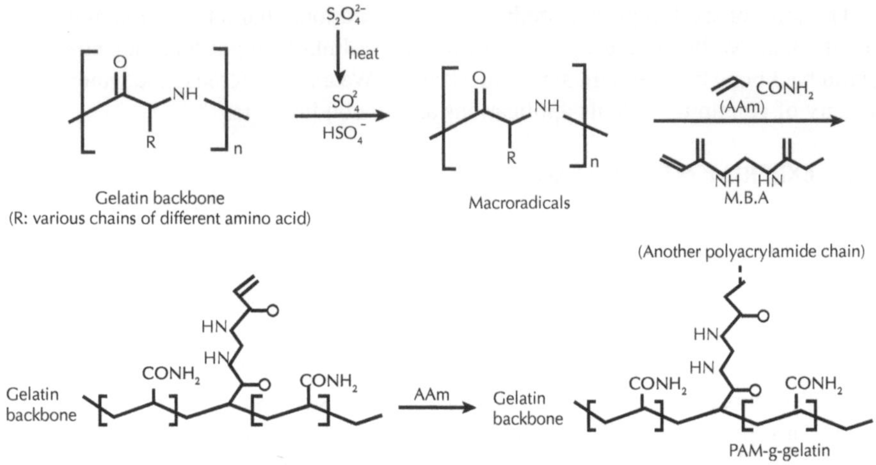

SCHEME 1   Proposed mechanistic pathway for synthesis of PAM g- gelatin hydrogel.

11.2.3   PREPARATION OF POLYACRYLAMIDE PAM (BLANK)

The same preparation method of PAM-g-gelatin hydrogel was used in the preparation of cross-linked PAM hydrogel in absence of gelatin under the same conditions.

Grafting percent (Gp %) and grafting efficiency (GE %) were calculated as follows [17] :

$$Gp\ \% = [(W1 - W0)/W0] \times 100 \qquad (1)$$

$$GE\ \% = ((Wt.\ of\ grafted\ PAM/Total\ Wt.\ of\ PAM)) \times 100 \qquad (2)$$

Where W1 is the weight of grafted copolymer hydrogel (PAM g- gelatin) and W0 is the weight of native polymer (Gelatin).

Percent weight conversion (WC %) of monomers (AM) into polymeric hydrogel (PAM) was determined from mass measurements [18]. Using the following expression:

$$WC\ \% = (Total\ Wt.\ of\ PAM\ /\ Total\ mass\ of\ AM\ in\ the\ feed\ mixture) \times 100 \quad (3)$$

11.2.4   WATER UPTAKE (SWELLING) EXPERIMENTS

The progress of the swelling process was monitored gravimetrically as described by other workers [19,20]. In a typical swelling experiment, a pre-weighed piece of sample (0.1 g) was immersed in an aqueous reservoir using distilled water (pH 7.2) and allowed to swell for a definite time period. At least three swelling measurements were performed for each sample and the mean values are reported. The swollen piece was taken out at predetermined time pressed in between two filter papers to remove excess water and weighed. Another sample (1 gm) was taken and allowed to swell for a definite time (5 hr) then weighted, the swollen sample was dried at 60°C for 12 hr., this method was repeated (reswelling) about10 times of drying. The variation of the swelling degree and weight loss of hydrogels during all drying times were noticed.

The percentage degree of swelling SD [%] of hydrogel [21], can be determined as a function of time as following;

$$SD\ (\%) = ((M_t–M_0)/M_0) \times 100 \qquad (4)$$

Where M_t weight of the swollen hydrogel sample at time t, and M_0 is the weight of the xerogel sample.

11.2.5 SWELLING IN SANDY SOIL EXPERIMENTS

DETERMINATION OF FLOW RATE OF WATER AND SWELLING DEGREE OF HYDROGEL IN SANDY SOIL

Two plastic measuring cylinders (1000ml, height 40cm, and area 39.57 cm^2) with small holes at the bottom were obtained, a filter paper was placed up the bottom for each cylinder [22], 500g of sand (particle size 500μm – 1mm) (height 15 cm) were placed in the first cylinder. 5 g of dry sample was mixed with 495 g of sand (total height 15 cm) with the same particle size were placed in the second cylinder, Tap water with height (10cm) was added up the mixture at the same time, the flow rate of water from up to down in the two cylinders was compared. Also the swelling degree of hydrogel was determined after saturation time (6 hr) [23,24]. Water releasing rate from sand and sand mixed with hydrogel during 10 days also determined. This method was also repeated with adding 15 cm from sand layers to the cylinders (total height 30 cm).

$$\text{Flow Rate} = \frac{\text{Volume of Flow Water } (cm^3)}{\text{Area of Cylinder } (cm^2) \times \text{Time of Flowing } (min)} \qquad (5)$$

DETERMINATION OF WATER LOSS OF SWOLLEN HYDROGEL UNDER EFFECT OF TEMPERATURE

35 gm of swollen hydrogel were mixed with sandy soil(100 gm) (particle size 500μm – 1 mm) in a plastic measuring cylinder and then covered by 5, 10, 20, and 30 cm of sand layers and weight the contents, then dried at 50°C for (1:5)days. The weight loss of swollen hydrogel with time will be determined as follow:

$$\text{Water loss \%} = ((W1-W2) / W1) \qquad \times 100 \qquad (6)$$

Where W1 and W2 are weight of mixture before and after drying respectively.

11.2.5　GRAFTING VERIFICATION

Physicochemical characteristics of synthesized PAM, gelatin, and PAM-g-gelatin copolymers were studied using Fourier Transform Infrared Spectrophotometer (Shimadzu FTIR-8400 S, Japan) and Thermogravimetric Analyzer (Shimadzu TGA -50, Japan). Morphological characteristics were followed using analytical Scanning Electron Microscope (SEM) (Joel JSM 6360 LA, Japan).

11.3　DISCUSSION AND RESULTS

11.3.1　GRAFTING PROCESS

EFFECT OF POLYMER (GELATIN) CONCENTRATION

Table (1) shows the effect of variation of gelatin in the range of (2–10%) on the grafting percent, grafting efficiency and weight conversion. It was clear from results that the grafting process was influenced by the amount of gelatin, where with increasing of gelatin concentration up to 6% the grafting percent and grafting efficiency were increased, and then tends to decreases slightly with further increase of gelatin concentration, while weight conversion remains constant. These results may be attributed to that with increasing of gelatin concentration the substrate chains will increases, where the initiated sites on the gelatin backbone will consequently increases and a large number of monomer units will be attached to it, then the grafting percent and grafting efficiency will increases. However, at gelatin concentration higher than 6%, the viscosity of mixture will increase and the free radicals will be hindered and the monomer cannot attached to the active sites on gelatin backbone, and then the grafting percent and grafting efficiency will decrease.

TABLE 1　. Effect of gelatin concentration on the percentage of grafting, grafting efficiency at constant conditions (5%AM, 0.05%MBA, and 0.05% APS/0.03FAS at 35°C, overnight)

Gelatin%	GP%	GE%	WC%
2	65	28.57	90.29
4	67.5	61.4	90.29
6	71.66	94.2	90.29
8	53.12	93.2	90.29
10	38	83.33	90.29

The effect of gelatin concentration on the swelling degree of hydrogel was studied as shown in Figure (1). It's clear from results that the swelling degree was increased with increasing of gelatin concentration up to 6% and then tends to decrease with increasing gelatin concentration beyond 6%. Increasing of swelling degree can be explained by the fact that gelatin is a natural water soluble polymer, where its contain a lot of hydrophilic groups which impart hydrophilicity to the molecule, thus increase affinity of water molecules to penetrate in to the gel and swells the macromolecular chains, thus resulting in a greater swelling of hydrogel. However, at much higher concentration of gelatin (beyond 6%) the density of network chains increases so much that both the diffusion rate of water molecules and relaxation rate of macromolecular chains are reduced resulting decreasing the swelling degree of hydrogel.

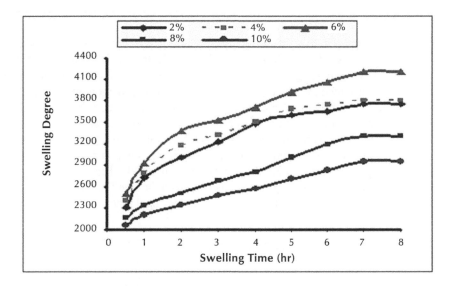

FIGURE 1 Effect of variation in gelatin concentration on the percentage of swelling degree (SD%). [AM] 5%, [MBA] 0.05%, [APS] 0.05%/[FAS]0.03%, at 35°C, overnight,

EFFECT OF MONOMER CONCENTRATION (AM)

The effect of acrylamide concentration on the grafting percent, grafting efficiency and total weight conversion was investigated. Table (2) shows that increasing the monomer concentration (AM) clearly increased the percentage of grafting and grafting efficiency. Maximum grafting percent and grafting efficiency were obtained at 5% PAM, also maximum weight conversion of monomer into PAM at the same conditions was observed at concentration 5%. These observations may be attributed to that increasing monomer concentration up to 5% facilates the diffusibility of monomer towards the initiated sites on the gelatin chains, which consequently increases the grafting yield.

However at monomer concentrations higher than 5%, phase separation was observed and the mixture was not homogenous, thus grafting process can not determined.

TABLE 2   Effect of monomer (AM) concentration on the percentage of grafting, grafting efficiency at constant conditions (6%Gelatin, 0.05%MBA, 0.05% APS/0.03%FAS at 35°C, Overnight)

AM %	GP%	GE%	**WC%**
2	16.66	80	60.97
3	36.66	89.79	80.32
4	55	93.2	87.4
5	71.66	94.2	90.29

Figures 2 (a) and (b) shows the effect of acrylamide concentration on behavior of the swelling degree gelatin-g-PAM and PAM hydrogels. It was clear from results that the swelling degree of hydrogel increased with increasing of AM concentration up to 5%, this can be attributed to that acrylamide being a hydrophilic monomer is expected to enhance the hydrophilicity of the hydrogel network when used in increasing concentrations in the reaction mixture. However beyond 5% phase separation occurred and this may be due to increasing density of AM monomers in the reaction mixture leading to separation from gelatin backbone. Also, it was observed from results that the swelling degree of gelatin-g-PAM hydrogel was slightly higher than PAM hydrogel, this indicate that the grafting process was enhanced the swelling process of hydrogel due to increasing the hydrophilic groups in the hydrogel matrix.

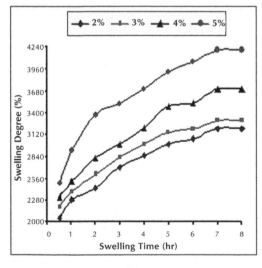

(a)

FIGURE 2   *(Continued)*

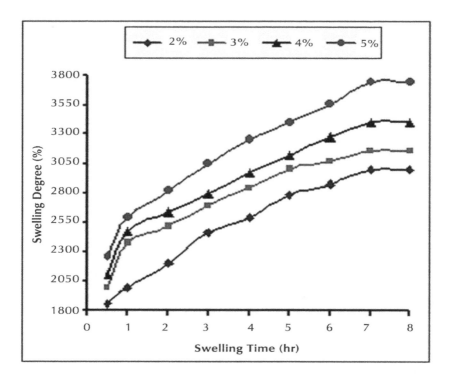

FIGURE 2   Effect of variation in acrylamide concentration (AM) on the percentage of swelling degree (SD %). [Gelatin] 6%, [MBA] 0.05%, [APS] 0.05% /[FAS]0.03%, at 35°C, overnight, In case of (a) gelatin-g-PAM, and (b) PAM hydrogels.

EFFECT OF CROSSLINKER CONCENTRATION (MBA)

The effect of methylene bis-acrylamide on the grafting yield, grafting efficiency and weight conversion was studied with concentration range (0.03:0.2%) as shown in Table (3). It was clear from results that with increasing the concentration of MBA up to 0.05% the grafting percent and grafting efficiency were increased then tends to decrease slightly with further increase of MBA concentration, but further increasing of MBA concentration beyond 0.2% lead to phase separation. These results may be attributed to that with increasing MBA concentration up to 0.05%, the weight conversion of monomer into polyacrylamide (PAM) will increase, and hence the grafting percent and grafting efficiency depends on the amount of PAM, then they will increases as MBA concentration increases up to 0.05%. While concentrations higher than 0.05% lead to decreasing weight conversion then the grafting yield and grafting efficiency decreased consequently. Also further increase of MBA concentration beyond 0.05% may be leads to formation of more dense three dimensional hydrogel structure which limiting the diffusion of AM monomer and acting in favor of PAM homopolymers formation. This reflects consequently on the drop of both grafting percentage and efficiency.

TABLE 3 Effect of methylene bis-acrylamide (MBA) concentration on the percentage of grafting, grafting efficiency and weight conversion at constant conditions (6%Gelatin, 5%AM, 0.05%APS/0.03% at 35°C, overnight.

MBA %	GP%	GE%	WC%
0.03	66	92.59	90.29
0.05	71.66	94.2	90.29
0.1	64.5	92.5	76.6
0.15	62.5	91.13	74.42
0.2	61.66	90.43	73.52

Crosslink's have to be present in a hydrogel in order to prevent dissolution of the hydrophilic polymer chains in an aqueous environment. The crosslinked nature of hydrogels makes them insoluble in water. Figures 3 (a) and (b) shows the influence of the crosslinking agent in the range (0.03:0.2%) on the swelling degree of PAM g- gelatin and PAM hydrogels. It was clear that maximum swelling degree observed at concentration (0.05%) then tends to decreases, this may be attributed to that on increasing the concentration of MBA beyond (0.05%) in the reaction mixture, the number of crosslink's increases in the gel network. This obviously leads to a slow diffusion of water molecules into the network and restricted relaxation of network chains in the hydrogel thus resulting in a fall in the swelling degree of the network. In fact with concentrations less than 0.03% no gel is prepared and slimy gel is formed due to that the amount of crosslinker was not enough to form the network of hydrogel. Also from figures it was shown that the grafted hydrogel have high swelling degree than PAM hydrogel which indicate that the grafted process increased the swelling degree of hydrogel due to increasing of hydrophilic groups in the hydrogel.

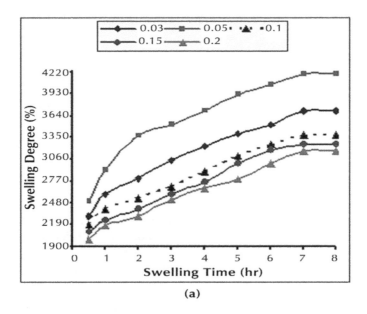

(a)

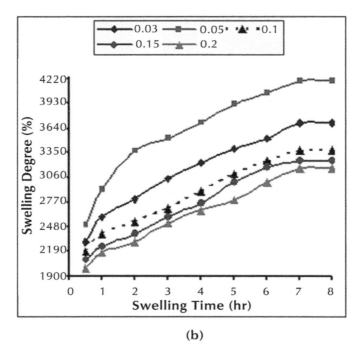

(b)

FIGURE 3 Effect of variation in crosslinker concentration (MBA) on the percentage of swelling degree (SD %). [Gelatin] 6%, [AM] 5%, [APS] 0.05%/[FAS] 0.03%, at 35°C, overnight, In case of (a) gelatin-g-PAM, and (b) PAM hydrogels.

EFFECT OF INITIATOR CONCENTRATIONS (APS/FAS)

Table (4) shows the effect of variation of ammonium persulphate (APS) concentration in the range of (0.025:0.1%) and ferrous ammonium sulphate concentration in the range (0.015:0.06%) on the studied grafting parameters. It was clear that the maximum grafting percent and grafting efficiency observed at concentration of (0.05% APS/FAS 0.03%), and decreased with higher concentrations. These results may be attributed to that with increasing initiator concentration up to (0.05%APS/0.03%FAS) will increasing number of free radicals, also active sites on the substrate will increase consequently, then a great number of monomer units will participate in the grafting process which lead to increasing of grafting yield and grafting efficiency. At higher concentrations (beyond 0.05% APS/0.03% FAS) the weight conversion nearly constant. With further increasing of initiator concentrations the number of produced radicals will be more increased and this lead to terminating step *via* bimolecular collision, which leading to decrease molecular weight of the resulting crosslinked PAM, thus shortening the macromolecular chains which leading to decreasing grafting yield and grafting efficiency.

TABLE 4   Effect of redox initiator (APS/FAS) concentrations on the percentage of grafting, grafting efficiency and weight conversion at constant conditions (6%Gelatin, 5%AM, and 0.05% MBA at 35°C overnight).

APS% / FAS%	GP%	GE%	WC%
0.025/0.030	70.8	94	89.5
0.050/0.030	71.66	94.2	90.29
0.075/0.030	70	93.33	89.1
0.100/0.030	66.6	90.9	87.1
0.050/0.0150	63.33	88.37	85.14
0.050/0.045	68.33	93.18	87.12
0.050/0.060	66.6	91.95	86.13

Figure 4 (a) and (b) shows the effect initiator (APS/FAS) on the swelling degree of hydrogels. The initiator has been varied in the feed mixtures in the concentration range (0.025:0.100)/(0.015:0.06%). It was clear from results that the swelling degree increase with decreasing of initiator concentration in the feed mixture, where higher swelling degree was obtained at concentration (0.05/0.03%) then tends to decrease with further increasing of initiator concentrations these results can be attributed to that the number of hydrophilic groups produced from grafting process was increased and reached to maximum where higher grafting yield (71.66%) observed at concentrations (0.05/0.03%). At concentrations more than (0.05/0.03%) the number of produced radicals will increase and this lead to terminating step *via* bimolecular collision resulting in enhanced crosslink density resulting decrease of the swelling degree.

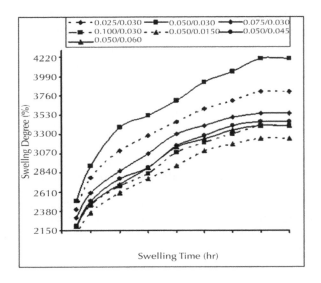

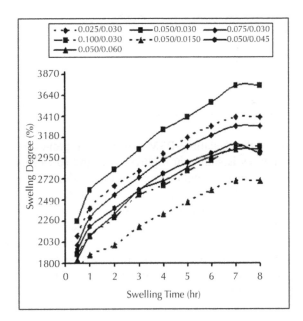

FIGURE 4 Effect of variation in initiator concentration (APS/FAS) on the percentage of swelling degree (SD %). [Gelatin] 6%, [AM] 5%, [MBA] 0.05%, at 35°C, overnight, In case of (a) gelatin-g-PAM, and (b) PAM hydrogels.

EFFECT OF REACTION TEMPERATURE

The effect of the variation of the reaction temperature (25–45°C) on the grafting parameters was illustrated in Table (5). It was clear from results that the grafting percentage, grafting efficiency also weight conversion increased with increasing of temperature up to 35°C then tends to decreases with further increasing temperature. These results can be attributed to that increasing temperature may be lead to increase diffusion of AM from the solution phase to the swellable gelatin phase, increasing rate of thermal dissociation of initiator, resulting in increasing the solubility of the monomer also formation and propagation of grafted chains. The net effect of all such factors leads to high grafting with increasing the polymerization temperature. While higher temperatures beyond 35°C may be lead to termination step occurring rapidly, then the grafting percentage, grafting efficiency also weight conversion will be decreased.

TABLE 5   Effect of grafting temperature on the percentage of grafting, grafting efficiency and weight conversion at constant conditions 6%Gelatin, 5%AM, 0.05%MBA, 0.05%APS/0.03%FAS, Overnight)

Grafting Temperature(°C)	GP%	GE%	WC%
25	52.25	86.53	72
30	65.16	90.3	85.74
40	71.66	94.2	90.29
45	61.83	95.087	83.16
50	53	85.48	73.66

The swelling degree of the (Gelatin-g-PAM and PAM) hydrogels prepared with various reaction temperatures is shown in Figures 5 (a) and (b). From results it's clear that the swelling degree increase with increasing of reaction temperature up to 35°C then tends to decrease with further increasing temperature, these results may be attributed to that higher temperatures favor the rate of diffusion of the monomers to the gelatin macroradicals as well as increase the kinetic energy of radical centers. The temperatures higher than the optimum value (35°C), however, lead to low-swelling superabsorbents. This swelling loss may be attributed to oxidative degradation of gelatin chains by sulfate radical-anions, resulting in decreased molecular weight and decreased the swelling degree [25].

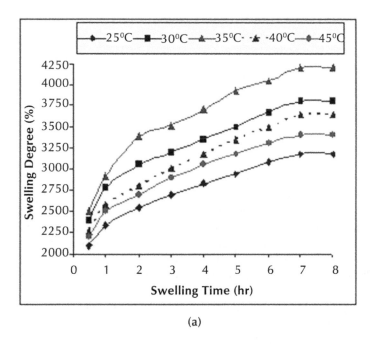

(a)

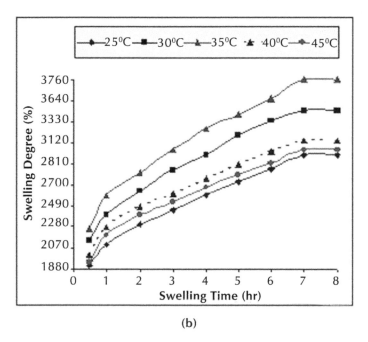

(b)

FIGURE 5 Effect of variation in reaction temperature on the percentage of swelling degree (SD %). [Gelatin] 6%, [AM] 5%, [MBA] 0.05%, [APS] 0.05% /[FAS]0.03%, overnight, In case of (a) gelatin-g-PAM, and (b) PAM hydrogels.

11.3.2 MATERIALS CHARACTERIZATION

INFRARED SPECTROPHOTOMETRIC ANALYSIS (FTIR)

The IR spectra of gelatin, PAM, and PAM g- gelatin respectively are shown in Figure 6 ((a)–(c)). From the IR spectra of gelatin, it was evident that it shows a broad absorption band at 3431cm-1, due to the stretching frequency of the -OH and NH$_2$ groups. The band at 2925 cm^{-1} is due to C–H stretching vibration. The presence of a strong absorption band at 1649 cm^{-1}confirms the presence of C=O group. The bands around 1446 and 1332 cm^{-1} are assigned to CH$_2$ scissoring and -OH bending vibration, respectively. In the case of PAM a broad absorption band at 3433 cm^{-1} is for the N–H stretching frequency of the NH$_2$ group. Two strong bands around 1693 and 1639 cm^{-1} are due to amide-I (C=O stretching) and amide-II (NH bending). The bands around 1400 and 2927cm^{-1} are for the C-N and C-H stretching vibrations. Other bands at 1454 and 1323 cm^{-1} are attributed to CH$_2$ scissoring and CH$_2$ twisting. The IR spectra of PAM g- gelatin, the presence of a broad absorption band ranged at 3438 cm^{-1} is due to the overlap of -OH and NH$_2$ stretching bands of gelatin and PAM. A band at 1641 cm^{-1} is due to amide-I (C=O stretching) and amide-II band of PAM overlap with each other and lead to a broad band at 1554cm^{-1}. The bands around 1382 and 2923 cm^{-1} are for the C-N and C-H stretching vibrations. Other band at 1456 cm^{-1} can be attributed to CH$_2$ scissoring.

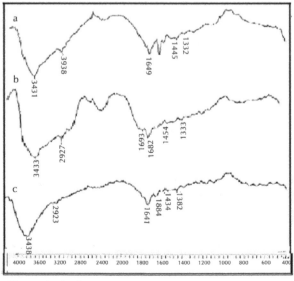

Wavenumber (cm^{-1})

FIGURE 6   FTIR of (a) Gelatin, (b) PAM, and (c) PAM-g-Gelatin.

THERMOGRAVIMETRIC ANALYSIS (TGA)

Figure (7), represent the thermal degradation of gelatin, PAM and PAM g- gelatin hydrogel respectively. In case of gelatin, the initial weight loss can be attributed to presence of small amount of moisture in the sample. The rate of weight loss was increased with increasing temperature. In case of PAM, the initial humidity weight loss followed by a continuous weight loss with increasing temperature, the degradation after that is due to the loss of NH_2 group in the form of ammonia. In case of degradation of PAM g- gelatin, it takes place more closely to gelatin, which indicated to low grafting yield. The data from the TGA was summarized in table (6). It is clear that the grafting of PAM chains onto gelatin backbone enhanced the thermal stability of the gelatin which was reflected on the increasing of half weight temperature (T_{50}).

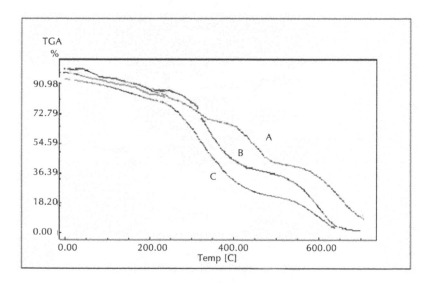

FIGURE 7   The TGA of (A) PAM, (B) Gelatin, and (C) PAM-g-Gelatin.

TABLE 7   Data showing TGA of gelatin, PAM and PAM g- gelatin hydrogels

		Weight loss %	
Sample	T_{50} °C	0-248°C	248-430°C
Gelatin	377	13.44	53.16
Gelatin-g-PAM	393	12.551	53
PAM	432	16.67	39.38

SCANNING ELECTRON MICROSCOPE (SEM)

The scanning electron microscope of gelatin, PAM, and graft copolymer (Gelatin-g-PAM, GP%71.66) was shown in figure (8). It was clear that the morphological structure of gelatin was differed than gelatin-g-PAM, where surface morphology of gelatin before grafting shows a smooth surface, which has been changed to surface curly as layers after grafting. PAM morphology was also changed drastically when grafted onto gelatin backbone.

(a)

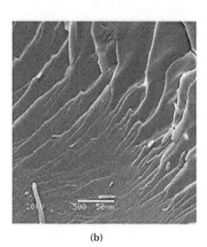

(b)

FIGURE 8 *(Continued)*

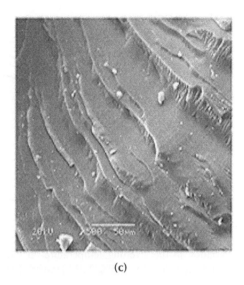

(c)

FIGURE 8   The SEM of (a) Gelatin, (b) PAM, and (c) PAM.-g-Gelatin.

11.3.3   PAM-G-GELATIN HYDROGEL EVALUATION

Effect of variation different operational conditions such as swelling temperature, re-swelling Ability and pH on the swelling process has been investigated. In addition, effect of hydrogels particle size on the swelling behavior has also been investigated.

EFFECT OF THE SWELLING MEDIUM TEMPERATURE

The swelling has a significant effect on the swelling and deswelling kinetics of hydrogels. The effect of temperature of the swelling medium on the swelling degree of PAM-g-Gelatin hydrogel was studied by variation in temperature of swelling medium in the range of (25–40°C) as shown in Figure (9). It is clear from results that the swelling degree was increased with increasing of temperature up to 30°C then, tends to decrease with further increase of temperature. These results can be attributed to that with increasing temperature up to 30°C, the expansion and flexibility of gel network will increase and this lead to increasing penetration of water molecules into gel network, resulting an increasing of swelling degree. While beyond 30°Cthe swelling degree decreased and this may be due to that the gel network was malformed and loss its mechanical properties with increasing of swelling time at higher temperatures.

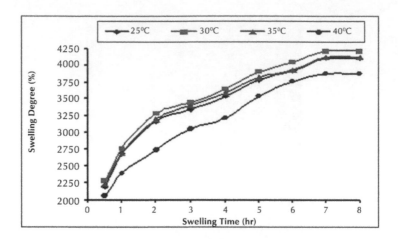

FIGURE 9   Effect of variation in the swelling temperature on the percentage of swelling degree (SD %) of hydrogel [Gelatin] 6%, [AM] 5%,[MBA] 0.05%, [APS] 0.05% /[FAS]0.03%, at 35°C overnight.

EFFECT OF RESWELLING ABILITY ON THE SWELLING DEGREE

The effect of reswelling ability (using distilled water pH7.2) of hydrogel after drying many times (10 times) was studied as shown in Figure (10). It was clear from results that the swelling degree and weight of sample decrease versus increasing reswelling and drying times (up to 10 times), in which the weight loss of dry sample (1 g) after reswelling and drying 10 time (0.64 g) was 36% from its started weight.

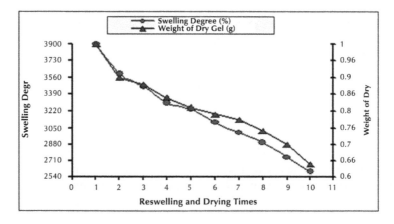

FIGURE 10   Effect of reswelling ability of dry gel and the swelling degree of hydrogel, [Gelatin] 6%, [AM] 5%,[MBA] 0.05%, [APS] 0.05% /[FAS]0.03% ,at 35°C overnight.

EFFECT OF PH MEDIUM ON THE SWELLING DEGREE OF HYDROGEL

As an environmental controlling factor, pH has a significant effect on the swelling degree of PAM-g-Gelatin hydrogel .The swelling degree of hydrogel is affected by the heterogeneous nature of functional groups of amino acids of gelatin that are protonated/hydrolyzed on interaction with the swelling medium [26]. Figure (11) show the effect of variation in pH of the swelling medium in the range of (1–11). It was clear from results that the prepared hydrogel showed a linear increase in the swelling degree in pH4 and this may be attributed to partial hydrolysis of the amide group, thus, an increased volume of network gives rise to greater voids within the gel and, therefore, the water sorption increases leading to an increase of the swelling degree of the hydrogel. On the other hand, crosslinked gelatin has an isoelectric pH (PI) in the range of 4.7 to 5.1, below the PI value the gelatin chains remain protonated. As a result, the chains contain NH_3^+ ions, and the cationic repulsion between them could be responsible for the high swelling degree [27].

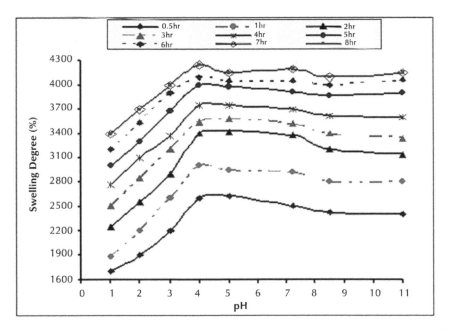

FIGURE 11   Effect of variation in pH of the swelling medium on the percentage of swelling degree (SD%) of hydrogel [Gelatin] 6%, [AM] 5%,[MBA] 0.05%, [APS] 0.05%/[FAS]0.03%,at 35°C overnight.

EFFECT OF PARTICLE SIZE OF HYDROGEL

The effect of variation in particle size on the swelling degree of (Gelatin-g-PAM) hydrogel was observed as shown in Figure (12) in the range of (4 mm–250 μm). With a lower the particle size at (1 mm–500 μm), a higher swelling degree was observed. The increasing of the swelling degree would be expected from the increase in surface area with decreasing particle size of hydrogel [28]. Additionally, the ultimate degree of swelling increased as the particle size becomes smaller. This is attributed to more water molecules being held in the volume between the particles. However the swelling degree was decreased at particle size smaller than (1 mm–500 μm) may be due to that the network of hydrogel began to destroyed and malformed by crushing and its swelling behavior was decreased.

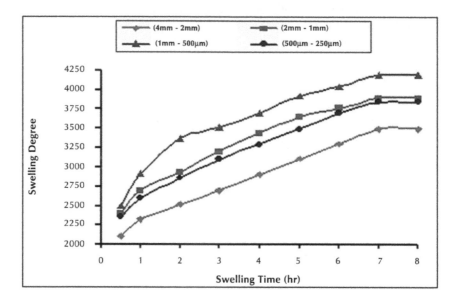

FIGURE 12   Effect of variation in particle size on the percentage of swelling degree (SD %) of hydrogel [Gelatin] 6%, [AM] 5%,[MBA] 0.05%, [APS] 0.05% /[FAS]0.03%, at 35°C overnight.

11.3.4   SWELLING OF (GELATIN-G-PAM) HYDROGEL IN SANDY SOIL

Table (8) show that the water flow in sand was decreased when hydrogel mixed with sand soil, these results may be attributed to that the hydrogel retained amount of water and then reduced the flow rate of water. Also it was clear from results that the swelling degree of hydrogel and the flow rate of water increased with decreasing the height of sand soil after saturation of sand soil pores with water (after 7 hr) as shown in Table (9), these results can be attributed to that increasing sand layers will increase the pressure occurring on the hydrogel particles resulting release of water from hydrogel, also hinder the dry gel to swell leading to decreasing the water uptake and the swelling degree of hydrogel.

Table 8   Data showing variation of water flow in case of sand soil and sand soil mixed with hydrogel [Gelatin] 6%, [AM] 5%, [MBA] 0.05%, [APS] 0.05/[FAS]0.03% at 35°C ,Overnight. (Using tap water pH 7.2)

Time (hr)	Water Flow in Sand Soil (ml)		Water Flow in Sand Soil mixed with Hydrogel (ml)	
	Height 15 cm Sand	Height 30 cm Sand	Height 15 cm Sand	Height 30 cm Sand
1	3600	2000	3390	2150
2	7000	4100	6680	2790
3	8800	5100	7530	3500
4	9900	5900	8400	4150
5	10850	6500	9200	4750
6	11800	7100	10000	5380
7	12800	7700	10800	5940
8	13800	8300	11600	6500
9	14800	8900	12400	7060
10	15800	8500	13200	7620

TABLE 9   Data Showing variation of flow rate of water and swelling degree of hydrogel mixed with hydrogel [Gelatin] 6%, [AM] 5%, [MBA] 0.05%, [APS] 0.05/[FAS]0.03% at 35°C ,Overnight. (Using tap water pH 7.2)

Flow Rate of Water in Sand Soil (cm/min)		Flow Rate of Water in Sand Soil mixed with Hydrogel (cm/min)		Swelling Degree(SD%) of Hydrogel in Sand Soil	
Sand Height (15cm)	Sand Height (30cm)	Sand Height (15cm)	Sand Height (30cm)	Sand Height (15cm)	Sand Height (30cm)
0.421	0.252	0.336	0.235	2300	1700

Table (10) show that the water loss from hydrogel which covered with different sand layers heights increase with increasing of drying time, these results can be attributed to that with increasing drying times soil-water content rapidly changes as water drains under gravitational forces and temperature. Also from results it was found that the prepared hydrogel from gelatin had lost about 50% from its water contents after 5 days from drying at 50°C which indicate that it have good mechanical properties.

TABLE 10   Data showing variation of water loss of the swollen hydrogel mixed with sand soil under effect of temperature 50°C. hydrogel compositions [Gelatin] 6%, [AM] 5%, [MBA] 0.05%, [APS] 0.05/[FAS]0.03% at 35°C, overnight. (Using tap water pH 7.2)

Time	Water loss	
(days)	Sand Height (30cm)	Sand Height (15cm)
1	12.2	16.5
2	20.6	23.3
3	28.9	33.3
4	36.15	41.66
5	41.67	50

11.4   CONCLUSION

From this study, it was found that the grafting of naturally protein such as gelatin by synthetic polymers as PAM will increase of the water uptake capacity of this natural polymer to have water swellable hydrogel. The water holding capacity of grafted hydrogel was found depend on the concentrations of the feed compositions of the hydrogel. Characterizations by FT-IR, TGA, and SEM provide further proof of grafting process on the backbone of gelatin. The hydrogel particle size was found also of determining effect where maximum swelling was observed with particles size ranged from 500 μm to 1mm. The prepared cross-linked PAM-g-gelatin hydrogel shows good thermal stability, high water absorbency and high pH, which recommended our preparation for biotechnological and agricultural applications. Finally grafted copolymer hydrogel have been evaluated using swelling and flow rate of water in sandy soil mixed with hydrogels experiments at optimum conditions. The retention of water by the hydrogel graft copolymer in sandy soil has been used as monitor to evaluate its applicability as soil conditioners.

KEYWORDS

- **Grafting verification**
- **Polyacrylamide**
- **Scanning electron microscope**
- **Thermogravimetric analysis**
- **Water-swellable hydrogel**

ACKNOWLEDGMENT

The authors acknowledge the help of Prof. M. R. Abdel Fatah, Professor of soil chemistry- arid land institute- Scientific research and technological applications city- Alexandria (Egypt), in performing the soil filtration experiments.

REFERENCES

1. Buchholz, F. L. and Graham, A. T. *Modern Superabsorbent Polymer Technology*, Wiley, New York (1997).
2. Peppas, L. B. and R. S. Harland In: Absorbent Polymer Technology. Elsevier, Amsterdam (1990).
3. Po, R. *J Macromol Sci Rev Macromol Chem Phys*, **34**, 607 (1994).
4. HoffmanIn, A. S. *Polymeric Materials Encyclopedia*. J. C. Salamone (Ed.), CRC. Press, Boca Raton, Florida, **5** (1996).
5. Mohy Eldin., M. S., El-Sherif, H. M., Soliman, E. A., Elzatahry, A. A., and Omer, A. M. *Journal of Applied Polymer Science.*, **122**, 469–479 (2011).
6. Burillo, G. and Ogawa, T. *Radiation Phys. Chem.*, (1981), 18, 1143-1147.
7. R. M.Hedrick, D. T. Mowry, Soil Sci., (1952), 73, 427-441.
8. R.Azzam, Commun. Soil Sci. Plant Analysis, (1980), 11, 767-834.
9. A.Huttermann, M. Zommorodi, K. Reise, Soil and Tillage Research.; (1990), 50, 295-304.
10. A. M. Omer., Preparation and Characterization of Graft Copolymer Hydrogels to be used as Soil Conditioners, MS.C. Al-Azhar University, Egypt, (2008).
11. M.S. Johnson, C.J. Veltkamp, Journal of the Science of Food and Agriculture, (1985), 36, 789-793.
12. R.P. Singh, Plenum Press, New York, NY,. p. 227 1995.
13. C.L. Swarson, R.L. Shogren, G.F. Fanta, J. Environ. Polym. Degrad. 1 (1993) 155–166.
14. B.R. Nayak, R.P. Singh, J. Appl. Polym. Sci. 81 (2001) 1776–1785.
15. S.K. Rath, R.P. Singh, J. Appl. Polym. Sci. 66 (1997) 1721–1729.
16. M.S. Mohy Eldin, A.M. Omer, E.A. Soliman, E.A. Hassan, J. Desalination and Water Treatment 51 (2013) 3196–3206.
17. M.S.Mohy Eldin, Preparation and Characterization of Some Cellulosic Grafted Membranes,MSC.(1995) .
18. O.Guven , M.Sen, Polymer, (1991), 32, 2491-2495.
19. B. D.Ratner, D. F.William, ed., CRS Press, Inc., Florida, (1981), vol 2, pp. 145-175.
20. L.F. Gudeman, N.A. Peppas, J. Appl. Polym. Sci. 55 (1995) 919–928.

21. K.Aleksandar, A. Borivoj, B. Aleksandar, J.Jelena. J. Serb. Chem. Soc. (2007), 72 (11) 1139–1153.

21. J. Akhter. Plant soil environ. (2004), 50(10); 463-469.

22. M.S. Johnson, J. Sci. Food Agric.; 1984b, 35, 1196-1200.

23. D.C. Bowman, R.Y. Evans, HortScience(1991).; 26(8), 1063-1065.

24. A. Hebeish, J. Cuthrie, "Chemistry and Technology of Cellulosic Copolymers", p. 46, (1981).

25. S.Ghanshyam Journal Applied Polymer Science,vol.90,3856-3871(2003).

26. B.Krishna, Journal Applied Polymer Science,vol.82,217-227 (2001).

27. P.Ali, journal of e-polymers 57 (2006).

CHAPTER 12

THE DYNAMICS OF BACTERIA AND PATHOGENIC FUNGI IN SOIL MICROBIOCENOSIS UNDER THE INFLUENCE OF BIOPREPARATIONS USE DURING POTATO CULTIVATION

V. V. BORODAI

CONTENTS

12.1 INTRODUCTION

The use of biological preparations such as Phytotsid and Planryz contributes the increase of the general number of soil bacteria population to 36.1% in compared with control, the number of saprophyte microflora and causes 1.2–1.8 times reduction in the number of soil fungi – Fusarium sp. and Alternaria sp. The Shannon ecological index of species biodiversity was lower at the application of Rovral Akvaflo than the biopreparation use. The decrease of species biodiversity was observed as well as strengthening the dominance of some species (dark pigmentation in fungi).

The biological means of plants protection is relevant direction which has got the wide scientifically innovative development. These methods are alternative to chemical ones which have negative impact on the biological components of agrophytocenosis [1,22-24,34,37,38].

Potato is one of the main and basic commercial crops cultivated in the world. Lately the distribution and increasing injuriousness of viral diseases, bacteriosises and mycosises of potato is observed. Emergence of new potato breeds and expansion of fungicide types testifies the need of seed-growing technologies improvement, application of essentially new systems of potato seed protection against pathogens.

Unlike chemicals for plant protection, the main part of biological preparations is presented microorganisms that take part in transformation of practically dissoluble compounds (for example phosphorus compounds), promote the formation of available nutrients in plant rhizosphere, produce the physiologically active substances (hormones, vitamins, amino acids, and so on.), contribute the induction of systemic plants resistance [1,3,8,15,16,20,23,24,30,33,34]. In addition, the substances with antibiotic and fungitoxic action are among metabolites of microorganisms and suppress the pathogens development. The active strains of microorganisms (biopreparations producers) do not cause the long-term genetic effects of human organism comparing to artificial chemically synthesized preparations.

In recent years, the investigations devoted to the studying of the effectiveness of biological preparations are carried out primarily for the potato treatment before planting and its protection against diseases in the vegetation period. For example, the treatment of potato by Hitozar P1 + Tekto in Karelia combined with biological system of plants protection induces the reduction of the scab development by 61%, 3 times decrease in *Rhizoctonia* stem canker cases and 8 times decrease in potato late blight cases; and insures the harvest increase by almost 23% [10,14,34]. In Russia the Baktofit and Planryz preparations use against the blackleg pathogen (tomato growing in greenhouses) and against late blight pathogen during the vegetation period was really biologically effective, and the calculation of the economical effectiveness showed the economic feasibility of this treatment [8].

The significant progress was made in the development of the potential biological preparations (application of the bacteria- and yeast antagonists) for the control of the post harvest damages and formation of the fungal toxins in fruits and vegetables [6]. Moreover, the effectiveness of preparation based on the *Gliocladium virens* bacteria-antagonists was examined against the scab pathogen *Rhizoctonia solani* [7], and bio-fungicide Lihnorin (based on *Trichoderma harzianum*) - against scab and Rhizoctonia stem canker [26].

In Germany, the experiments with isolated potato leaves and plants in pots were made. It was found that application of the commercial preparation Serenade (on the basis of *Bacillus subtilis* metabolites) inhibited the late blight pathogen infection in potato leaves, but it was less effective than application of preparations based on copper [32]. According to German scientists investigations, the bacterisation of 2 bacteria strains - *Pseudomonas fluorescens* L 13-6-12 and *Serratia plymuthica* 3 Re 4-18 against *Rhizoctonia solani* limited the negative impact of scab disease on potato. The antagonist *Pseudomonas fluorescens* L 13-6-12 depressed the potato diseases development better than others, but only partially [6].

According to Russia scientist's investigations, the treatment of potato by biological preparations induced the inhibition of Rhizoctonia stem canker and scab pathogens development in the storage period. In comparison with control variant, Ryzoplan caused 2–2.5 times reduction the spread of the late blight of potato, scab, Rhizoctonia stem canker and alternariosis in the tubers [34].

In Russia, the investigations devoted to the studying of the impact of potato treatment before growing with biopreparations (Phytosporyn, Planryz, Baktophosfin, Azovit, Integral) and liquid fertilizers (RUSP) on the harvest and tubers quality of the medium-early potato breed Nevsky showed that it ensured the plants resistance to diseases, which usually developed in the plants apex and tubers (Rhizoctonia stem canker development decreased by 8–35%) [34].

The effects of biological preparations and growth regulators (Ryzoplan, Cherkaz, Humate A, Agat 25K, Bioplant) on the potato resistance (breed Nevsky) to viroid infection and on harvest were studied in Russia [26].

The application of biological preparations—Ekstrasol-55 and Agat-25 provided the potatoes harvest increase by 15–22%, which increased the harvest of seed tubers on 20–30% (standard fraction) and was important in the faster reproduction of the new and perspective potatoes breeds [33].

A significant effect on the potato harvest and its resistance to fungi pathogens was observed after bio-fertilizers application (Azotobacterin, Phosphobakteryn, Kremniyebakteryn): potato harvest increased in the range from 1.2 to 2.5 times depending on the breed, the soil enrichment by the main nutritional elements in the available for plants form was observed, the inhibition of pathogenic microflora development (scab pathogen) was determined, the nitrite and nitrate content in crops was examined, the maturing of potato tubers accelerated [30]. The application of the bio-fertilizer Binoram in Russia (based on *Pseudomonas sp.*) decreased the Rhizoctonia stem canker lesions by 8–16 times [33] and bio-fertilizers Bioplan-Complex (which includes the nitrogen-fixing bacteria *Klebsiella planticola*)—accelerated the emergence of quality, development and ultimately - productivity of potato [15]. The treatment of potato tubers by bio-fertilizers in the spring based on of *Klebsiella sp.* Did not affect the harvest, but had influence on its structure - improved the commodity potatoes quality, the starch content increased, and during the storage technical withdrawal decreased by 2.2–2.3%, natural weight loss and total losses also decreased [34].

The biopreparation Supresivit based on spores from the fungus *Trichoderma harzianum* was applied as a dressing mixed with mineral fertilizers: NPK, LAV (ammonium nitrate with limestone) and DASA (ammonium nitrate and ammonium sulphate) [11]. It was indicated that *Trichoderma harzianum* suppresses pathogenic fungi at the concentration 0.5 g of Supresivit per 1 kg of the fertilizer and higher. The plants from treated plots had lower infestation - decrease by about 5–15% superficial infestation (potato—blight fungus, and so on). Simultaneously the effect on higher yields was observed.

In the Institute of Molecular Biology and Genetics of Ukraine, the protective effects of the *Pseudomonas sp.* bacteria strains IMBG 163 and *Pseudomonas sp.* IMBG 287 were examined at the time of their introduction into higher plants, which is the basis for development of the biotechnological methods of potato protection (*Solanum tuberozum* L.), against the complex soil infection at different stages of seed and commercial production receiving. It was shown that *Pseudomonas sp.* strains IMBG 163 and *Pseudomonas sp.* IMBG 287 are perspective components of biopreparations for the protection of potato breeds sensitive to pathogenic microorganisms during the production of ecologically marketable tubers [15].

The investigation of biological preparations efficacy for the treatment of potato tubers before planting and protection it from diseases during the growing season is important in the recent decades. In Canada, for seed potatoes treatment against scab, the most effective were strains EF-76 *Streptomyces melanosporofaciense* + Khitosan [25] and the strains of bacteria - antagonists, which were isolated from the soil under potato growing and showed the effectiveness against scab pathogen –*Streptomyces scabili* (after analysis of the 16S-RNA gene it was classified as *Bacillus sp. sunhua*) [9]. The current research in Tunisia showed that isolates of *Bacillus sp.* suppressed the development of dry tubers rot caused by various types of *Fusarium in vitro* and *in vivo* conditions [21]. In biological methods of the control of dry tubers rot development the application of the antagonist cells concentration and mixtures of two types were also studied [27].

Chrysomal and globerin complexes containing antibiotics from polyether and heptaene aromatic macrolide groups have been isolated from the mycelium of *Streptomyces chrysomallus* R-21 and *Streptomyces globisporus* L-242 strains, respectively [29]. Physicochemical characteristics of these complexes have been investigated, including UV and NMR spectra, chromatographic parameters, and so on. The fungicidal and antiviral activity of the studied complexes has been demonstrated on tomato plants.

Soil antagonistic streptomycetes are particularly suitable for the biological control; they were proved to be highly efficient in reducing the incidence of fungal pathogens [3]. Streptomycetes isolated from the podzolic soils were evaluated for the biosuppression of fungal populations. Seventeen strains of streptomycetes (out of the 279 isolates total) were found to be strongly antagonistic to fungal pathogens *in vitro* and were selected for further experiments *in situ*. The full protection of plants against *Fusarium* spp. was obtained with the *Streptomyces hygroscopicus* strain K49.

At the same time, the studies at the Institute of Microbiology and Virology NASU showed the high antagonistic activity and simulative activity of *Bacillus sp.* isolates against bacterial diseases of tomatoes, which resulted in resistance for two years [8].

In the investigations of the Ukrainian scientists, the bacteria *Bacillus amyloliquefaciens* was observed for the lupine treatment. The bacteria can move in the plant and produce the complex of the biologically active substances. The emergence of the first disease signs occurred later (in 2–2.5 weeks) compared with the control and reduction of plants lesions during the growing season was observed. It was also observed that biopreparation has the stimulating properties—the improvement of the growth indicators and development of plants [13,18,19].

Plants were sprayed and watered alternately using Biosept 33 SL [12]. The mycological analysis showed that Biosept 33 SL influenced on the reduction of *Fusarium oxysporum* colony number and in part inhibited alternariosis on sweet pepper plants. Biosept 33 SL did not decrease the number of *Fusarium equiseti* and *Colletotrichum coccodes* on sweet pepper plants. The bio-preparation affected the *Trichoderma* spp. growth on roots and stem base at sweet pepper.

The aim of three-year field investigations in Poland was to evaluate the effect of bio-preparations Polyversum (B.A.S. *Pythium oligandrum*) and Biochikol 020 PC (chitosan) applied on infected tubers by *Rhizoctonia solani* sclerots during vegetation period. As a standard fungicide Vitavax 2000 FS (karboxin and thiuram) was used [35]. During potato vegetation period all applied preparations affected both the lower tuber infestation level by *R. solani* sclerots and the lower tuber infestation percent by these pathogens. According to the results obtained from conducted *in vitro* examinations, it was found that Polyversum and Biochikol 020 PC bio-preparations significantly (in comparison with control) reduced level and percent of tuber infestation by *R. solani*. Application of chemical standard preparation karboxin and thiuram mixture as a dressing had the best influence on inhibition of tubers infestation by *R. solani*. Among all the tested preparations under *in vitro* conditions the most effective in reduction *R. solani* mycelium linear growth turned out to be Polyversum biopreparation. *In vitro* response of the tested pathogen depended on the type of preparation and its concentration. Also, based on the results it was found that the preparations under examination significantly inhibited top leaf and tuber infestation by *Phytophthora infestans*. Moreover, according to the results obtained from *in vitro* tests, a significant effect from the Vitavax 2000 FS and Polyversum preparations and from the highest concentration (2%) of Biochikol 020 PC preparation on the percentage of inhibition of *P. infestans* mycelium linear growth was observed (in comparison to the control) [17].

Bacteria isolated from soil rhizosphere samples of healthy Malian indigenous trees were screened for their antagonistic effect against this pathogen [2]. Three actinomycetes isolates (*Streptomyces* spp. RoN, G1P, and N1F) were the most effective microbioagents in suppressing the growth of the pathogen. The biological control essay showed the possibility of controlling potato soft rot by these three actinomycetes isolates under conservation conditions. These treatments significantly decreased soft rot compared with the untreated potato tuber slices. The microbiological control results of this study suggest that the actinomycetes isolates RoN, G1P and and J1N are effective microbioagents in controlling soft rot of potato and could be considered as promising alternative to chemical products.

In Russia the biopreparation Planryz is widely used for potato treatment, but in Ukraine Planryz BT (based on the bacterial strain *Pseudomonas fluorescence* AR-33, with concentration 2.5×10^9 cells/ml, NV—1.5–2.0 l/ha), despite of being recorded in the "List of pesticides and agrochemicals permitted for application in Ukraine "[36], is recommended only for grain, corn and on vineyards. In Ukraine Planryz is applied for vegetables and potato on private plots, for example, in Lviv region where biological laboratory of Planryz production is situated.

In Ukraine the research of influence of biological preparations on soil microflora, the changes of qualitative and quantitative microbiotes structure during potato cultivation have fragmentary character and are actual.

The purpose of research was studying of the impact of biological preparation on the level of population dynamics of a soil microbiocenosis at the time of potato cultivation.

12.2   MATERIALS AND METHODS

Research was carried out on experimental plots of UAAS Institute for Potato Research (soil - light or medium loamy for sod-podzolic soil) where potato (varieties Oberig and Skarbnitsa) was grown. The potato tubers were treated by biological preparations (Phytotsid and Planryz BT) and fungicide (Rovral Akvaflo) before landing, and later - plants in the end of the period of flowering [34]. The experimental variants:

1.   Control (water treatment),
2.   Biological control——Phytotsid treatment (on the basis of bacteria *Bacillus subtilis*) titer $1–9 \times 10^9$ CFU/sm^3, 2 l/hectares,
3.   Biological control——Planryz treatment (*Pseudomonas fluorescence* strain AP-33, titer 2.5×10^9 cells/ml, consumption rate- 1.5–2.0 l/hectares), and
4.   Chemical control——Rovral Akvaflo treatment, suspension concentrate –0.4 l/ton.

The object of research was biodiversity of soil microorganisms at the potato cultivation with application of different treatment methods. During the period of beginning of potato flowering the soil samples were selected according to standard techniques [3–5]. The microbiological investigations were carried out for studying the quantitative and qualitative structure of bacteria and fungi in soil microflora with the help of its cultivation on selective nutrient media of Zvyagintsev and Capeka. Calculation of colonies and studying of morphological, cultural properties of cultivated strains was made according to the methods, described in works [5,31,35,37]. The results were expressed as the number of colony forming units in 1 g of the soils (CFU/g). For estimation of the variety of microorganisms in soil Shannon diversity index (H) and the Pielou's evenness index were counted [31,34,35,37]. Mathematical processing of experimental data was made according to B. Dospehov [4].

12.3   DISCUSSION AND RESULTS

The studying of soil microbiocenosis structure during potato cultivation (before landing of the crop) resulted in determination of the total number of microorganisms which ranged within 153.4–259.2 thousands/g of soil in all variants of investigation. Among phytopathogenic micromycetes *Fusarium sp.* (pathogenic agent of dry rot of potato that causes considerable losses of the potato during the period of storage) and *Al-*

ternaria sp. (agent of dry spottiness of potato) were the most widespread –0.4–1.6 thousands/g of soil and 2.4–3.7 thousands/g of soil respectively. Among saprophytes such representatives as *Penicillium spp.*, *Rhizopus spp.*, and *Trichoderma spp* were found. The amount of bacteria exceeded the quantity of fungi and actinomycetes by 100–1000 times. Dominating prevalence of non-sporeforming bacteria (71%) was the general prominent feature observed in all variants of experiment.

The obtained results are in accordance with the data of other scientist concerning increase of the microorganisms quantity in the first half of potato vegetation because of the soil temperature rising, intensity of mineralization processes and, at the same time, mostly because of expense of root exudate [1,23,24,34].

It was established that application of biological preparations promoted the total increase of saprophyte bacteria in soil in comparison with control by 13.0–36.1% at cultivation of a potato variety Skarbnitsa and by 4.5–24.6% of Oberig variety (Table 1 and 2).

As a result of potato treatment by biological preparations in comparison with control the qualitative structure of soil microflora was also changed with domination of *Pseudomonas*, *Bacillus*, and *Micrococcus* bacteria. Among saprophyte fungi *Penicillium* and *Trichoderma* were the most numerous.

The reason of biological preparations impact on suppression of phytopathogens development could be explained by activation of saprophyte soil microflora and its antagonistic action against phytopathogenic fungi that are the pathogenic agents of rots [1,22-24,34,37,38]. By numerous studies it has been proved that the level of specific variety of soil microbiocenoses can be considered one of the most important criteria of stable soil fertility [1,23,24,37].

During the investigation 62 morphotypes of bacteria and 15 morphotypes of fungi were isolated as the pure culture and described. According to the Shannon index the variety of soil bacteria and fungi was the highest in soil where potato was treated by biological preparations –1.89–2.22 and 1.82–1.89 in comparison with control –1.87–2.10 and 1.52–1.67. At the application of chemical preparation Rovral Akvaflo reduction of the specific kinds of microorganisms was observed, in addition to the domination of dark-painted mycomycetes —*Alternaria sp.*, *Cladosporium sp.*, *Phoma sp.*, *Doratomyces sp.*, and chromogenic bacteria. In comparison with other variants of experiment the Shannon index in this case was 1.46–1.88 against 2.04–2.24, and the Pielou's index –0.63–0.79 against 0.83–0.91.

The regularity of phytopathogenic fungi microflora changes in the soil at treatment of various varieties of potato by biological and chemical preparations was established. So, at the treatment by preparations Phytotsid and Planriz the number of pathogenic agents such as *Alternaria sp.* and *Fusarium sp.* decreased, but during application of Rovral Akvaflo – increased. At the same time the considerable decreasing of the total number of saprophyte microflora (Figure 1) was observed.

TABLE 1   The quantitative and qualitative composition of bacterial microflora during the potato treatment by chemical and biological preparations in vegetation period

The variant of experiment	The total number of colonies, x10⁴ CFU/g of soil	Shannon diversity index	Pielou's evenness index
	Skarbnitsa variety		
Control (water treatment)	42.55 ± 4.04	2.10	1.01
Biological control (Phytotsid)	48.08 ± 5641	2.22	1.07
Planryz	57.91 ± 6.35	2.24	1.15
Chemical control (Rovral Akvaflo)	34.85 ± 8.39	1.88	0.85
	Oberig variety		
Control (water treatment)	49.34 ± 5.93	1.87	0.96
Biological control (Phytotsid)	51.56 ± 2.39	1.89	0.97
Planryz	61.49 ± 3.42	2.04	1.05
Chemical control (Rovral Akvaflo)	44.03 ± 3.91	1.46	0.75

TABLE 2   The quantitative and qualitative composition of fungi microflora during the potato treatment by chemical and biological preparations in vegetation period

The variant of experiment	The total number of colonies, x 10⁴ CFU/g of soil	Shannon diversity index	Pielou's evenness index
	Skarbnitsa variety		
Control (water treatment)	43.56 ± 3.36	1.52	0.78
Biological control (Phytotsid)	59.33 ± 3.05	1.82	0.88
Planryz	75.18 ± 2.59	1.89	0.91
Chemical control (Rovral Akvaflo)	81.0 ± 2.77	1.22	0.63
	Oberig variety		
Control (water treatment)	88,62±2,3	1.67	0.86
Biological control (Phytotsid)	130,42±5,7	1.82	0.83
Planryz	94.75 ± 3.27	1.85	0.89
Chemical control (Rovral Akvaflo)	65.54 ± 4.32	1.53	0.79

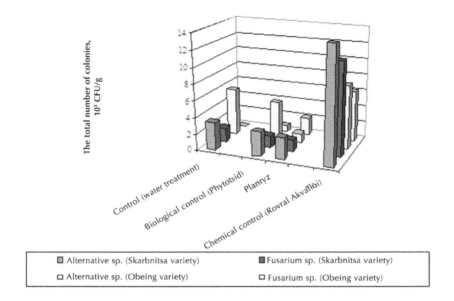

☑ Alternative sp. (Skarbnitsa variety)	■ Fusarium sp. (Skarbnitsa variety)
☐ Alternative sp. (Obeing variety)	☐ Fusarium sp. (Obeing variety)

FIGURE 1   The impact of potato treatment by preparations during the cultivation on qualitative structure of phytopathogenic micromycetes.

In the control variant the number of pathogenic agents —*Alternaria sp.* and *Fusarium sp.* was 1.7–5.7 thousand/g, in the variant of chemical control –6.2–14.0 thousand/g, Phytotsid –0.8–4.7 thousand/g, Planriz –1.1–2.7 thousand/g.

It is known that the increase of pesticide loading leads to reduction of number of all ecologic-trophic groups of microorganisms, the ratio between them is considerably changed which is resulted in disturbance of functional links in agrosystem and decreasing of the biological activity of the soil [1,23,24,34,37]. The application of the organophosphoric insecticide chlorpyrifos has led to decreasing of total number of bacteria and actynomycetes in the soil [28]. Essential changes in structure of saprotrophic microorganisms was also observed at the application of chemical means of protection against fungi phytopathogens of potato such as Ridomil gold MC and Kuprikol in comparison with biological preparations Trihodermin and Agrohit [20]. That is, suppression of the autochthonic microflora could be accompanied with increasing of the number of phytopathogenic microorganisms, narrowing species variety and occurrence of new dominant ones.

Analysis of the fungi and bacteria species structure which were isolated from the soil in the various variants of experiment, showed the essential difference in microbic complex structure and changes of dominant morphotypes (Figure 2–5). At the treatment of the potato by biological preparations 18 various morphotypes of bacteria (treatment by fungicide – 10) have been isolated and identified. The application of fungicide has led to changes of soil microflora qualitative structure that further can lead to the soil toxicosis and decreasing of plant productivity.

The analysis of tubers lesion by phytopathogens after their storage has shown that the quantity of pathogenic agents after potato treatment by Planriz has decreased compared with control from 3.7–8.6 thousand/g soils to 0.6–2.2 thousand/g (Figure 6).

It is necessary to mention that at the treatment of potato tubers before landing and plants during the vegetation period by biological preparations such as Phytotsid and Planriz the crop increased in 1.2–1.3 times.

Thus, the application of Phytotsid and Planriz promoted the total increase in saprophyte bacteria number in soil in comparison with control by 13.0–36.1% at cultivation of potato breed Skarbnitsa and by 4.5–24.6% of Oberig variety. In addition, the quantity of phytopathogenic fungi such as *Fusarium* and *Alternaria* has decreased in 1.2–1.8 times that is probably related to the increase of the number of saprophyte microorganisms which are capable to effective competition with phytopathogens. Reduction of quantitative and qualitative structure of soil microflora has led to changes in microbiocenoses composition of soil and occurrence of new dominant morphotypes.

The quantitative and species structure of soil bacteria and fungi was the highest in soil where potato was treated by biological preparations —the Shannon index was accordingly 1.89–2.22 and 1.82–1.89 compared with control 1.87–2.10 and 1.52–1.67. At the application of chemical preparation Rovral Akvaflo reduction of the specific variety of microorganisms was observed, besides the domination of dark-painted mycomyces —*Alternaria sp.*, *Cladosporium sp.*, *Phoma sp.*, *Doratomyces sp.*, and chromogenic bacteria was represented. In comparison with other variants of experiment the Shannon index in this case was 1.46–1.88 against 2.04–2.24, and an index of uniformity of Pielou –0.63–0.79 against 0.83–0.91.

Prelanding treatment of potato tubers and the subsequent one in vegetation period by biological preparations promoted the decreasing of pathogenic agent's population density in the soil and increasing of new crop tubers resistance to pathogenic agents.

Thus, application of biological preparations Phytotsid and Planriz for tubers treatment before landing, during vegetation and before putting on storage are effective methods of reduction of tubers lesions (caused by pathogenic agents) and improvement of potato quality.

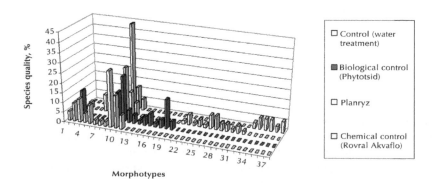

FIGURE 2 The qualitative structure of bacterial microflora in soil at the beginning of potato flowering, % (Skarbnitsa variety).

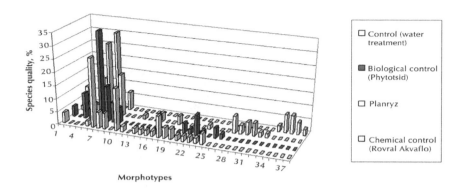

FIGURE 3 The qualitative structure of bacterial microflora in soil at the beginning of potato flowering, % (Oberig variety).

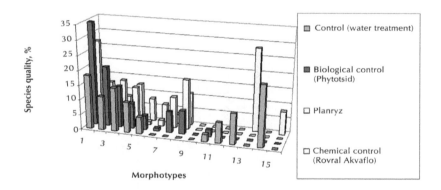

FIGURE 4 The qualitative structure of mycomycetes microflora in soil at the beginning of potato flowering, % (Skarbnitsa variety).

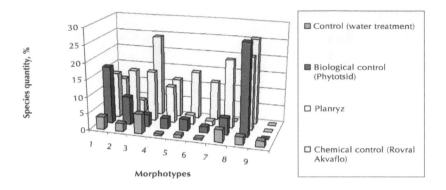

FIGURE 5   The qualitative structure of mycomycetes microflora in soil at the beginning of potato flowering, % (Oberig variety).

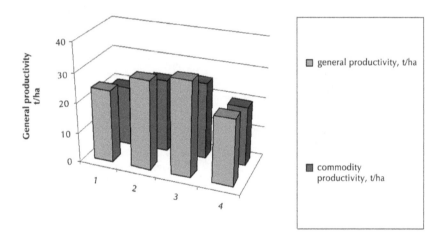

FIGURE 6   The general and commodity productivity of potato at the application of biological preparations (1—control (water treatment), 2—biological control—Phytotsid, 3—treatment of Planriz), and 4—chemical control—Rovral Akvaflo)

KEYWORDS

- **Biopreparations**
- **Phytopathogenic fungi**
- **Potato**
- **Soil microbiocenosis**

ACKNOWLEDGMENT

We thank the colleagues from Department of Phytopathogenic Bacteria, Institute of microbiology and virology of D.K.Zabolotnoho, NASU, especially Zitkevich N. V. and Gnatuk T.T., for their guidance, helpful suggestions, advices, encouragement, and assistance.

REFERENCES

1. Andreyuk K.I., Iutynska G.A., Antypchuk A.F.: The functioning of soil microbial communities under the anthropogenic stress. Oberegy, Kyiv, 2001.(in Russian).
2. Babana A., Hamed B., Fasse S., Kadia M., Diakaridia T., Amadou D.:Brit. Microbiol. Res. J.,1(3), 41-48 (2011).
3. Degtyareva E.,Vinogradova K., Aleksandrova A., Filonenko V., Kozhevin P.: Vest.Mosk. Univer., Pochv., 2, 22–26 (2009). (in Russian).
4. Dospehov B.A.: The methodology of field experiment (with the basics of statistical processing of results). Ahropromyzdat, Moscow, 1985.
5. Gerhard et al. (Eds): Methods of general bacteriology. Mir, Moscow, 1983.535 p. (in Russian).
6. Grosch R., Lottman J. and oth.: Gusunde Pflanz., 57(8), 199- 205(2005).
7. Grebenisan J.: Bul.Univ.Agr., 62, 392 (2006).
8. Gvozdyak R.I., Chernenko E.P., Moroz S.M., Yakovleva L.M.: Quar. and Plant Prot. J.,12, 15-17(2007).
9. Han J.S. and oth.: Sunhua. J.Appl.Microbiol., 99(1), 213(2005).
10. Hsieh S., Huang Ruizhi. Method for producing working reagent and for biological prevention or control : Заявка 1312263 ЕПВ, МПК⁷ A01N 63/00 /: Yuen Foong Yu Paper MFC Co, Ltd — № 00954262.2 (2006);
11. Hýsek J., Vach M., Broňová J., Sychrová E., Civínová M., Neděelník J., Hrubý J.: Arch.of Phytopath. And Plant Prot., 35 (2), 115 – 124(2002).
12. Jamiołkowska A.: El. J. of Pol. Agricult. Univ., 12 (3),13 (2009).
13. Korniychuk M.S. et al.: Quaran. and Plant Prot. J., 7, 6-8 (2008). (in Russian).
14. Kotova Z.P., Kuznetsova L.A.: Mat. of the rep. on the world conf.dev.to 75 anniv. of the boil. Depart. of MSU named after Lomonosov, Moscow, 86-88 (2006). (in Russian).
15. Kovalchuk M.V., Ryazantsev V.B., Kostiuk I.I., Kozyrovska N.O.: Bulletin agr. Science, 7, 43-45 9(2005). (in Russian).
16. Kurzawinska H., Mazur S.: Scien. works of the Lithuan.ins.of horticult. and Lithuan.univ. of agricult., 27(2), 419-425 (2008).
17. Kurzawinska H., Mazur S.: Folia horticulturae Ann., 21(2), 13-23(2009).
18. Kyprushkyna E.I.: The stor. and proc. of the agric. Prod., 7, 24-27(2009). (in Russian).
19. Kyprushkyna E.I., Kolodyaznaya V.S., Chebotar V.K.: Plants prot. J., 3, 17-24 (2003). (in Russian).
20. Kulikov S.N., Alymova F.K., Zakharova N.G. et al: App. Bioch. and microb., 42(1), 86-92(2006). (in Russian).
21. Mejda Daami-Remadi and oth.: Plant Path. J., 5(3), 283-290(2006).
22. Odum E. : Basic Ecology. Saunders College Publishing, Philadelphia, 1983.
23. PATYKA N.V., KRUGLOV Y., Omelyanets T.G.: Agroecol.J.l (1), 60-65(2009). (in Russian).
24. Patyka V.P..: Agroecol. J., l (2), 21-24 (2005). (in Russian).
25. Perevast K. and oth.: Biocontrol, 51(4), 533-546 (2006).

26. Pryshchepa L.I.: The world org. of boil. Prot. system. Inform. bulletin, 36, 146-155(2007). (in Russian).
27. Schisler D., Slininger P., Bothast R.: Phytopath., 87(2),177-183. (1997).
28. *SHAN M., FANG H., WANG* X., Feng B., Chu XQ, Yu YL .: J. Environ Sci. 18(1), 4–5 (2006).
29. SHENIN YU.D., NOVIKOVA I.I., SUIKA UARKAJA P.V.: Applied Bioch. and Microb., 46 (9), 854-864 (2010).
30. Sokolova M.G. et al.: Agrochem., 6, 62-67(2008). (in Russian).
31. Some new methods of the quantitive accounting of soil microorganisms and studying of its properties: Ju.M. Voznyakovsky, 1982. 52 p. (in Russian).
32. Stephan D., Schmitt A., Martins Carvalho S., Seddon B., Koch E.: Eur. J. Plant Pathol., 112(3), 235-246(2005).
33. Tektonydy I.P., Myhalyn S.E.: Potat. and Veg.J., 7, 17-18(2006). (in Russian).
34. Tikhonovich I.A., Kozhemyakov A.P., Chebotar V.K. et al.: The biopreparations applica- tion in agriculture. Russian Agricultural Academy, Moscow, 2005. (in Russian).
35. The methodology of soil microbiology and biochemestry: Zvyahyntsev D.G. MSU, Mos- cow, 1991. (in Russian).
36. The list of pesticides and agrochemicals permitted for application in Ukraine. Young West Media, Kyiv, 2008. (in Ukrainian).
37. Voznyakovskaya Y.M.: The coll. Art. of ANIAA micr., 57, 21-31(1988). (in Russian).
38. Zhuchenko A.A.: The ecological genetics of plants and the problems of agrosphere (theory and practice). Ahrorus Publishing, Moscow, 2004. (in Russian).

CHAPTER 13

IRRADIATION OF FRUITS, VEGETABLES, AND SPICES FOR BETTER PRESERVATION AND QUALITY

MD. WASIM SIDDIQUI, VASUDHA BANSAL, and A. B. SHARANGI

CONTENTS

13.1   INTRODUCTION

Foods such as fruits and vegetables are highly perishable and most of them are not available throughout the year, which needs their preservation to consume in all the seasons along avoiding their wastage. Specific type of treatment is essential for preservation. Method of preservation requires retardation of the growth of microbes which are leading cause of food spoilage without deteriorating the nutrients. Foods, safe from the microbes are not only said to be preserved, their shelf life in terms of nutrients is also primary. Preservation cannot improve the original quality of food but can retain the original nutrients along optimised ripeness of the foods can be achieved.

Food irradiation is one of the processes of food preservation which renders the usage of radioisotopes in the form of ionizing radiations (Wilkinson, 2005). It is considered as nonthermal processing technique as no application of heat is involved and devastating effects on food characteristics is limited [19]. There are wide applications of irradiation technology so as to prolong the shelf life of perishable commodities as fruits, vegetables, sea food, and organ meat, retardation of sprouting in vegetables like potatoes, onions, destruction of surface microbes, and sterilization of food packages (Molins, 2001). Preserving food using irradiation possesses an immense potential to cut down the world's food shortage and produce safe consumables (Olson, 2004).

Irradiation is also considered as energy efficient when compared with conventional heat treatments. This processing technology can replace the use of food preservatives, additives and surfactant fumigants for enhancing the shelf life of food stuffs as fresh fruits and vegetables (Osterholm, 2004). Irradiation is an effective surface preservative technique with non tedious procedures (Scott Smith, 2004). Various food borne diseases come up on contaminated surfaces as presence of larvae or insect eggs, improper handling conditions during harvesting and transportation invite various pathogenic bacteria which therefore impair the sensory quality of fresh fruits and vegetables (Koopmans, 2004). The process involves exposing the food materials to permitted doses of radiations for edibles. Foods are exposed to prescribed level of ionizing radiations This technique is also referred as cold pasteurization process where no significant changes caused to state of food, their aromatic compounds with no matter if the food is being dried or frozen. Processed products such as jams, jellies, sugar syrups, juices, sausages, dehydrated powders, and so on are prepared from whole fruits and vegetables. Therefore, processing of refined products is a second step which requires safe and disinfected raw material. Irradiation can prove an optimum candidate for rendering fresh products to get processed (Tauxe, 2001).

13.1.1   FOOD IRRADIATION

Food Irradiation is defined as the potential technique for treating the microbes by exposing them to ionizing radiation using radioisotope as cobalt-60 and cesium-137, and so on where machines using electricity are being used for generating X-rays (Suresh, 2005). It has been declared as a safe process by U.S Food and Drug Administration. Application of targeting can be direct or indirect. When any moving charged particle strikes with microbial cell or biological tissue, its metabolic functions get metamorphosed or destroyed. The other type is when other material as water comprising of microbial cells lead to ionization by which highly reactive hydrogen radicals are formed due to the excitation of cells in the water and therefore radicals impart effect of ionizing radiations.

The wide uses of food irradiation are as follows:

1. Sterilization of microbes to inhibit their multiplication on the surface of food stuffs in order to avoid the usage of chemical compounds as ethylene dioxide.
2. Imparting shelf life to sea food, poultry and meat.
3. Inhibition of sprouting in tubers and bulbs as potatoes, onions. Garlic, and so on.
4. Enhancing safety to cereals by killing the insects.
5. Delaying of ripening in fruits as apple, mango, papaya, and banana.
6. Disinfection of packaging material.

By controlling all the above factors, the outbreak of diseases would be prevented. There are certain doses being fixed for their suitable action for example 0.03–0.015 kGy used for inhibiting sprouting where as upto 10 kGy viable for spices and dried vegetables (Loaharanu, 2003). The factors behind appropriate dosage is to avoid the change in physical and biochemical properties of food as flavor, aroma, taste, texture, taste, nutritional value, and so on. Energy giving food groups as carbohydrates, fats, proteins do not get affected with major changes during application of radiation. Sensitive nutrients in the forms of vitamins as C, thiamine (B1), riboflavin (B2), pyridoxine (B6), cyanocobalamin (B12), and so on gets affected with the radiation but not to an adverse extent. Fat soluble vitamins as Vitamin A, E, K, and D are found to be stable at low doses.

There are certain types of radiations that are produced by electron accelerators as Gamma rays which are produces by two types of radioactive sources as cobalt-6 or cesium-137 and X-rays, produced by a machine (Loaharanu, 1994). These sources are regarded as safe and radioactivity does not get induced into the food material. In current scenario more than 40 countries are having regulation of irradiation for over more than 100 food stuffs. International agencies as Food and Agriculture Organization (FAO) and World Health Organization (WHO) has added it to a recommended and certified list of irradiated food is safe for consumption (Thayer, 2004).

The techniques of measuring thermoluminescence and chemiluminescence intensities are used to detect irradiated spices. Heide and Bogl studied on twenty-nine different irradiated spices (10 kGy) [23]. With the exception of garlic, onions, white and black pepper, the irradiated spices can be identified more than 6 months after irradiation. Some, curcuma, juniper berries, basil, chillis, paprika and celery, show increased luminescence intensities for a year or longer after irradiation. Electron paramagnetic resonance and thermoluminescence signals induced by gamma irradiation in some herbs, spices and fruits were also systematically studied by Raffi et al (2000). Gamma-radiation of powdered tumeric, black pepper, dry mustard, cinnamon and paprika, at dosage of 30 kGy, at 30°C, produced free radicals detectable by EPR. The intensities of the EPR signals were proportional to the total irradiation dose up to a saturation dose. The signal decayed most significantly in the first four post-irradiation days, but could be detected even after 34 days. The paramagnetic signal was unaffected by the presence of oxygen, but decayed rapidly in water. Similar observations were made for spray-dried fruit powder (Yang et al, 2007).

13.1.2 PROBLEMS OF PERISHABLES

Unlike spices, fruits and vegetables are considered as highly perishable agricultural produce because of being made up of live tissues to which improper storage and transportation conditions lead to their spoilage (Sivapalasingam, 2004). To the matter of concern, fruits, and vegetables are the utmost important commodities all over the world as they are the key elements for balancing the healthy diet by providing ample amount of vitamins, micro-nutrients that play a major role as various co-factors in metabolic activities.

Many factors can pose threat to them. Presence of pathogenic bacterial load, vulnerable storage conditions as temperature and humidity, contaminated vessels, are the driving forces that call for effective preservation. Immediate surface treatment can make the perishables free from lethal deterioration. Particularly fruits and vegetables require prolong shelf life during transportation to sold under global markets and irradiation occupies the most appropriate place for preservation treatment (Sivapalasingam, 2004). To be attractive in order to have sensory appeal, they need to be looking fresh in terms of color as well. Fruits are active sources of pectins. Pectins are responsible for maintaining the physicochemical properties of fruits as brix, turbidity, and so on. Non-conducive conditions activate the enzyme pectin methyl esterase that degrades the pectin and quality of fruits is lost, others as phenoloxidase, polygalacturonase can degrade the color of fresh fruits. This treatment retards the activity of enzymes at very higher doses. Fruits and vegetables consist of heat labile nutrients so in order to reduce the urgent risk of spoilage, highly suitable preservation technique can prevent the damage along not posing danger to sensitive nutrients.

13.1.3 FACTORS POSING THREAT TO PERISHABLES

Environmental conditions in terms of physical factors as uncontrolled temperature, humidity, oxygen availability are directly linked to the deterioration of fruits and vegetables. Fast damaging actions are due to tremendous metabolic activities of live tissues of perishables that make the fruits and vegetables vulnerable to be attacked by microbes, insects and pests (Koopmans, 2004).

Second factor is injuries or bruises which can either be physiological or mechanical. Any minor cut may be during harvesting as improper plucking of fruits or careless storage will rupture their make them prone to attack of organisms or being eaten up by birds and consequently their cell membrane gets permeabilised for uncontrolled gaseous exchange. Operational post-harvest tasks as peeling, cutting, trimming, polishing, and spraying, contaminated vessels are also equally responsible for degradation of soft lining tissues. These threats will not only lead to the wastage of superior resources but also the cost, labor, use of land will go in vain.

Third cause is the inadequate selling or marketing of harvested goods. Most of the harvested fruits and vegetables do not get well transported due to lack of storage space, refrigeration facilities and usage of conventional machines of cutting, picking rolling which ultimately severe losses. When the root cause problem would be curbed, an end can be put to the wastage. This can be done by operating careful post harvest operations, sensitive handling and sterile packaging. In order to combat the above threats food irradiation serves has a significant role to play.

13.1.4   ROLE OF IRRADIATION IN PRESERVATION OF FRUITS, VEGETABLES, AND SPICES

Preservation of fruits and vegetables using irradiation process falls under the category of low and medium dose requirement. This is due to the fact of the fragile tissue covering on perishables. 0.06–1.0 kGy is considered as the effective dose range for preserving fruits while 1–10 kGy for vegetables. Treatment of spices with irradiation doses of 10 kGy has proved to extend shelf life without causing significant changes in sensory or chemical quality [3].

- Inhibition of sprouting of potatoes, onions garlic and ginger are treated in the range of 0.06–0.2 kGy.
- Insect disinfection grains as fresh and dried fruits are irradiated in the range of 0.15–1.0 kGy.
- Parasitic infection is prevented by giving the treatment at 0.3–0.1 kGy to the fresh fruits.
- Delaying of ripening in fresh fruits is done at 0.5–1.0 kGy.
- Extension of shelf life to several days to fruits and vegetables is done at 1.0–3.0 kGy.
- Inactivation of pathogenic bacterial load of dried vegetables is given at 1.0–7.0 kGy.
- Improving the extraction as juice yield from fruits and vegetables is done at 3.0–7.0 kGy.

Irradiation is the only technique that can render preserving treatment to raw and frozen fruits and vegetables which we keep in refrigeration storage for non-seasonal perishables. The major advantage is the property of being unaltered changes to aroma, taste and texture where no conventional treatment has the potential of keeping them safe for long term storage and consumption. That is why the developed countries as United States, European ventures are turning 70% raw materials to the processed food stuffs. However, due to economic constraints in developing countries, technology is much far behind to be adopted at commercial purpose. Irradiation cannot come over the mishandling which is being done at initial level of harvesting. It can only be beneficial when well handled food is kept at proper vessels and further irradiated food must be packaged properly to ensure its safety from re-contamination.

KEY BENEFITS OF IRRADIATION TO FRUITS AND VEGETABLES ARE AS FOLLOWS

Improper storage conditions as increased temperature and humidity give rise to sprouting in vegetables and fully grown vegetables left no more effective for marketing.

Irradiation is indispensable for inhibiting sprouting. Moreover, very low dose is required for rendering the treatment where as loads of doses of chemicals are used for fulfilling the same purpose. All tubers as potato, sweet potato, roots as carrots, turnips, radishes and bulbs as onion, garlic are included.

Characteristics of food quality can be enhanced using irradiation. Over ripening of fruits can be inhibited with low doses of radiations for example tender fruits as banana, papaya, guavas and apples. While hard covering fruits as all berries for example raspberries, cherries, strawberries and blueberries are effectively stored after irradiation.

Adding variety to the shelf by making non-seasonal fruits and vegetables by making them available in frozen state.

Pathogenic parasites found on the surface can cause deadly diseases as pork usually infected with fungi worms and can be lethal when consumed raw. But irradiated pork is safe to consume and worms cannot survive as their life cycles get disrupted.

Most of the spices are currently treated with ionizing radiation to eliminate microbial contamination. It was unambiguously confirmed that the treatment with ionising energy is more effective against bacteria than the thermal treatment, and does not leave chemical residues in the food product (Loaharanu 1994; Thayer *et al*. 1996; Olson 1998). Thus, ethylene oxide and methyl bromide treatments can be effectively replaced by food irradiation, which in fact is less harmful to the spices than heat sterilization, which implicates the loss of thermolabile aromatic volatiles and/or causes additional thermally induced changes (for example thermal decomposition or production of thermally induced radicals). Chmielewski and Migdal also opined that radiation decontamination of medicinal plants and spices is a safe and very effective method. The losses of the biologically active substances are lower than in the case of other decontamination methods [8].

13.1.5   EFFECTS ON NUTRITIONAL VALUE

The foremost concern is nutrients for every processing. The awareness of the safety of process is the key matter to commercialization producers and consumers. Other methods of preservation as canning or freezing are known to cause significant effect on the texture and as well to the nutrients which are difficult to avoid. As matter related to fruits and vegetables, they contain heat sensitive vitamins which are essential to preserve. There are some factors which need to keep in consideration regarding the parameters of irradiation as dose of irradiation, temperature, type of food to irradiated, packaging material of irradiated food, and so on. The energy content of the foods in terms of carbohydrates, proteins and fats are far have less effects on their structure and composition but vitamins as vitamin C, B1, E, and K are relatively sensitive. Studies have reported retention of thiamine in irradiated foods as compared to canned foods.

From nutritional point of view doses up to 1 kGy are absolutely safe for irradiating fruits and vegetables where as some loses may get occurred on medium doses as 1–10 kGy, foods are being treated in presence of air while irradiation above 10 kGy, losses of vitamins as thiamine can be significant. Muller and Diehl 1996 reported the minimum loses of folate in the vegetables during processing of irradiation where as more than 50% of loses observed in conventional cooking and heating. Micronutrients as folate has an important role in metabolic activities and conservation of these sensitive nutrients are utmost desirable.

In case of bulbs and tubers potato and garlic are the most consumed vegetables all over the world. Sprouting in potatoes is the major concern during storage where irradiation has significantly inhibited the sprouting and enhanced the shelf life to more than 6 months. The losses of vitamin c are evaluated during the storage life and it was observed that high content of vitamin c are preserved in irradiated potatoes as compare to control (Evangelista, 2000 and Moy, 2002) [14].

In regard to content of vitamin c found in garlic was studied for about 10 months after being irradiated at 0.05 kGy gamma radiation and content found to be preserved after several months where as some loses are observed which are similar to loses occurred in control (Kilcast, 1994).

In case of perishables it is suggested to follow the fumigation process at varied temperatures in order to avoid the insects or pests to prevent their entry to them. But the deleterious effects of fumigation cannot be ignored to public and environment and this process takes long time as well. This problem has successfully over by the usage of food irradiation less than doses of 1.0 kGy and infections caused by pests has significantly controlled (Patil et al, 2004). Effect of irradiation on the storage of capsicum, zucchnis, and cumcumber were evaluated at 0.75 kGy and stored at 1–7°C for the period of 1 month and less significant changes were observed on their nutrient composition (Mitcheel, 1992). Doses of irradiation of about 1 KGy has been optimized for frsh cut lettuce and after being treated to be stored at refrigeration temperature and observed that vitamin c content was much better preserved at 1 kGy dose in relation non-irradiated lettuce.

Fruits like mango and papaya usually get involved with the problem of fungi where 0.5–0.95 kGy gamma irradiations has prevented their spoilage and conserved the content of vitamin c as well. Since higher doses of irradiation have large chances for vitamin c gets destroyed but safe use of irradiation has overcome from this problem (Lacroix et al., 1990). Fruits are good sources of riboflavin; niacin and thiamine studies reported that doses upto 2 kGy do not cause nutrition losses. China has a popular fruit called lyceum and evaluated for effects of irradiation at the doses of 2–14 kGy and observed no degradation of riboflavin and thiamine where as content of vitamin c is lost (Wen et al., 2006).

Vitamin C is an essential vitamin found in citrus fruits and irradiation upto lower doses has non-significant changes on it. Similarly, study on the storage of vitamin C in oranges were conducted and stored at 0°C for the period of 3 months and no damage was observed where losses were found in oranges after 1 month at the storage of 15°C (Fan and Mattheis, 2001) [15]. In addition to fruits, fruit juices are also immense sources of vitamin C, however studies reported its loss 0f around 70% on severe doses (2.5–10 kGy) of irradiation of gamma rays. The appropriate dose for retention of ascorbic acid was reported as 3 kGy in carrot and kale juice (Song et al., 2007).

Irradiation processing is a boon for imparting safety to fruits and vegetables and keeping their nutritional quality intact as compared to their fresh values. Therefore, irradiation by gamma rays, X-rays or accelerated electrons under controlled conditions does not make the food radioactive and make the food readily available in its original form to far flung areas.

The only major drawback of food irradiation is its limited use on restoring the shelf life of fresh foods as enzymes are present in them and in order to inactivate the enzymatic reactions, large doses of radiations need to be given which in turn make the food stuff unpalatable. Immediate treatment is commendable as number of advantages offered by irradiation and number of multi-functions can be performed at later changes with less tedious efforts and is proved fruitful. Thus, this method should be used because of its low cost and efficiency compared to other conventional methods.

13.1.6   EFFECTS OF IRRADIATION ON FRESH-CUT QUALITY

In the last years, consumers have shown increased trends of consuming fresh cut fruits and vegetables due to changes in the trends of lifestyle and consumer preferences for natural products in replace of synthetic chemicals, preservatives, and colorants, and so on. Ready to eat, healthy, and safe food is the demand all over the world. Minimally processed fresh cut fruits and vegetables have short shelf life and in addition, they provide favorable conditions for micro-organisms to grow and multiply because of their surface area exposed and moisture present.

Firmness on suitability of cherry tomatoes was observed by treatment at 0–3 kGy of gamma radiations. 1.0 kgy dose found to be suitable for microbial sterilization to carrots without affecting the sensory quality. Packed mixed salad is sold on a large scale by food services and rendering microbial and sensory quality to it is a major contribution. 1.15 kGy is sufficient to maintain the quality with minimum pathogenic load and 4 log reduction of L. Monocytogenes was found. 0.5–1.0 kGy found to sterilize the turnip and brought 4–5 reduction in bacterial load (Thayar, 1999 and Zhang, 1996). No difference was observed on the surface color of the turnip.

Similarly irradiation of coriander at 1 kGy observed in inhibition of bacterial load and no changes in color and texture was observed and enhanced the shelf life to 2 days as compared to non-irradiated coriander. Shelf life of water melon increased to 4 days on application of 0.5 to 1 kGy irradiation with only slight darker coloration was observed on the first day while later no detectable changes were observed. To be precise, the optimum doses would be 1 kGy for lettuce, turnip, watermelon and melon and only 0.5 kGy for coriander and mint (IAEA, 2001).

13.1.7   EFFECTS ON SENSORY QUALITY

Irradiation processing has significant effect on the organoleptic properties of fruits and vegetables. It has been observed that not all the fruits and vegetables are viable for irradiation processing. As their aromas are so delicate and any high intensity treatment can result in serious losses. For example, fruits like avocado are very much sensitive to irradiation treatment. In treatment of strawberries, the retention of its texture, color and flavor are irradiation doses dependent and that is why not more than 2 kGy of dose is not recommended for strawberries (Yu, 1995). Similarly, seedless grapes were also studied for their preservation through irradiation and no changes were observed in their color, firmness and sensory quality (Thompson, 1995). Role of irradiation on sweet cherries were reported by Drake and Neven in 1997 and 0.90 kGy of treatment was found suitable with no adverse effects in terms of titrable acidity, soluble solid content and flavor were observed [11].

Antioxidant rich fruits as blueberries were affected till 3 kGy of the gamma radiations and surface bloomness was intact (Miller, 1995). Depending on the dose of irradiation, citrus fruits shown to have effects on their discoloration dose of 0.3 kGy with storage at lower temperature seemed to have good effected and worked as antifungal (Oufedjikh, 2000). It terms of sensory quality, fruit juices are not considered as suitable for treatment as delicate flavors are fragile and do not respond against irradiation. However concentrate of orange was treated at 2.5 kGy and stored at 0–5°C refrigeration temperature (Spoto, 1997). 0.6 kGy dose of irradiation worked well for treating apple at 2–5°C and shelf life was enhanced to 6 months with no changes on firmness and color was observed [5](Bhushan and Thomas, 1998).

Lescano in 1993 reported the prevention of discoloration of asparagus with lower doses of treatment and sensory quality was remained intact. Vegetables like mushroom which is consumed in large numbers all over the world was observed for its texture after being stored in polystyrene trays covered with PVC film with 3 kGy dose of irradiation and stored at 10°C for 17 days against control and no change in weight and acidity was seen (Narvaiz, 1994). In other study, only 10% of losses were seen in firmness in cut lettuce with treatment of 0.35 kGy of dose and no change in color, appearance and flavor was observed (Prakash et al., 2000). Therefore, promising results were reported on the organoleptic properties of fruits and vegetables and no significant changes were neither on their appearance and firmness, nor on their flavor and texture.

13.1.8   EFFECTS ON EXTENSION OF SHELF LIFE

The ultimate goal of any processing treatment is to enhance the shelf life with quality near to fresh. As all the fruits and vegetables are not available throughout the year and every region cannot have all the conditions to grow each variety. Here role of enhanced life comes into play where stored foods stuffs can be transported to far distant places and provide different kinds of variety as well.

Shelf life can be enhanced with conducive conditions of harvesting and storage along inhibiting the microbial load. Food irradiation is one the suitable processes covering all the required parameters. Depending upon the commodity, number of doses shelf life is enhanced from days to weeks. For example cut lettuce has a shelf life of 14–18 days where treated lettuce with dose 0.35 kGy can restored the quality up to 22 days (Prakash et al., 2000). Similarly, grated carrots can be stored a fresh for the period of 10 days at 10°C by treating at lower dose of 0.5 kGy [6](Boisseau, 1991).

Effects on shelf life were observed on button mushrooms where 2 kGy of dose increased the shelf life to 10 days at 10°C. radiation can significantly enhance the shelf life provided storage conditions were given as in case strawberries their shelf life was increased to 4 days by applying 1–2 kGy of doses while keeping them to at 1–5°C (IAEA, 1994)[22].

Fruitful results were reported on the extension of shelf life with retained quality. Irradiation has many parameters to evaluate their effects on number of doses and its co-relation with storage temperature.

13.2   IRRADIATION OF SPICES- FACTS AND IMPACTS

13.2.1   IRRADIATION TO REMOVE MICROBIAL POPULATION

Spices, even when used in small quantities, pose a potential source of microbial pollution for foods to which they were added. This unique food ingredient, originated from several countries, where harvest and storage conditions are inadequately controlled with respect to food hygiene. Thus there is a chance of exposure to a high level of natural contamination by mesophylic, sporogenic, and asporogenic bacteria, hyphomycetes, and faecal coliforms (Bendini *et al*. 1998). Most spices become seriously contaminated by air- and soil-borne bacteria, fungi, and insects when dried in the open air. Microorganisms of public health significance such as *Salmonella, Escherichia coli, Clostridium perfringens, Bacillus cereus*, and toxigenic moulds can also be present. Bacterial plate counts of one to 100 million per gram of spice are usual [4] (Bendini *et al*. 1998).

A wide range of bacterial and fungal species has been identified in spices. Bacterial counts in this products vary from 10^4–10^7 per gram. The most contaminated spices are generally black pepper, turmeric, paprika, chili, and thyme. Bacteria like *Bacillus cereus* and *Clostridium perfringens* are frequently found in spices, but usually in low numbers. However, in extreme cases, *B. cereus* counts up to 10^5 CFU/ g. Several other *Bacillus* spp., that are opportunistic pathogens, are also more frequently isolated from spices. Since their spores may survive cooking, ingredients harbouring these spores must be considered as a potential health hazard. *Salmonella* has been found, although infrequently, in a variety of spices. However, its presence is of great concern, when spices are used in food that are consumed raw or when the spices are added to food after cooking. Mold counts of spices and herbs may reach to 10^5 CFU/ g level, and also a high incidence of toxigenic molds has been found.

The mold genre *Penicillium* and *Aspergilus* are common, and hence, mycotoxins may be present. Peppermint showed bacterial loads that ranged from 10^5 to 10^7 CFU/ g fresh weight. The application of Good Hygienic Practices (GHP) and the Hazard Analysis of Critical Control Points (HACCP) concepts are of importance in the field of dried ingredients and should consider the whole process, including production in the countries of origin. However, importers, lack direct control in exporting countries and this diminishes CCP opportunities. Therefore, the need for microbial decontamination treatements for spices and herbs is an important question.

Irradiation of spices basically aims to reduce the microbial population, ensuring their hygienic quality. Generally, there are three energy levels viz., low doses (up to 1 kGy); medium doses (1–10 kGy) - the improvement of hygienic quality of spices (10 kGy) and high doses (above 10 kGy).According to Codex Alimentarius General Standard for irradiated foods, the uppermost level delivered to a food should not exceed 10 kGy, except when necessary absolutely [10]. Limitation of FDA for culinary herbs, seeds, spices, vegetable seasonings and blends of aromatic vegetable substances has not to exceed 30 kGy [9](Code of Federal Regulation, 2006).

Until the early 1980s, the most widely used method to destroy microorganisms in dried food ingredients was fumigation with ethylene oxide. A number of alternative technologies have been developed for decontaminating dried food ingredients. However, none of them match the applicability of treatment with ionizing radiation, because of their low antimicrobial efficiency, changes in flavor and color, loss of functional properties. Irradiation has a very strong antimicrobial effect. Radiation doses of 3–10 kGy reduce the total aerobic viable cell counts even in highly contaminated spices and other dry ingredients to bellow 10^3–10^4 CFU/g. Recontamination can be prevented because irradiation can be applied to products as a terminal treatment, in their final packaging.

13.2.2   IRRADIATION ON ORGANOLEPTIC CHANGES AND ON VOLATILE COMPOUNDS

The organoleptic characteristics of some spices are affected by irradiation, but the extent of the effect and the dose level to cause a perceptible change vary with the product. The environmental and storage condition of the spice prior to irradiation may also affect the product's sensory characteristics. Ingredients that have been stored for a long time, or whose quality, flavor or aroma is not up to usual standards, may be more affected by the same irradiation dose than better quality spices. The sensory properties of most spices are well maintained between 7.5–15 kGy compared to that of ethylene oxide (EtO) gas treatment (Marcotte 2001).Irradiation-decontaminated spices are superior to conventional heat-treated ones in aromatic quality [21].Various works have been done by researchers in this direction throughout the globe. The methods include gas chromatography with flame-ionization detection (GC/FID), gas chromatography-mass spectrometry (GC/MS), sensory evaluation, GC-olfactometry, and so on. Emma *et al* compared two methods of decontamination [13]. Powdered black pepper was irradiated with different recommended doses of gamma rays (5 and 10 kGy, respectively) and treated with microwaves for different periods (20, 40, and 75 s). Gamma irradiation was found to be a safe and suitable technique for black paper decontamination. However, Sadecka *et al* (2005a) found a significant increase in monoterpenes under thermal treatment of black pepper (using 130°C hot dry steam for 4 min, internal temperature of the treated berries was 95°C).

13.2.3   IRRADIATION ON ANTIOXIDANT ACTIVITY OF SPICES

Irradiation with a ^{60}Co source, Gammacell 220 (Many spices, generally used to flavor dishes, are an excellent source of phenolic compounds which have been reported to show good antioxidant activity (Rice-Evans et al., 1996 and Zheng and Wang, 2001). The influence of irradiation process on antioxidant activity of seven dessert spices (anise, cinnamon, ginger, licorice, mint, nutmeg, and vanilla) was evaluated by Murcia et al (2004). The result in decreasing order of antioxidant capacity was cinnamon ≅ propyl gallate >mint >anise >BHA >licorice ≅ vanilla >ginger >nutmeg >BHT. Irradiated samples did not show significant antioxidant activity with respect to the non-irradiated samples (1, 3, 5, and 10 kGy) in the assays used. However, in another study gamma (A.E.C.L.) using 0, 5, 10, 15, 20, and 25 kGy at room temperature in cinnamon had no effect on the antioxidant potential of the cinnamon compounds (Kitazuru et al, 2004). Gamma -irradiation (10KGy) had also no effect on the antioxidant property of turmeric extracts as investigated by Chatterjee et al [7].

Antioxidant properties of clove (*Syzygium aromaticum*) and ginger (*Zingiber officinale*) irradiated by doses of 5, 10, 20, and 30 kGy, respectively, were studied by Suhaj and Horvathova (2007). Irradiation did not have much influence on contents of phenolic compounds in spices immediately after radiation treatment, but their contents increased during the ginger storage, mostly when irradiated at 20 kGy.

In the following tables (Table 1 and 2) the threshold doses for flavor or aroma changes and the effects of irradiation on microbial contamination in spices has been given.

TABLE 1   Threshold Dose (T.D.) for flavor or aroma changes in spices/products

Threshold dose in kGy for sensory change	Products with threshold at that dose
5< T.D.< 10	Cardamom, chives, coriander, cumin, fenugreek, garlic powder, ginger, lemon peel, marjoram, mustard seed, orange peel, paprika, white pepper, and turmeric
>10	Basil, caraway, cayenne pepper, celery seed, charlock, cinnamon, curry, dill fennel, marjoram, mustard seed, nutmeg, onion powder, paprika, black pepper, white pepper, red pepper, sage, thyme, and turmeric
>15	allspice, cloves, juniper, paprika, and black pepper

Source: Sterigenics Food Safety (2013)

TABLE 2   Irradiation effects on microbial contamination in spices/products

Spices or its Product	Dose (kGy)	Comments	Reference
Chili, coriander, cumin, and turmeric	0–10	a dose of 10 kGy reduced bacterial counts to below detectible levels, dose of 5 kGy eliminated fungi and colilforms	[1]
Fresh dry ground ginger	10	radiation treatment resulted in a decrease in aerobic microbial populations from 108 to 101 CFU/g	[2]

TABLE 2   *(Continued)*

Allspice, Greek oregano, black pepper, garlic powder, Egyptian basil, thyme, Mexican oregano, domestic paprika, Spanish paprika, celery seed, and crushed red pepper	10 except paprika and red pepper at 6.5	radiation treatment reduced standard plate counts to less than 3000 yeast, mold and coliforms reduced to non-significant levels	[12]
Black pepper, fennel, anise, coriander, and turmeric	5.0–10.0	irradiation an effective decontamination treatment at both levels; losses of flavor components observed	[16]
Marjoram, ginger rhizomes, and hot pepper	5–30	a dose of 10–20 kGy was required to reduce microorganisms to below a detectable level (103/g, coliforms (found only in hot pepper) eliminated by a dose of 5 kGy; generally molds and coliforms eliminated with lower irradiation doses than were bacteria; doses of 16–20 kGy would likely be necessary to achieve sterility (< 10/g)	[18]
Paprika, and mixed seasoning (marjoram, ground pepper ground paprika)		a dose of 3–4 kGy led to a 2–3 log cycle reduction in viable cell count, total sterility was obtained with doses of 15–20 kGy, depending on initial bioburden	Farkas et al., 1973
Black pepper, thyme, anise seed, curry, poultry seasoning, pickling spices, cardamom, cumin and cream of tartar	1–9	suggested minimum doses to achieve commercial sterility (< 10/g) were: black pepper (13 kGy), thyme (13 kGy), anise seed (10 kGy), curry (7.3 kGy), poultry seasoning (6 kGy), pickling spices (7kGy), cardamom (9.4 kGy), cumin (12 kGy) and cream of tartar (4 kGy) based on initial bioburdens and resistance of contaminating microorganisms	[20]

TABLE 2   *(Continued)*

17 different spices		doses of 5–15 kGy and 4–10 kGy were required to reduce total aerobic bacteria and spore-forming bacteria, respectively, to below detectable levels; 4–10 kGy doses also eliminated coliforms; when untreated spices were stored for 1–3 months at 30-35°C and >80% humidity, mold counts increased up to 108 per gram in many powdered spices; samples treated with 4 kGy did not show any signs of mold contamination	Ito et al., 1985
Paprika, black pepper, spice mixture and onion powder	5–10	5 kGy dose reduced microbial counts by 2–3 orders in reducing microbial counts; most microbes eliminated by a dose of 15 kGy	Kiss, 1982
Paprika, black pepper	0–4	log of mesophilic aerobe counts decreased linearly with increasing irradiation dose	Kiss and Farkas, 1981
Ginger, turmeric, cayenne pepper, onion powder, garlic powder	10 and 30	doses of 30 kGy resulted in sterilization of the samples	Lescano et al., 1991
Paprika (granulated, added oil and fine grind types)	6.5	microbial populations reduced to below recommended (ICMSF) permissible levels by 6.5 kGy dose	Llorente et al., 1986
White and black pepper (whole and powdered), turmeric (whole and powdered), rosemary (whole and powdered), basil (powdered)		doses of 12–15 kGy reduced total aerobic bacteria to non-detectable level; doses <10 kGy reduced spore-forming bacteria to less than 103 per gram; 4–10 kGy eliminated coliforms; 4 kGy doses prevented proliferation of molds over a 1–2 month storage period	Muhamad et al., 1986
Black pepper, red chili, turmeric, coriander, curry powder	5–12.5	a dose of 10 kGy destroyed all micro-organisms in prepackaged samples; molds eliminated by 5 kGy; no effects on spice quality were found	Muna-siri et al., 1987
White pepper, turmeric, paprika and ginger	5	found large reductions in total aerobic plate count, coliforms, Bacillus spp and yeasts after irradiation treatment; average reduction was 1.75 logs	Niemand and duPlessis, 1986

TABLE 2   *(Continued)*

Ashanti pepper (whole and ground berries)	2.5–10	2.5 kGy dose reduced fungal and bacterial loads by 2 log cycles; dose of 7.5 kGy eliminated fungi and reduced total viable bacteria by more than 4 log cycles	Onyenekwe et al., 1997
Red chili pepper (whole dry and ground)	2.5–10	the fungal count reduced by 2 log cycles after a dose of 2.5 kGy and eliminated by 7.5 kGy; bacteria destroyed by a dose of 10 kGy	Onyenekwe and Ogbadu, 1995
Black pepper, red chili (C. annum), turmeric (whole and ground)	10	examined by six different labs; 3 of 6 labs reported no colony forming units (CFUs) after irradiation treatment while the other 3 found levels of 0–90 CFUs; in the few cases where E. coli and B. cereus were found in untreated samples, they were completely eliminated by doses of 10 kGy	Sharma et al., 1989
Black pepper, nutmeg, mace, cinnamon, cardamom, cloves	5–10	doses of 7.5–10 kGy were in sterilizing or adequately reducing bacterial contamination in spice samples; fungal contamination eliminated at doses lower than 5 kGy	Sharma et al., 1984
Ground black and white pepper	10, 17, and 20	10 kGy dose effective in obtaining zero plate counts for total bacteria, yeast mold and coliforms without affecting quality; no effect on volatile oil or piperine content	Shigemura et al., 1991
Onion powder average	0–15	minimum dose of 9 kGy resulted in a count reduction to below 2.5 x 104, even when initial contamination was very high; results not dependent on spice quality or type of packaging; a dose of 15 kGy reduced aerobic plate counts to 50 per gram	Silberstein et al., 1979
Turmeric, black pepper, chili	1–10	dose of 4–5 kGy is effective in eliminating most microorganisms in spices, particularly the spore formers	Singh et al., 1988

TABLE 2 *(Continued)*

Paprika	5–11	dose of 5 kGy resulted in a reduction in mesophilic aerobic cell count of 2.5 orders of magnitude (identical to that achieved with 600g m--3, 25°C, 6 hr EO treatment; higher doses led to even greater reductions; coliforms and E. coli I were eliminated; mold count not affected by EO treatment, but was significantly reduced by 5 kGy dose of irradiation	Szabad and Kiss, 1979
White pepper, nutmeg and ginger	5–45	doses of 5 kGy effectively reduced numbers of microorganisms in samples while 15.8 kGy sterilized them; actual reductions depended on initial bioburden	Tjaberg et al., 1972
Cinnamon, ginger (ground), fennel,fenugreek	2–10	doses of 6 kGy for cinnamon and ginger, 6–10 kGy for fennel and 6–8 kGy for fenugreek gave microbial effects similar to those achieved with commercial EO fumigation methods	Toofanian and Stegeman, 1988
Black pepper, paprika, thyme, marjoram, mustard powder, curry powder, spice mixtures for various meat products		doses of 7.5 kGy led to a 104–105 reduction in viable organisms in a number of spices	Wetzel et al., 1985

Source: Sterigenics Food Safety (2013)

13.3 CONCLUSION

Available studies show than 50% of the countries are engaged in applications of food irradiation and producing fruitful results. The majority of the world is facing the problem of shortage food supply and scanty methods for advanced storage. This scenario is in current need of peculiar treatment as irradiation where large number of primary losses can be eliminated and procurement will be increased. An overwhelming data from all over the world left no doubt over the safety of irradiated foods in terms of nutritive value, appearance, sensory quality, and free from contamination. Large numbers of food borne diseases are due to the insanitary conditions during handling from fields, storage in contaminated vessels and irradiation has opened the wide safety profile for variety of foods. Especially the perishables where large experimentation cannot be opted due to their fragile skin and irradiation proving as boon for their treatment in their raw stage and alleviating the multiple losses, along rendering healthy food to the consumers and combating the large number of pathogenic bacteria which are very difficult to dealt with the conventional treatments where other losses are more prominent than food safety.

Presently more than 30 countries have successfully applied the irradiation technology. Majorly we have United States, Canada, Belgium, France, Japan, and South Africa. An important step will have to commercial it on a fast pace that is labeling the irradiation processed foods. So that consumers can themselves know about the food stuff they are going to consume and this will create truthfulness of the process and question of misleading will be eradicated. The unique feature of irradiation is of providing on spot sanitary treatment against microbes where no foreign species can enter to food. FDA has thorough supported the food irradiation as scientific technology for safety of masses. Earlier major concern is for final processed product where as now-a-days raw materials are being given priority. Therefore, irradiation has proved as extremely beneficial for enhancing the shelf life of fruits and vegetables by 4–5 times. Gamma radiation has totally been employed for inhibiting sprout formation in potato crop along killing of pests on early stage.

In addition, studies of marketing on irradiated foods have given an overwhelming response that consumers are more willing to buy the irradiated stuffs as information given on the labeling driving a boost for better knowledge and safety in consumption. Therefore education to the retailers as well as the consumers is required to take the applications of irradiation to an advanced level.

Major benefit is in the case of fruit and vegetables, where less than 1 kGy doses of radiation can keep the organisms at bay, along that flavor and aroma is near to fresh. Celery is the major food item for transporting to shipping services. Chances of contamination gets doubled in fresh green leaves as spinach and broccoli which are always demanded as fresh as provided from farms, here irradiation has showcased the potential and shippers and consumers are enthusiastic to go for irradiated greens. A code of good irradiation practice for the control of pathogens and other micro-flora in spices has been developed by the International Consultative Group on Food Irradiation (ICGFI) under the aegis of FAO, IAEA, and WHO. Importantly, irradiation is neither used for the preservation of these ingredients (obtained through proper drying, packaging, and storage) nor to correct quality deficiencies. The code details pre-irradiation treatment, packaging requirements, the irradiation treatments and the dose requirement for radiation disinfection together with threshold doses that cause organoleptic changes (http://www.iaea.org/icgfi/documents/5spices.htm). Therefore much data of relevance of food safety is there for making it industrialized and studies on sensory evaluation is needed in form of acceptance to consumers. Organoleptic properties are last but powerful to comment on the regular consumption of irradiated food as fruits and vegetables. There are still many challenges as regulatory approval, consumer acceptance and pre-market studies for implanting on a large commercial scale.

13.4 FUTURE PROSPECTS

In coming future, irradiation is surely expected to cover all the parts of the world. Food irradiation has shown a massive growth during the past decade, though poultry has been irradiated since 1990's but the food stuffs as fruits and vegetables, frozen foods, red meat has engulfed the whole process and commercialization of the processed products along worldwide acceptance of consumers has made it to grow so rapid. US markets are targeting tropical fruits to get treated in order to make them available fresh to the consumers as guavas, leeches, longans, mango from India. In addition of giving variety to consumers, it is also imparting healthy food with enhanced shelf life. As most of the foods get wasted by the time they reach to consumers. These losses can only covered up by rendering irradiation treatment. This method has opened a trade in fresh and dried foods.

Any technology can only be accepted if it has the characteristics of safety, nutritional friendly, acceptance of consumers, and perseverance of sensory quality, cost effective over other existing technologies. Irradiation has all the qualities and great future to grow ahead. It is a technique of wholesomeness including combined function of microbiological action, secured nutritional value, unaffected flavor and texture. Currently, spices and herbs are the main targets of irradiation. The effect is exceptional that in Japan itself, 8,000 tones of potatoes are sold in Japanese market for every year (Masakazu Furuta).

Ionizing radiation has more advantages to fruits and vegetables as it is more efficacious than conventional thermal treatment where as irradiation has lot more to perform as major is the delaying in the ripening and decontamination of vegetable seasoning, updating the quality of dried foods. The commercial prospects of irradiated fruits and vegetables are very promising. The reason of being fruitful is that fresh fruits and vegetables are the significant sources of pathogens and lethal contaminations and due to fragile nature of the tissues of perishables, hard treatment cannot be given.

Therefore, irradiation is the most suitable technique for making the perishables to be consumed afresh and tones of losses can be converted to consumable stuff. Moreover, replacement of fumigation and synthetic chemicals is the attractive part which has strong advantage. Interesting part is the lowest doses of irradiation are required to combat all the disadvantages and dosage is minimal for perishables. Precise values have been summarized by James H. Moy that 0.15 kGy is for the disinfection from microbes, 0.02–0.75 kGy is for inhibiting sprouting along delaying in ripening, while 10–30 kGy for vegetable decontamination and 0.30–5 kgy for dried foods.

The safety of irradiation has been clearly accepted as effective technology and regulatory authority has established on global basis. Consumers' choice is the final preference for taking the technology to the market. In coming years, irradiation will empower the existing processing technologies. Irradiation is therefore providing safety and health along minimising the losses on large front and emerging as economical processing as well.

KEYWORDS

- Food irradiation
- Foremost concern
- Good hygienic practices
- Irradiation
- Irradiation processing

REFERENCES

1. Alam MK, Choudhury N, Chowdhury NA, et al: Decontamination of spices by gamma radiation. *Lett Appl Microbiol* 14:199-202, 1992.
2. Andrews LS, Grodner RM, Chung HY, et al: Chemical and microbial quality of irradiated ground ginger. *J Food Sci* 60:829-32, 1995.
3. Andrews, L.S.; Ahmedna, M.; Grodner, R.M.; Liuzzo, J.A.; Murano, P.S.; Murano, E.A.; Rao, R.M.; Shane, S. and Wilson, P.W.(1998) Food preservation using ionizing radiation. *Rev Environ Contam Toxicol.*, 154:1-53.
4. Bendini A., Galina Toschi T., Lercker G. (1998): Influence of gamma irradiation and microwaves on the linear unsaturated hydrocarbon fraction in spices. *Zeitschrift fur Lebensmittel Untersuchung und Forschung A*, 207: 214–218.
5. Bhushan B, Thomas P. (1998). Quality of apples following gamma irradiation and cold storage. *Int. J. Food Sci. Nutrit.*, 49, 485-92.
6. Boisseau P, Jungas C, Libert MF. (1991). Effets des traitements combines sur les qualities microbiologiques et organoleptiques des vegetaux frais peu transformes.XVIIIth International Congress of Refrigeration. Montreal. August 10-17, 1991.
7. Chatterjee, S.; Desai, S.R. P. and Thomas,P.(1999). Effect of γ-irradiation on the antioxidant activity of turmeric (**Curcuma longa L.**) extracts. *Food Research International*,32(7):487-490.
8. Chmielewski, A.G. and Migdal,W. (2005). Radiation decontamination of herbs and spices. *Nukleonika*, 50(4):179-184.
9. Code of Federal Regulation (2006). Irradiation in the production, processing and handling of food. Revised as of April 1, 2006. *Code of Federal Regulations*. Title 21 - Food and Drugs, Vol. 3, Part 179, pp. 451-456.
10. Codex Stan (2003) Codex general standard for irradiated foods. In: *Codex Alimentarius*, Vol. 1, Section 8. Rome: FAO/WHO, 1992, pp. 311-315.
11. Drake S R, Neven L G. (1997). Quality response of 'Bing' and 'Rainier' sweet cherries to low dose electron beam irradiation. *J. Food Process Preserv.*, 21, 345-51.
12. Eiss MI: Irradiation of spices and herbs. National Symposium on the Ionizing Energy Treatment of Foods; Sydney (Australia); 5-6 Oct 1983. *Food Technol Aust* 36:362-66,370, 1984.
13. Emam O.A., Farag S.A., Aziz N.H. (1995): Comparative effects of gamma and microwave irradiation on the quality of black pepper. *Zeitschrift fur Lebensmittel Untersuchung und Forschung*, 201: 557–561.
14. Evangelista. J. (2000). Alimentos Irradiados. In: Alimentos - um estudo abrangente. São Paulo: Editora Athencu, 135 -169.

15. Fan X, Mattheis, J P. (2001). 1-Methylcyclopropene and storage temperature influence responses of "Gala" apple fruit to gamma irradiation. *Postharvest Biol. Technol.*, 23, 143-151.

16. Farag RS, El Khawas KHAM: Influence of gamma-irradiation and microwaves on the antioxidant property of some essential oils. *Advances in Food Sciences* 18:107-12, 1996.

17. Farag, SEDA, Aziz NH, Ali AM: Comparing the effects of washing, thermal treatments and gamma irradiation on quality of spices. *Nahrung* 40:32-36, 1996.

18. Farag, SEDA, Aziz NH, Attia ESA: Effect of irradiation on the microbiological status and flavouring materials of selected spices. *Zeitschrift Fuer Lebensmittel-Untersuchung Und-Forschung* 201:283-88, 1995.

19. FDA. (1997). Irradiation in the production, processing, and handling of food. Food and Drug Administration. Fed. Reg. 62(232), 64107-64121.

20. Grecz N, Al Harithy R, Jaw R: Radiation sterilization of spices for hospital food services and patient care. *J Food Safety* 7:241-55, 1986.

21. Furuta,M.(2004). Current status of information transfer activity on food irradiation and consumer attitudes in Japan. *Radiation Physics and Chemistry*, 71(1-2):501-504.

22. Gautam S, Thomas P, Sharma A. (1998). Gamma irradiation effect on shelf-life, texture, polyphenol oxidase and microflora of mushroom (Agaricus bisporus). *Int J Food Sci Nutrit.*, 49, 5-6.

23. *Heide, L. and Bögl, W.(2007). Identification of irradiated spices with thermo-and chemiluminescence measurements.* International Journal of Food Science & Technology, *22(2):93-103.[DOI: 10.1111/j.1365-2621.1987.tb00463.x]*

24. IAEA (2013) Food and Environmental Protection. http://www.iaea.org/icgfi/documents/5spices.htm (accessed on 10.06.2013).

25. IAEA ICGFI. (1994). Irradiation of strawberries. A compilation of technical data for its authorization and control. FAO/IAEA/WHO IAEA-TECDOC-779, 34.

26. International Atomic Energy Agency (IAEA). (2001) Report of the first FAO/IAEA Research Coordination Meeting on the Use of Irradiation to Ensure Hygienic Quality of Fresh, Pre-cut, Fruits and Vegetables and other Minimally Processed Food of Plant Origin, 5-9 November 2001, Rio de Janiero, Brazil.

27. Ito H,Watanabe H, Bagiawati S, et al: Distribution of microorganisms in spices and their decontamination by gamma irradiation. *Poster Presentation. Food Irradiation Processing. Proceedings International Symposium on Food Irradiation Processing. IAEA/FAO. Washington, D.C. March 4-8, 1985*: 171, 1985.

28. James H, Moy. (1989). Quality factors of fruits and vegetables. Chapter 25, 328-336, American Chemical Society Publishers.

29. Kilcast D. (1994). Effect of irradiation on vitamins. *Food Chem.*, 49, 157-164.

30. Kiss I: Reduction of microbial contamination of spices by irradiation. *Proceeding of the European Meeting of Meat Research Workers; No. 28, Vol. I*, 322-25, 1982.

31. Kiss I, Farkas J: Combined effect of gamma irradiation and heat treatment on microflora of spices. In: *Combination Processes in Food Irradiation. Proceedings of an International Symposium Jointly Organized by the IAEA and FAO. Colombo, Sri Lanka. Nov. 24-28, 1980*: 107-13, 1981.

32. Kitazuru, E.R.; Moreira, A.V.B.; Mancini-Filho, Delincée ,J. H. and Villavicencio, A.L.C.H.(2004). Effects of irradiation on natural antioxidants of cinnamon (**Cinnamomum zeylanicum** N.). *Radiation Physics and Chemistry*,71(1-20:39-41.

33. Koopmans M, and Duizer E. (2004). Foodborne viruses: an emerging problem. *J. Food Microbiol.*, 90, 23-4.

34. Lacroix M, Bernard L, Jobin M, Milot S and Gagnon M. (1990). Effect of irradiation on the biochemical and organoleptic changes during the ripening of papaya and mango fruits. *International Journal of Radiation Applications and Instrumentation. Part C. Radiation Physics and Chemistry*, 35, 296-300.

35. Lescano G, Narvaiz P, Kairiyama E (1993). Gamma irradiation of asparagus (Asparagus officinalis, var. Argenteuil). *Lebensm Wiss Technol* 26:411-16.

36. Lescano G, Narvaiz P, Kairiyama E: Sterilization of spices and vegetable seasoning by gamma radiation. *Acta Alimentaria* 20:233- 42, 1991.

37. Llorente Franco S, Gimenez JL, Martinez Sanchez F, et al: Effectiveness of ethylene oxide and gamma irradiation on the microbiological population of three types of paprika. *J Food Sci* 51:1571-72,1574, 1986.

38. Loaharanu P. (1994). Status and prospects of food irradiation. *Food Technology*, 48, 124-130

39. Loaharanu P. (2003). Irradiated Foods. Ed 5. Publishers American Council on Science and Health, pp. 10.

40. Marcotte M (2001) Effect of irradiation on spices, herbs and seasonings – comparison with ethylene oxide fumigation. http://www.food-irradiation.com/Spices.htm

41. Masakazu Furuta. (2011). Present status and future prospects of food irradiation. Journal of Urban living and Health Association, 55(1), 23-33.

42. Miller W R, McDonald R E, Smittle B J. (1995). Quality of 'Sharpblue' blueberries after electron beam irradiation. *HortSci* 30, 306-8.

43. Mitchell G E, McLauchlan R L, Isaacs A R, Williams D J, and Nottingham S M. (1992). Effect of low dose irradiation on composition of tropical fruits and vegetables. *J. Food Comp. Anal.*, 5, 291-311.

44. Molins R. (2001). Food Irradiation: Principles and Applications. New York: Wiley Interscience, 2001.

45. Moy J H and Wong L. (2002). The efficacy and progress in using radiation as a quarantine treatment of tropical fruits - A case study in Hawaii. *Radiat. Phys. Chem.*, 63, 397-401.

46. Muhamad LJ, Ito H,Watanabe H, et al: Distribution of microorganisms in spices and their decontamination by gammairradiation. *Agric Biol Chem* 50:347-55, 1986.

47. Müller H and Diehl J F. (1996). Effect of Ionizing Radiation on Folates in Food. *Lebens-Wiss. U-Technol*, 29, 187-199.

48. Munasiri MA, Parte MN, Ghanekar AS, et al: Sterilization of ground prepacked Indian spices by gamma irradiation. *J Food Sci* 52:823-24, 1987.

49. Murcia,M.A.; Egea, I.; Romojaro, F.; Parras, P.; Jiménez, A. M. and Martínez-Tomé,M. (2004). Antioxidant Evaluation in Dessert Spices Compared with Common Food Additives. Influence of Irradiation Procedure. *J. Agric. Food Chem.*, 52 (7): 1872–1881. [DOI: 10.1021/jf0303114]

50. Narvaiz P. (1994) Some physicochemical measurements on mushrooms (Agaricus campestris) irradiated to extend shelf-life. Lebensm Wiss Technol 27, 7-10.

51. Niemand JG, du Plessis TA: The decontamination of dry and dehydrated agricultural commodities by radurization. *Acta Hortic* 194:301-8, 1986.

52. Olson D.G. (1998): Irradiation of food. *Food Technology*, 52: 56–62.

53. Olson D O. (2004). Food Irradiation Future Still Bright. *Food Technology* 58 (7), 112.

54. Onyenekwe PC, Ogbadu GH, Hashimoto S: The effect of gamma radiation on the microflora and essential oil of Ashanti pepper (*Piper guineense*) berries. *Postharvest Biology and Technology* 10:161-67, 1997.

55. Onyenekwe PC, Ogbadu GH: Radiation sterilization of red chili pepper (*Capsicum frutescens*). *Journal of Food Biochemistry* 19:121- 37, 1995.

56. Osterholm M T and Morgan A P.(2004). The Role of Irradiation in Food Safety. *New England Journal of Medicine, April 29th.*

57. Oufedjikh H, Mahrouz M, Amiot M J. (2000). Effect of gamma-irradiation on phenolic compounds and phenylalanine ammonia-lyaseactivity during storage in relation to peel injury fro peel of Citrus clementina Hort. ex. Tanaka. *J Agric Food Chem* 48, 559-65.

58. Patil B S, Vanamala J and Hallman G. (2004). Irradiation and storage influence on bioactive components and quality of early and late season "Rio Red" grapefruit (*Citrus paradisi* Macf.). *Postharvest Biol. Technol.*, 34, 53-64.

59. Prakash A, Guner A R, Caporaso F. (2000). Effects of low-dose gamma irradiation on the shelf life and quality characteristics of cut romaine lettuce packaged under modified atmosphere. *J Food Sci* 65, 549-53.

60. Raffi, J. ; Yordanov, N.D.; Chabane, L.; Douifi, S.; Gancheva, V. and Ivanova S.(2000). Identification of irradiation treatment of aromatic herbs, spices and fruits by electron paramagnetic resonance and thermoluminescence. *Spectrochimica Acta Part A: Molecular and Biomolecular Spectroscopy*, 56(2):409-416. [http://dx.doi.org/10.1016/S1386-1425(99)00252-8]

61. Rice-Evans, C.; Miller, N. and Paganga, G. (1996). Structure–antioxidant activity relationships of flavonoids and phenolic acids. *Free Radical Biology and Medicine*, 20: 933–956.

62. Sadecka J., Kolek E., Peťka J., Suhaj M. (2005a): Influence of two sterilization ways on the volatiles of black pepper (*Piper nigrum* L.). *Chemicke Listy*, 99: 335–338.

63. Scott Smith J and Pillai S. (2004). Irradiation and Food Safety. *Food Technology,* 58 (11), 48-55.

64. Sharma A, Padwal-Desai SR, Nair PM: Assessment of microbiological quality of some gamma irradiated Indian spices. *J Food Sci* 54:489-90, 1989.

65. Sharma A, Ghanekar AS, Padwal-Desai SR, et al: Microbiological status and antifungal properties of irradiated spices. *J Agric Food Chem* 32:1061-63, 1984.

66. Shigemura R, Gerdes DL, Hall WR: Effect of gamma processing on prepackaged black and white pepper (*Piper nigrum L.*). *Lebensm Wiss Technol* 24:135-38, 1991.

67. Silberstein O, Galetto W, Henzi W: Irradiation of onion powder: effect on microbiology. *J Food Sci* 44:975-76, 981, 1979.

68. Singh L., Mohan MS, Padwal-Desai SR, et al: The use of gamma irradiation for improving microbiological qualities of spices. *J Food Sci Technol, India* 25:357-60, 1988.

69. Sivapalasingam S, Friedman C R, Cohen L and Tauxe R V. (2004). Fresh produce: A growing cause of outbreaks of foodborne illness in the United States, 1973 through 1997. *Journal of Food Protection*, 67(10), 2342-2353.

70. Song H P, Byun M , Jo C, Lee C-H, Kim K-S and Kim D-H. (2007). Effects of gamma irradiation on the microbiological, nutritional, and sensory properties of fresh vegetable juice. *Food Control*, 18, 5-10.

71. Spoto M H F, Domarco R E, Walder J M M. (1997). Sensory evaluation of orange juice concentrate as affected by irradiation and storage. *J Food Process Preserv.*, 21, 179-91.

72. Sterigenics Food Safety (2013) Irradiation of Herbs, Spices and Dry Ingredients.http://www.sterigenics.com/services/food_safety/food_irradiation__questions_answers.pdf (accessed on 06.06.2013)

73. Suhaj,M. and Horvathova,J. (2007). Changes in antioxidant activity induced by irradiation of clove (*Syzygium aromaticum*) and ginger (*Zingiber officinale*). *Journal of Food and Nutrition Research*, 46 (3): 112-122.

74. Suresh D, Pillai Leslie A, Braby, Lavergne. (2005). Electron Beam Technology for Food Irradiation. *The International Review of Food Science and Technology (Winter 2004/2005).*

An Official Publication of the International Union of Food Science and Technology (IU-FoST).

75.   Szabad J., Kiss I: Comparative studies on the sanitizing effects of ethylene oxide and of gamma radiation in ground paprika. *Acta Aliment* 8:383-95, 1979.

76.   Tauxe R V. (2001). Food Safety and Irradiation: Protecting from Foodborne Infections. *Centers for Disease Control and Prevention.*

77.   Thayer D W, Rajkowski K T. (1999). Development in irradiation of fresh fruits and vegetables. *Food Technol.*, 53, 62-65.

78.   Thayer D W. (2004). Irradiation of Food – Helping to Ensure Food Safety. *New England Journal of Medicine, April 29th. 2004.*

79.   Thomas P, Bhushan B, Joshi M R. (1995). Comparison of the effect of gamma irradiation,heat-radiation combination, and sulphur dioxidegenerating pads on decay and quality of grapes. *J Food Sci Technol.*, 32, 477-81, 1995.

80.   Thayer D.W., Josephson E.S., Brynjolfsson A., Giddings G.G. (1996): Radiation Pasteurization of Food. Council for Agricultural Science and Technology – CAST. *Ames*, Apr. 1996: 7.

81.   Tjaberg TB., Underdal B, Lunde G: The effect of ionizing radiation on the microbiological content and volatile constituents of spices. *J Appl Bacteriol* 35:473-78, 1972.

82.   Toofanian F, Stegman H: Comparative effect of ethylene oxide and gamma irradiation on the chemical, sensory and microbial quality of ginger, cinnamon, fennel and fenugreek. *Acta Aliment* 17:271-81, 1988.

83.   Wen H W, Chung H P, Chou F I, Lin I H and Hsieh P C. (2006). Effect of gamma irradiation on microbial descontamination and chemical and sensory characteristic of lycium fruit. *Radiat. Phys. Chem.*, 75, 596-603.

84.   Wetzel, K.; Huebner, G. and Baer, M.(1985).Irradiation of onion, spices and enzyme solutions in the German Democratic Republic. *Food Irradiation Processing, pp.35-46.*International Atomic Energy Agency, Vienna.

85.   Wilkinson V M, and Gould G W. (2005). Food Irradiation: The Computype Media, New Delhi.

86.   Yang, G. C.; Mossoba, M. M.; Merin, U.and. Rosenthal,I.(2007). An EPR study of free radicals generated by gamma-radiation of dried spices and spray-dried fruit powders. *Journal of Food Quality,*10 (4): 87-94. [DOI: 10.1111/j.1745-4557.1987.tb00820.x]

87.   Yu L, Reitmeier C A, Gleason M L. (1995). Quality of electron beam irradiatcd strawberries. *J Food Sci.,* 60, 1084-87.

88.   Zhang S, Farber J M. (1996). The effects of various disinfectants against Listeria monocytogenes on fresh-cut vegetables. *Food Microbiology,* 13, 311-321.

89.   Zheng, W.and Wang, S. (2001).Antioxidant activity and phenolic composition in selected herbs. *Journal of Agricultural and Food Chemistry*, 49: 5165–5170.

CHAPTER 14

ANTIOXIDANT PROPERTIES OF VARIOUS ALCOHOL DRINKS

N. N. SAZHINA, A. E. ORDYAN, and V. M. MISIN

CONTENTS

14.1   INTRODUCTION

The total content of phenol type antioxidants and their activity with respect to oxygen and its radicals are measured by two operative electrochemical methods: ammetric and voltammetric in samples more than 100 various alcohol drinks. Results of measurements show good (80–90 %) correlation of these methods for dry red and white wines. Use of indicated methods allows revealing forged alcoholic production.

For a long time salutary influence of dry grape wines, cognacs, and other qualitative alcoholic drinks on human health is known, certainly at their moderate consumption. It is connected, basically, with presence in them of natural antioxidants (AO), the main things from which are polyphenols. According to some researches [1,2], polyphenols of grapes, in particular bioflavonoids, play a main role in display of "the French paradox", consisting that among the regularly consuming red dry grape wines population, lower susceptibility to cardiovascular and oncological diseases [1,2] is revealed. Therefore, the sufficient attention is given to research antioxidant activity of various wines and their components *in vitro* and *in vivo*, despite a considerable quantity of already available works is paid now. Authors of the book [3] result the wide review of these works in which structure of different alcoholic drinks, the total antioxidant content and their biological activity are investigated by various methods. Basically, these methods are chemical methods and liquid chromatography methods [4,5]. The aim of the present work was research of the total antioxidant activity of various alcoholic drinks by two electrochemical methods: ammetric and voltammetric and comparison of the received results.

14.2   EXPERIMENTAL DETAIL

Objects of research were dry, semidry, dessert, fortified red and white wines, cognacs, liquors, and infusions received from manufactures or their distributors at an exhibition in Moscow in 2009.

Used ammetric method allows defining the total phenol type antioxidant content in investigated samples [3,6]. The essence of this method consists in measurement of the electric current arising at oxidation of investigated substance on a surface of a working electrode at certain values of electric potential (0–1.3). At such values of potential there is an oxidation only OH-groups of natural phenolic antioxidants (R-OH). The electrochemical oxidation proceeding under scheme R-OH $\rightarrow$ R-O$^{\bullet}$ + e^{-} + H^{+}, can be used as model at measurement of free radical absorption activity. Capture of free radicals is carried out according to reaction R-OH $\rightarrow$ R–O$^{\bullet}$ + H$^{\bullet}$. Both reactions include rupture of same communication O-H. In this case ability to capture of free radicals by flavanoids or other polyphenols can be measured by oxidability of these substractions on a working electrode of ammetric detector [7]. The registered signal (the area under a current curve) is compared to a signal received in the same conditions for the sample of comparison with known concentration. Gallic acid (GA) was used as such sample in the present work. Measurement error of the total phenol type antioxidant content taking into account reproducibility of results was 10%. Measurement time of one sample makes 10–15 min.

The voltammetric method uses the process of oxygen electroreduction (ER O_2) as modeling reaction. It proceeds at the working mercury film electrode (MFE) in several stages with formation of the reactive oxygen species (ROS), such as O_2^- and HO_2 [8]:

$$O_2 + e^- \rightleftarrows O_2^{\cdot-} \tag{1}$$

$$O_2^{\cdot-} + H^+ \rightleftarrows HO_2^{\cdot} \tag{2}$$

$$HO_2^{\cdot} + H^+ + e^- \rightleftarrows H_2O_2 \tag{3}$$

$$H_2O_2 + 2H^+ + 2e^- \rightleftarrows 2H_2O \tag{4}$$

For determination of total antioxidant activity it is offered to use the first wave of ER O_2 corresponding to stages (1)–(3) when on a surface of a mercury film electrode active oxygen radicals and hydrogen peroxide are formed. It should be noted that antioxidants of various natures were divided into four groups according to their mechanisms of interaction with oxygen and its radicals (Table 1) [8,9].

TABLE 1 Groups of biological active substances (BAS) divided according mechanisms of interaction with oxygen and its radicals

N Group	One group	Two group	Three group	Four group
Substance names	Catalyze, phtalocyanines of metals, and humic acids.	Phenol nature substances, vitamins A, E, C, B, and flavonoids.	N, S, Se-containing substances, amines, and amino acids.	Superoxide dismutase (SOD), porphyry metals, and cytochrome C
Influence on ER O_2 process	Increase in current of ER O_2, potential shift in negative area	Decrease of current of ER O_2, potential shift in positive area	Decrease of current of ER O_2, potential shift in negative area	Increase in current of ER O_2, potential shift in positive area

TABLE 1 *(Continued)*

The prospective electrode mechanism.	EC* mechanism with the following reaction of hydrogen peroxide disproportion and partial regeneration of molecular oxygen.	EC mechanism with the following chemical reaction of interaction of AO with active oxygen radicals	CEC mechanism with chemical reactions of interaction of AO with oxygen and its active radicals	EC* mechanism with catalytic oxygen reduction *via* formation of intermediate complex.

*The note: E—electrode stage of process and C—chemical reaction.

Kinetics criterion K is used as an antioxidant activity criterion of the investigated substances. It reflects quantity of oxygen and their radicals which have reacted with AO (or their mixes) in a minute:

For the second and third groups of antioxidants: $K = \frac{C_0}{t}(1 - \frac{I}{I_0})$, µmol/l×min,   (5)

For the firth and four groups of antioxidants: $K = \frac{C_0}{t}(1 - \frac{I_0}{I})$, µmol/l×min,   (6)

where I and I_0—limited values of the current of oxygen electroreduction, accordingly, at presence I and at absence I_0 of AO in the supporting electrolyte, C_0—initial concentration of oxygen in the supporting electrolyte (µmol/l), that is solubility of oxygen in supporting electrolyte under normal conditions, and t—time of interaction of antioxidants with oxygen and its radicals, min.

To implement this method automated device "AOA" (Ltd. "Polyant", Tomsk, Russia) was used [9]. A phosphate buffer solution of 10 mL with known initial concentration of molecular oxygen was used as the background electrolyte, in which doses of the test samples (50–150 µl) were added. That concentration of investigated wine samples in volume of a buffer solution was corresponded to concentration of the samples entered into a cell of the ammetric device; wines were diluted in 50–300 times before introduction in an ammetric cell. The voltammetric method has good sensitivity and is simple and cheap. However, as in any electrochemical method of this type, the scatter of instrument readings at given measurement conditions is rather high (up to a factor of 1.5–2.0). Therefore each sample was tested 4–5 times, and results were averaged. The maximum standart deviation of kinetic criterion K from average value for all investigated samples was 30 %.

14.3   DISCUSSION AND RESULTS

Table 2 shows results of measurements of the total phenol type antioxidant content in units of GA concentration (C, mg/L GA), obtained by ammetric, and kinetic criterion

(K, μmol/L min), characterizing the total antioxidant activity with respect to oxygen and its radicals, for more than 100 samples of alcoholic drinks.

TABLE 2 Results of measurements of the total phenol type antioxidant content (C, mg/L GA), and kinetic criterion (K, μmol/L min)

№ of sample	The company, country	Name, type of wine, wine age.	C, mg/l GA	K, μmol/l☐min
1	Navarra, Spain	Palacio de OTAZU, red semi-dry, 2001	328.1	1.20
2	Navarra, Spain	Palacio de OTAZU, red semi-dry, 2003	302.3	1.21
3	Navarra, Spain	Palacio de OTAZU, red semi-dry, 2006	277.7	1.05
4	Lancelot, Italy	Castello del Sono, red semi-dry, 2009	261.1	1.07
5	Launcelot, France	Cabernet Sauvignon, red semi-dry, 2008	254.6	0.90
6	Legenda Kryma, Moldova	Sauvignon de Purcari, dry white, 2005	82.5	0.57
7	Batitu Classic, Chile	Sauvignon Blanc, a dry white, 2009	74.2	0.58
8	Vega Libre, Spain	Malbeo, dry white, 2008	83.5	0.61
9	Batitu Classic, Chile	Chardonnay, dry white, 2008	122.5	0.85
10	Batitu Reserv, Chile	Cabernet Sauvignon, dry red, 2007	245.5	0.77
11	Batitu Classic, Chile	Cabernet Sauvignon, dry red, 2008	217.4	0.80
12	Batitu Reserve, Chile	Merlot, dry red, 2007	236.0	1.09
13	Negru De Purcaru, Moldova	Cabernet Sauvignon, dry red, 2003	209.2	0.95
14	Batitu Classic, Chile	Merlot, dry red, 2008	206.2	1.03
15	Inkerman, Crimea	Aligote, dry white, 2006	86.0	0.57
16	Legenda Kryma	Aligote, dry white, 2006	127.3	0.88
17	Koktybel, Crimea	Aligote, dry white, 2006	83.8	0.47
18	Inkerman, Crimea	Cabernet Kachinske, dry red, 2006	199.9	0.97

TABLE 2   *(Continued)*

19	Inkerman, Crimea	Cabernet Coptobe, dry red, 2009	312.5	0.95
20	Legenda Kryma	Cabernet, dry red wine, 2009	254.8	0.88
21	Legenda Kryma	Muscat Sater, white semi-sweet, 2009	115.1	0.6
22	Legenda Kryma	Muscat Sater, semi-sweet red, 2009	153.3	0.91
23	Koktybel, Crimea	Monte Rouge, semi-sweet red, 2008	206.5	0.95
24	Cherny Polkovnik, Crimea	Coupage, red special, 1998	192.4	0.71
25	Koktybel, Crimea	Cahors, red special, 2008	174.3	0.84
26	Cherny Doktor, Crimea	Solnechnaya Dolina, red special, 2003	196.1	0.98
27	Parteniyskaya Dolina, Crimea	Tauris, red port, 2009	99.2	1.19
28	Magarach, Crimea	Chernovoy Magarach, red port, 2005	180.1	1.48
29	Krymskoe Shampanskoe Sevastopol Winery, Crimea	semi-sweet champagne , 2009	155.9	0.62
30	Sevastopol Winery, Crimea	Sparkling white muscat, 2009	84.4	0.52
31	Artemevkoe Shampanskoe, Crimea	Brut champagne white, 2009	92.1	0.52
32	Novy Svet, Crimea	Brut champagne, 2005	132.9	0.82
33	Novy Svet, Crimea	Brut champagne white, 2006	90.7	0.50
34	J. Bouchon, Chile	Sauvignon Blanc, dry white, 2009	60.9	0.32
35	Grand, France	Chateau Grand Xilon, dry white, 2008	98.8	0.65
36	Vin De Bordeux, France	Chateau Grand Xilon, dry red, 2005	257.7	0.95
37	Krevansky combine Ararat, Armenia	Noah, cognac, 2002	29.2	0.91
38	Hardy, France	V.S.O.P., Cognac, 2002	59.6	0.51
39	Chaten Le Don, France	Bordeaux, dry red, 2006	324.9	1.04

TABLE 2   *(Continued)*

40	Sadylly, Azerbaijan	dry white (coupage)	83.8	0.52
41	Matrasa, Azerbaijan	dry red (coupage)	244.0	0.94
42	Baku, Azerbaijan	semi-sweet red (coupage)	251.2	0.78
43	Ganuzh, Azerbaijan	semi-sweet red (coupage)	238.1	0.80
44	Baku, Azerbaijan	Cognac, 2001	18.3	0.79
45	Moscva, Azerbaijan	Cognac, 1998	26.3	0.76
46	Shirvan, Azerbaijan	Cognac, 1995	45.3	0.96
47	Apsheron, Azerbaijan	Ordinary cognac, 2005	5.5	0.33
48	Port wine Adam, Azerbaijan	Port wine, 2008	63.6	1.29
49	Martini Vermouth, Azerbaijan	Vermouth white, 2009	28.7	0.50
50	Dar Bogov, Amtel	Ginseng infusion, 2009	5.3	0.22
51	Vilage Saint Deni, France	Blanc Mouellux, white semi-sweet, 2009	58.7	0.41
52	Maison De La Fer, France	Blanc Mouellux, white semi-sweet, 2009	69.8	0.50
53	Baron Du Rua, France	Blanc Mouellux, white semi-sweet, 2009	57.3	0.39
54	Vilage Saint Deni, France	Rouge Mouellux, semi-sweet red, 2009	185.8	-
55	Maison De La Fer, France	Rouge Mouellux, semi-sweet red, 2009	160.0	0.75
56	Baron Du Rua, France	Rouge Mouellux, semi-sweet red, 2009	196.2	-
57	Severnaya Zvezda, St. Petersburg	Cognac, 2006	23.1	0.70
58	Severnaya Zvezda, St. Petersburg	Cognac, 2004	33.2	0.75
59	Generalissimus, St. Petersburg	Cognac, 2006	19.3	0.70
60	Chateau Du Razo, France	Sauvignon, dry white, 2007	75.8	0.53
61	Russkaya Loza, Krasnodar	Sauvignon, dry white, 2007	103.6	0.65
62	Russkaya Loza, Krasnodar	Chardonnay, dry white, 2007	103.3	0.61

TABLE 2   *(Continued)*

63	Russkaya Loza, Krasnodar	Muscat, dry white, 2007	125.1	0.75
64	Russkaya Loza, Krasnodar	Cabernet dry red, 2007	271.0	0.91
65	Chateau Del Ain, France	Bordeaux, dry red, 2007	289.1	0.99
66	Carmen, Chile	Merlot, dry red, 2007	347.0	1.21
67	Carmen, Chile	Carmenere,dry red, 2006	321.6	1.01
68	Norton, Argentina	Malbec, dry red, 2004	274.0	0.88
69	Lo Tengo, Argentina	Malbec, dry red, 2007	277.2	1.23
70	Saint-Anac, France	Moellux Dor De Alix, white semi-sweet, 2007	86.5	0.60
71	Cocoa	Liquor, liqueur, 2009	7.6	0.41
72	Strawberries	Liquor, 2009	75.0	0.83
73	Mint	Liquor, 2009	6.2	0.43
74	Taso Real, Spain	Gresalino, white semi-sweet, 2008	93.1	0.50
75	El Toril, Mexico	Tequila, 2009	4.8	0.60
76	Vina Sutil, Chile	Chardonnay, dry white, 2009	104.9	0.66
77	Vismos, Moldova	White Agate, cognac, 2004	69.7	0.82
78	Barton & Guestier, France	Sauvignon Blanc, dry white, 2009	136.0	0.88
79	Barton & Guestier, France	Bordeaux dry white wine, 2008	105.5	0.67
80	Vina Sutil, Chile	Cabernet dry red wine, 2006	276.2	0.94
81	Barton & Guestier, France	Merlot, dry red, 2008	290.4	1.05
82	Chinese fruit, Rotor House, China	Fruit wine, plum, 2007	94.0	-
83	Algeria, Algeria	Medea, Rose, 2008	172.4	-
84	ONCV, Algeria	Rose de la Hitijia, Rose, 2009	172.5	-
85	Algeria, Algeria	Medea, dry red, 2008	287.0	1.01
86	Vinariatiganga, Moldova	Fiteasca Regala, dry white, 2007	86.9	0.59
87	Lambouri, Cyprus	Lambouri, dry white, 2007	102.7	0.91

TABLE 2   *(Continued)*

88	Stretto, Russia	Stratton, dry red, 2006 -	312.6	1.15
89	Lambouri, Cyprus	Maratheptico, dry red wine, 2006	323.3	1.06
90	Mavrud, Bulgaria	Fsenovgrad, dry red, 2004	350.3	1.08
91	Vicariya Tsyganka , Moldova	Cabernet Sauvignon, dry red wine, 2006	259.7	0.89
92	Carmello Pati, Argentina	Cabernet Sauvignon, dry red wine, 2002	317.9	1.02
93	Carmello Pati, Argentina	Cabernet Sauvignon, dry red wine, 1999	289.7	0.87
94	Tokaji, Hungary	Furmint, dry white, 2005	137.5	1.02
95	Tokaji, Hungary	Furmint, white semi-sweet 2003	155.6	0.71
96	Tokaji, Hungary	Furmint, white semi-sweet 2002	185.9	-
97	Tokaji, Hungary	Kuve white sweet, 2006	164.5	-
99	Tokaji, Hungary	Aszu 5 putonyami, white sweet, 2003	222.9	-
100	Tokaji, Hungary	Aszu 3 putonyami, white sweet, 2000	236.7	-
101	Dalvina, Macedonia	Vranec, dry red wine, 2008	412.0	1.50
102	Dalvina, Macedonia	Merlo , dry red wine, 2008	408.1	1.45
133	Dalvina, Macedonia	Cabernet Sauvignon, dry red, 2008	399.5	1.21
104	Dalvina, Macedonia	Rhein Risling, dry white, 2008	115.8	0.95
105	Dalvina, Macedonia	Astraton, dry white, 2008	90.8	0.53
106	Skovin, Macedonia	Makedonsko Belo, dry white, 2008	210.1	1.39
107	Skovin, Macedonia	Chardonnay, dry white, 2008	153.2	0.95
108	Skovin, Macedonia	Syrah Cabernet, dry red, 2007	427.1	1.50
109	Skovin, Macedonia	Kale, dry red wine, 2007	410.0	1.46
110	Skovin, Macedonia	Makedonsko Crveno, dry red, 2008	346.7	1.07
111	Zonte's Foot step, Australia	Cabernet, dry red, 2008	428.5	1.35

TABLE 2   *(Continued)*

112	Zonte's Foot step, Australia	Shiraz, dry red wine, 2008	415.0	1.28
113	Woop Woop, Australia	Shiraz, dry red wine, 2008	321.7	0.99
114	Ries, Australia	Merlot, red semi-dry, 2009	370.3	1.02
115	Farrese, Italy	Trebyano a'Arbutstso Farnese, dry white, 2008	128.9	0.89
116	Isla Negra, Chile	Cabernet Sauvignon Merlot, red semi-dry, 2009	410.4	1.32

For clarity, diagrams (Figure 1 and 2) presents comparative measuring data of the total AO content in some typical samples of drinks from Table 2 for dry, semi-dry, semi-sweet wines and champagnes (Figure 1), and fortified wines and cognacs (Figure 2).

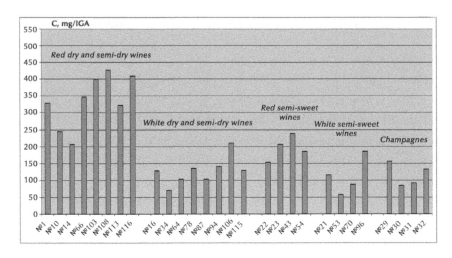

FIGURE 1   The total phenol type antioxidant content (C, mg/L GA) in samples of dry, semi-dry and semi-sweet wines.

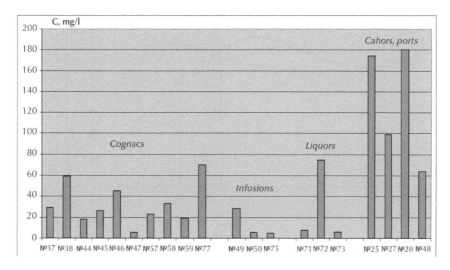

FIGURE 2   The total phenol type antioxidant content (C, mg/L GA) in samples of fortified wines, infusions, liquors, and cognacs.

Diagrams show that dry and semi-dry red wines have the highest values of the total phenol AO content (from 200 to 430 mg/L GA), that is essential more than dry and semi-dry white wines (from 50 to 200 mg/l GA). The phenolic content of red semi-sweet wines also dominate compared with semi-sweet white wines and champagnes. In work [10] the total content of phenols has been measured in dry red wines by the Folin-Ciocalteau method with GA as a standard. Values of the phenol content have appeared in 2–5 times more, than in the present work. Apparently, it is possible to explain that besides phenols, with a reactant of Folin-Ciocalteau can react other substances presenting in wines, such as restoring sugars, proteins and amino acids [11]. The ammetric method excludes it. Smaller values of the total phenol content in the red wines, received in the present experiments, can be explained, besides, by not total oxidation of these phenols at passage through ammetric cell [6].

Red port wines and cahors have the lead positions between fortified wines (Figure 2, sample №25, 28). The level of phenol compounds in cognacs was significantly lower (20–70 mg/L GA). At this level the cognac sample №47 is sharply behind (5.5 mg/l GA). A similar results for infusion of ginseng (№ 50), for a tequila (№ 75), cocoa and peppermint liqueurs (№ 71, № 73) were observed. The same minor values of C were recorded for 40% ethanol. Relatively low values of phenol AO content in these samples may indicate that they are probably falsified.

With regard to the total activity of drink AOs with respect to oxygen and its radicals, different samples show different mechanisms of its influence on the ER O$_2$ process and the reduction current form. Figure 3 shows a typical voltammograms (VA-grams) for all red dry, semi-dry, semi-sweet, special wines, including ports, and cahors.

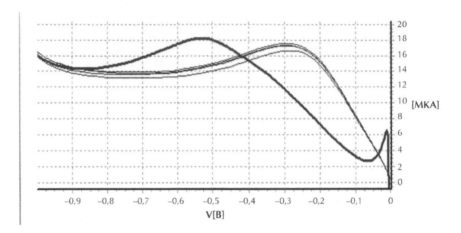

FIGURE 3 Typical VA-grams of ER O_2 current for red wines in absence of sample (the left curve) and presence (the right curves).

It is visible that the mechanism of action of the polyphenols containing in these wines, is characteristic for compounds of second groups in Table 1 and the values of potential shift of EV O_2 current limit increase with growth of the total phenol content. For this group of substances (R-OH) the following mechanism of interaction with the reactive oxygen species (ROS) is offered [8].

$$O_2 + e^- \rightleftarrows O_2^{\cdot-} + R\text{-}OH \rightleftarrows HO_2^{\cdot} + R\text{=}O \tag{7}$$

$$HO_2^{\cdot} + R\text{-}OH \rightleftarrows H_2O_2 + R\text{=}O \tag{8}$$

where R=O—oxidative form of the antioxidants (R-O˙). According to the theory of electrode processes, an indicator of thermodynamics of electrochemical processes is the potential of a half wave of a EV O_2 current which is described by the classical Nernst equation having for electrode process Equation (7) at temperature T = + 25C the following kind [9]:

$$E_{O2/O2}^{\cdot-} = E^{\circ}_{O2/O2}^{\cdot-} + 0,059 \cdot lg(C_{O2}/C_{O2}^{\cdot-}) \tag{9}$$

If superoxide anion ($O_2^{\cdot-}$) reacts in reaction Equation (7) with an antioxidant (R-OH), $C_{O2}^{\cdot-}$ decreases, the potential increases according to Equation (9) and the maximum of a EV O_2 current haves shift in positive area. The values of kinetic criterion K arise with increasing of the total phenol AO content, reaching values of 1.3–1.5 μmol/L min for samples of red wines (Table 2, №66, 101, 108). Figure 4 shows the correlation between K and C for 40 samples of dry and semi-dry red wines from Table 2. The correlation coefficient was r = 0.8212.

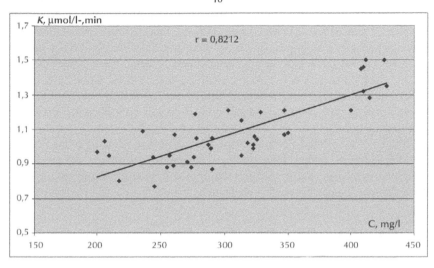

FIGURE 4 Correlation dependence between the kinetic criterion K and the total AO content C for dry and semi-dry red wines. (r = 0.8212)

It is known from numerous studies [3] of red grape wines, that in they more than 100 polyphenolic compounds, including flavonoids and polyphenols non-flavonoid nature were identified [12]. The main contribution to their AO activity make compounds of following groups: anthocyans, catechins, oligomeric procyanidins, flavonols, stilbenes, glucosides, and and so on. [3]. Anthocyans (anthocyanidins and anthocyanins) are the main pigments of grape and cause the red wine color [12]. They have a wide spectrum of biological activity for human increase elasticity of blood vessels and improve visual acuity [13] In addition, anthocyans affect capillary permeability, as well as the hematopoietic function of bone marrow [14] Catechins and procyanidins, whose content is more than 90% of the total polyphenols of grape and wine [15] are the most activity AO [16] Some catechins, for example epicatechin and epigallokatehin-gallat, can induce apoptosis in cancer cells [17] Procyanidins brake oxidation of low-viscosity lipoproteins in blood, preventing cardiovascular disorders [18] The high AO activity have in wine also phenolic acids, particularly gallic and caffeic and flavonols: quercetin, rutin, myricetin, reducing blood cholesterol level [19]. The above and other phenol compounds contribute mainly their part in high values of the total content of antioxidants and their activity for red wines [3].

Obtained in this study VA-grams of ER O_2 current for dry and semi-dry white wines are not much different from those for red wines (Figure 3), however, potential maximum shift of working current relatively to the background is slight, indicating a lower level of the total phenol AO content in white wines. The range of kinetic criterion values K for investigated white wines was from 0.4 to 1.0 μmol/l min. Correlation between the kinetic criterion and the total phenol AO content, built for 24 samples of dry and semi-dry white wines from Table 2, showed better correlation between two methods than for red wines, with r = 0.9061. Comparative analysis of white and red wines, made in [20,1], shows a smaller (5–10 times) total polyphenol content in white wines than in red wines. It is associated with the fact, that in fermentation of grape juice skins and seeds of grapes, rich by anthocyans, catechins, and other polyphenol compounds are included. During the fermentation of white wine, grape juice is used only, therefore much less active AO: catechins, epicatechins, gallic acid, quercetin, rutin, cyanidin, miricitin, and proanthocyanidins are found in white wines [20]. Inclusion of skins and grape seeds in production of white wines significantly increases their antioxidant activity [21].

With regard to fortified alcoholic drinks, in addition to reducing of the total phenol AO content (Figure 2), they show another character of VA-grams and mechanism of interaction with oxygen and its radicals. On Figure 5 typical VA-gram of a EV O_2 current for the investigated cognacs is presented.

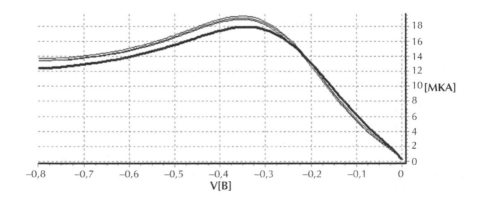

FIGURE 5 Typical VA-grams of ER O_2 current for cognacs in absence of sample (the bottom curve) and presence (the top curves).

Such behavior of a current is characteristic for compounds of first group in Table 1. For these compounds increase in current of ER O_2 results from influence of their catalytic properties on EV O_2 kinetics thanks to reaction of hydrogen peroxide disproportion and partial regeneration of O_2 Equation (10)–(12) [9]:

$$O_2 + e^- + H^+ \rightleftharpoons HO_2^{\cdot} \tag{10}$$

$$HO_2^{\cdot} + e^- + H^+ \rightleftharpoons H_2O_2 \tag{11}$$

$$2H_2O_2 \qquad 2H_2O + O_2 \tag{12}$$

Values of kinetic criterion K, calculated under the formula (6), have appeared for cognacs high enough (from 0.5 to 1.0 μmol/l min), despite of the low total phenol AO content (Figure 2). Any essential correlation between the AO content and values K for cognacs was not observed. For the sample of cognac №47 value K has made 0.33 μmol/l min and character of VA-grams did not correspond to typical VA-grams for cognacs (Figure 5), and reminded VA-gram for 40%-s' ethanol (Figure 6), on which background and working currents practically merge. The same basic kind VA-gram had ginseng infusion (№50), tequila (№75) and liquors №71 and №73. It once again gives the reason to suspect their falsification. Strawberry liquor (№72) behaved like wines, showing decrease of ER O_2 current and potential shift in positive area (Figure 3), also as vermouth (№49) and ports.

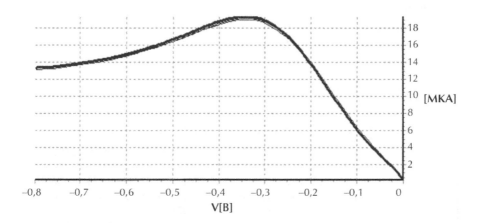

FIGURE 6   Typical VA-gram of ER O_2 current for 40%-s' ethanol in absence of sample (the top curve) and presence (the bottom curves).

Quality cognacs contain antioxidants, which are contained in brandy spirit and also tannins, a lignin, reducing sugars, oils, pitches, and enzymes getting to a ready product due to the extraction of them by spirit from wood of oak butts during "aging or endurance" of cognac. Cognac alcohol becomes golden color and is filled with woody-vanilla aromas due to the extraction of timber oksiaromatic acids, the main ones are lilac, ellagic, vanillic, as well as vanillin and lilac aldehyde [3]. Apparently, some enzymes of cognac and aromatic compounds in which include phthalocyanines, being effective catalysts of electrochemical reduction of oxygen, and determine the dominant mechanism of ER O_2 and character of VA-grams. Thanks to presence of tannins in cognacs, they help from gastro enteric frustration. Cognac also expands vessels of a brain and raises an organism tone. However medical properties of cognac are shown, if it is used only in small doses (30–50 ml).

14.4 CONCLUSION

Thus, use of two operative electrochemical methods allows you to determine quickly and relatively cheaply the total content of antioxidants and their activity with respect oxygen and its radicals in various grades of wines and cognacs and to identify low-quality production. The results of the present work received for more than 100 samples of alcoholic drinks, show rather high (80–90%) correlation of these methods for dry, semi-dry red and white wines. These methods can be applied for quality control of wines and other alcoholic drinks.

KEYWORDS

- Ammetric method
- Antioxidant content
- Electrochemical methods
- Gallic acid
- Voltammetric method

REFERENCES

1. Renaud, S. and De Longeril, M. Wine, alkogol, platelets, and the French paradox for coronary heart disease. Lancet, 339, 1523–1526 (1992).
2. Perez, D. D., Strobel, P., Foncea, R., Diez, M. S., Vasquez, L., Urquiaga, I., Castillo, O., Cuevas, A., San Martin, A., and Leignton, F. Wine, diet, antioxidant defenses, and oxidative damage. Ann. NY Acad. Sci., 957, 136–145 (2002).
3. Yashin, A. Ya., Yashin, Ya. I., Ryzhnev, V. Yu., and Chernousova, N. I. Natural antioxidants. The total content in foodstuff and their influence on health and human aging. Moscow, Translit, 92–103 (2009).
4. Dolores Rivero-Perez M., Muniz Pilar and Gonzalez Sanjose Maria L. Antioxidant profile of red wines evaluated by total antioxidant capacity, scavenger activity, and biomarkers of oxidative stress methodologies. J. Agric. Food Chem., 55, 3 8 (2007).
5. Alonso Angeles M., Dominguez Cristina, Guillen Dominico A., and Barroso Carmelo G. Determination of antioxidant power of red and white wines by a new electrochemical method and its correlation with polyphenolic content. J. Agric. Food Chem., 50, 3112 3115b (2002).
6. Yashin, A. Ya. Inject-flowing system with ammetric detector for selective definition of antioxidants in foodstuff and drinks. Russian chemical magazine, V. 52(2), 130 135 (2008).
7. Peyrat Maillard, M. N., Bonnely, S., and Berset, C. Determination of the antioxidant activity of phenolic compounds by coulometric detection. Talanta, V. 51 709 714 (2000).
8. Korotkova, E. I., Karbainov, Y. A., and Avramchik, O. A. Investigation of antioxidant and catalytic properties of some biological-active substances by voltammetry. Anal and Bioanal Chem., V. 375(1-3), 465 468 (2003).
9. Korotkova, E. I. Voltammetric method of determining the total AO activity in the objects of artificial and natural origin. Doctoral thesis Tomsk, (2009).

10. Roginsky, V., De Beer, D., Harbertson, J. F., Kilmartin, P. A., Barsukova, T., and Adams, D. O. The antioxidant activity of Californian red wines does not correlate with wine age. J. Sci. Food Agric., 2006, V. 86, 834 840.

11. Korenman, I. M. Method of organic compounds determination. Photometric analysis M., (1970).

12. De Villiers, A., Vanhoenacker, G., Majek, P., and Sondra, P. Determination of anthocyanins in wine by direct injection liquid chromatography-diode array detection-mass spectrometry and classification of wines using discriminant analysis. J. Chromat. A., V. 1054, 194 204 (2004).

13. Kong, J. M., Chia, L. S., Goh, N. K., Chia, T. F., and Brouillard, R. Analysis and biological activities of anthocyanins. Phytochemistry, 64(5), p. 923–933 (2003).

14. Lila, M. A. Anthocyanins and Human Health: An in vitro investigative approach. J. Biomedicine and Biotechnology, 5, p. 306 313 (2004).

15. Gachons, C. P., and Kennedy, J. A. Direct method for determining seed and skin proanthocyanidin extraction into red wine. J. Agric. Food Chem., V. 51, p. 5877 5881 (2003).

16. Bagchi, D., Bagchi, M., Stohs, S. J., Das, D. K., Ray, S. D., Kuszynski, C. A., Joshi, S. S., and Pruess, H. G. Free radicals and grape seed proanthocyanidn extract: importance in human health and disease prevention. Toxicology, V. 148, p. 187 197 (2000).

17. Ahmad, N., Gupta, S., and Mukhtar, H. Green tea polyphenol epigallocatechin-3-gallate differentially modulates nuclear factor kB in cancer cells versus normal cells. Archives of Biochemistry and Biophysics. V. 376, p. 338 346 (2000).

18. Bagchi, D., Sen, C. K., Ray, S. D., Dipak, K., Bagchi, M., Preuss, H. G., and Vinson, J. A. Molecular mechanisms of cardioprotection by a novel grape seed proanthocyanidin extract. Mutation Research, V. 523, p. 87 97 (2003).

19. Chen, C. K., and Pace-Asciak, C. R. Vasorelaxing activity of resveratrol and quercetin in isolated rat aorta. Gen. Pharmacol, V. 27, p. 363 366 (1996).

20. Simonetty, P., Pietta, P., and Testolin, G. Polyphenol content and total antioxidant potential of selected Italian Wines. J. Agric. Food Chem, V. 45, p. 1152 1155 (1997).

21. Fuhrman, B., Volkova, N., Soraski, A., and Aviram, M. White wine with red wine-like properties: increased extraction of grape skin polyphenols improves the antioxidant capacity of the derived white wine. J. Agric. Food Chem., V. 49, p. 3164 3168 (2001).

CHAPTER 15

A STUDY ON THE POTENTIAL OF OILSEEDS AS A SUSTAINABLE SOURCE OF OIL AND PROTEIN FOR AQUACULTURE FEED

CRYSTAL L. SNYDER, PAUL P. KOLODZIEJCZYK, XIAO QIU,
SALEH SHAH, E. CHRIS KAZALA, and RANDALL J. WESELAKE

CONTENTS

15.1 INTRODUCTION

Aquaculture is currently the world's fastest growing primary food production sector, but it is also a leading consumer of fish oil and fish meal from capture fisheries, which is a major concern for the industry's long-term sustainability. Oilseeds have been widely explored as an alternative, land-based source of oil and protein for aquaculture feeds, but their use has been limited because conventional seed oils do not contain very long chain ω-3 fatty acids such as eicosapentaenoic acid (EPA) and docosahexaenoic acid (DHA). Over the past decade, however, there has been considerable progress toward the genetic engineering of oilseeds to produce very long chain ω-3 fatty acids. Such modified oilseeds show promise as nutritional supplements for aquafeed applications. Advances in oilseed processing technology may also assist in the development of sustainable oilseed-based aquafeeds.

Aquaculture currently supplies almost half of the seafood for human and animal consumption [1], and as one of the world's fastest growing food production sectors, the industry is under pressure to address the sustainability of its supply chain, particularly with respect to its reliance on fish oil and fish meal from capture fisheries [2]. The growing concern over the decline of the world's capture fisheries has prompted considerable research into alternative aquafeed ingredients, especially those from land-based plants. Oilseeds, for example, have been widely explored as an alternative source of oil and protein in fish diets, but present a number of challenges that currently limit their use in aquafeeds. Unlike marine oils, which are enriched in very long chain (20-22 carbon) polyunsaturated fatty acids (VLCPUFA) such as eicosapentaenoic acid (EPA, 20:5ω-3) and docosahexaenoic acid (DHA 22:6ω-3), conventional seed oils are rich in 18-carbon fatty acids (mainly linoleic acid, 18:2ω-6 and α-linolenic acid, 18:3ω-3), which fish are unable to efficiently convert to VLCPUFA. In addition, animals lack the ability to convert omega-6 fatty acids to omega-3 fatty acids, making most high linoleic vegetable oils an inadequate substitute for fish oil. Maintaining high levels of ω-3 VLCPUFA in farmed fish is of paramount concern not only for optimal fish health, but also for the nutritional quality of the end products destined for human consumption. Indeed, the relatively low consumption of ω-3 VLCPUFA in the Western diet, together with high intake of ω-6 fatty acids, has been blamed in part for the growth of chronic, diet-related disease among Western populations [3].

While oilseeds offer many advantages for the sustainable production of aquafeed ingredients, engineering oilseeds to produce VLCPUFA has proven to be a formidable biotechnological challenge. Over the past decade, research groups around the world have explored various strategies for producing some or all of the fatty acids in the VLCPUFA biosynthetic pathway [4]. Almost all of these approaches require the insertion and coordinated expression of multiple transgenes, the effectiveness of which tends to vary depending on the nature of the genetic construct and of the host plant [5,6]. To date, substantial accumulations of various precursors leading to the formation of arachidonic acid (ARA, 20:4ω-6) and EPA have been obtained [4]; however, commercially-viable levels of DHA have yet to be achieved.

Although much of the commercial value of commodity oilseeds is in the oil fraction, the meal of many oilseeds, if processed appropriately, can be used as a high-quality protein source for animal feeds. Research is ongoing into the use and enhancement of oilseed meal for aquafeed, with many studies focusing on the enrichment of high-value bioactive components (e.g. carotenoids) or reduction of anti-nutritional factors [7]. At the same time, advances in oilseed processing technology are also helping to maximize the value of the oil and the meal for aquafeed and other applications.

This paper will review progress toward the production of ω-3 fatty acid-enriched seed oils for aquaculture feeds, their efficacy as aquafeed ingredients, and the role of processing technologies in maximizing the potential of oilseeds for aquaculture applications.

15.2   ENGINEERING OILSEEDS FOR THE SUSTAINABLE PRODUCTION OF Ω-3 FATTY ACIDS FOR AQUACULTURE

In animals, linoleic acid (LA, 18:2 ω-6) and α-linolenic acid (ALA, 18:3 ω-3) are considered to be the only dietary fatty acids that are essential for the production of VLCPUFA, including EPA and DHA. Animals possess a biosynthetic pathway for the further desaturation and elongation of linoleic and α-linolenic acids to form ARA and EPA, respectively (Figure 1A). However, the poor efficiency of this process, particularly the initial Δ6 desaturation step, can severely limit the accumulation of VLCPUFA in many individuals [8]. As a result, direct consumption of dietary VLCPUFA is recommended for optimal nutrition [9]. Even fish, which are generally considered to be excellent dietary sources of VLCPUFA for humans, must obtain their VLCPUFA from their diet, usually through bioaccumulation of VLCPUFA produced by marine algae and other microorganisms.

The VLCPUFA-producing pathways of these marine microorganisms have offered a diverse array of genetic tools for oilseed biotechnology. Unlike animals, these microorganisms possess multiple pathways for VLCPUFA biosynthesis. Some organisms rely on aerobic desaturation/elongation processes analogous to the pathway present in animals [10], while others use a polyketide synthase (PKS) to build VLCPUFA in a process similar to *de novo* fatty acid biosynthesis in plants [11]. A few higher plants, including borage and echium [12,13], also possess the critical Δ6 desaturation activity and naturally accumulate gamma-linolenic acid (GLA, 18:3ω-6) and stearidonic acid (SDA, 18:4 ω-3); however, these plants lack the enzyme activities required to convert GLA or SDA to downstream VLCPUFA. Nevertheless, these plants have served as an important source of genes supporting oilseed biotechnology efforts.

Indeed, the Δ6 desaturation pathway has been one of the most extensively studied in transgenic plants. Expression of recombinant desaturases from various sources has resulted in relatively high accumulation of GLA, and to a lesser extent, SDA. In *Brassica juncea*, expression of a Δ6 desaturase alone yielded up to 40% GLA and 8% SDA [14]. Other groups have used additional Δ12 or Δ15 desaturases to increase the supply of LA or ALA precursors, respectively [15,16]. Using a high linoleic safflower (*Carthamus tinctorius*) line coexpressing a Δ6 and Δ12 desaturase, Nykiforuk et al, [5] were able to obtain more than 70% GLA in the seed oil; these lines have recently been approved for commercialization as Sonova™ 400. Achieving high levels of SDA, particularly without significant coproduction of GLA, has been somewhat more challenging.

To this end, an ω-3-selective Δ6 desaturase was introduced into flaxseed in order to preferentially increase the accumulation of SDA [17]. SDA-enriched seed oils, having bypassed the limiting Δ6 desaturation step, may be particularly useful for aquafeed applications; however, the efficacy of these oils is still being tested in fish [13,18].

Conversion of GLA and SDA to ARA and EPA, respectively, requires the introduction of at least two additional genes encoding a Δ6 elongase and a Δ5 desaturase to complete the "conventional" Δ6 desaturation pathway (Figure 1B). This pathway was first reconstituted in tobacco (high LA background) and flax (high ALA background), resulting in approximately 30% GLA or SDA, respectively, but only up to about 5% total 20-carbon VLCPUFA [19]. It appeared that the Δ6 elongation step was not very efficient, even though the recombinant elongase was successfully expressed in developing seeds and exhibited *in vitro* enzyme activity. Analysis of the acyl-CoA pool of transgenic flaxseed, however, indicated that there were only trace levels of Δ6 desaturated PUFA in the acyl-CoA pool, where elongation occurs [19]. Since the Δ6 desaturase utilizes phospholipid substrates and the elongase requires acyl-CoA substrates [20], this so-called "substrate dichotomy" was suggested to be a major limiting factor for VLCPUFA formation in transgenic plants [19]. These observations were supported by another study [21] employing an "alternative" Δ9 elongation pathway (Figure 1B), in which LA and ALA are first elongated, and then desaturated twice to form ARA and EPA. In this pathway, there is no need to transfer of substrates from phospholipids to the acyl-CoA pool and then back again. When this pathway was introduced into Arabidopsis leaves, up to 7% ARA and 3% EPA was obtained, and the total levels of 20-carbon PUFA were greater than 20% [21].

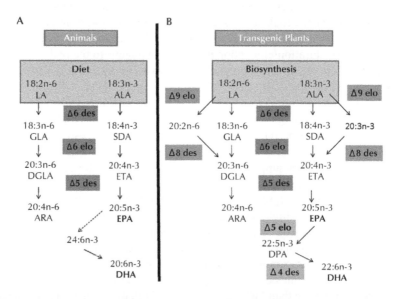

FIGURE 1 Pathways for very long chain polyunsaturated fatty acid (VLCPUFA) biosynthesis in animals and plants. A) In vertebrates, VLCPUFA are synthesized from the dietary essential fatty acids, linoleic acid (LA) and α-linolenic acid (ALA), by the Δ6-desaturation pathway, consisting

of a Δ6-desaturase Δ6-elongase, and a Δ5-desaturase, leading to the formation of arachidonic acid (ARA; ω-6 pathway) and eicosapentaenoic acid (EPA; ω-3 pathway). Docosahexaenoic acid (DHA) is synthesized from EPA following further elongation and desaturation to 24:6, which undergoes one round of β-oxidation to form DHA. In most vertebrates, however, the Δ6-desaturation step is limiting, resulting in poor rates of conversion to EPA and DHA. **B)** Two desaturation/elongation pathways have been tested extensively in transgenic plants. The Δ6-desaturation pathway (red) is analogous to the pathway present in animals, but uses LA and ALA generated through *de novo* biosynthesis. The Δ9-elongation pathway (blue) also leads to the formation of ARA and EPA, but elongation takes place first, and is followed by two desaturation steps. From EPA, at least two additional activities (purple) are required to generate DHA. (des: desaturase; elo: elongase. All fatty acid abbreviations are defined in the text. The notation n-3/n-6 and ω-3/ω-6 are used interchangeably.)

Other groups have tested a number of variations on the Δ6 desaturation pathway in efforts to overcome this problem of substrate dichotomy. Adding a Δ12 desaturase and an additional Δ6 elongase to the minimal three gene pathway resulted in increased elongation efficiency in transgenic *Brassica juncea* seeds, yielding up to 25% ARA [22]. The further introduction of an ω-3 desaturase, which converts ω-6 intermediates to their ω-3 counterparts, resulted in up to 11% EPA [22]. Zero erucic acid *Brassica carinata*, however, seems to be a more suitable host for EPA production, accumulating up to 20% EPA in lines expressing a five gene construct (Δ12 desaturase, Δ6 desaturase, Δ6 elongase, Δ5 desaturase, and an ω-3 desaturase) [6]. In zero-erucic acid *B. juncea*, the same construct yielded only 5% EPA[6], which highlights the importance of host species on the outcome of such transgenic experiments.

Some groups have also explored the use of acyl-CoA desaturases as another approach to overcoming substrate dichotomy in transgenic plants. In this strategy, Δ6 desaturation and elongation both occur in the acyl-CoA pool. Transgenic soybeans expressing an acyl-CoA/phospholipid desaturase from liverwort accumulated more than 19% total 20-carbon PUFA [23]. Although this increase comprised mostly ω-6 fatty acids, reflecting the high levels of linoleic acid precursor in soybean, this study demonstrated the effectiveness of acyl-CoA desaturases in overcoming substrate dichotomy. Another study using an ω-3-specific acyl-CoA desaturase showed a similar proof-of-concept in *Arabidopsis*, but achieved much lower total levels of VLCPUFA [24].

While it appears that the nature of the genetic construct, source of the transgenes, and selection of the host species all play a crucial role in determining the success of VLCPUFA-engineering experiments, recent advancements in our basic understanding of plant lipid biochemistry are also shedding insights into the biochemical basis for substrate dichotomy. The discovery of new enzymes involved in substrate trafficking [25,26] and the elucidation of a proposed "acyl-editing" pathway in developing oilseeds [27,28] will greatly assist ongoing efforts to develop a sustainable land-based source of VLCPUFA.

Although much of the research to date has focused on EPA as an essential first step in achieving high levels of VLCPUFA in transgenic plants, it is well recognized that DHA is equally important for aquaculture applications Several studies have shown proof-of-concept production of DHA in transgenic systems. Production of DHA has been technically more difficult, both in terms of overcoming substrate dichotomy and in handling the very large multi-gene constructs required for DHA production. Wu et al [22], for example, used a nine-gene construct to demonstrate proof-of-concept DHA synthesis in *Brassica juncea*. Recent methodological advancements such as the transient leaf-based assay described by Wood et al[29], promise to accelerate development of complex multi-gene constructs by facilitating rapid testing and interchangeability of independent genetic elements. This technique has already been successfully applied to the VLCPUFA pathway, resulting in up to 2.5% DHA in tobacco leaves and similar levels in stably-transformed *Arabidopsis* seeds [30]. Given the rapid pace of research in this area, it is likely only a matter to time before commercially viable levels of DHA are achieved in a crop plant [31-34].

15.3   EFFICACY OF Ω-3 VEGETABLE OILS IN FISH NUTRITION

Over the past few decades, a wide range of vegetable oils and vegetable oil blends have been explored as possible alternatives to fish oil for aquafeed applications (see [35,36]for recent reviews). All vertebrates, including fish, have certain dietary requirements for essential fatty acids (EFA). In fish, dietary EFA requirements differ according to species; freshwater and diadromous fish may only require 18-carbon PUFA (eg. LA or ALA), while marine species require ω-3 VLCPUFA (eg. EPA and DHA) [13]. A deficiency in EFA can lead to adverse effects on growth and reproduction and in severe cases, could result in increased mortality [37]. The diversity of vegetable oils available and the differences in the dietary needs of cultivated fish species have led to a variety of different strategies for the incorporation of vegetable oils into aquaculture feeds. Some of these studies will be reviewed here.

Although flaxseed is a relatively minor commodity oilseed in terms of total production, it has received a great deal of attention for aquafeed applications due to its high levels of ALA (50-60%). One of the earliest fish oil replacement studies compared flax oil, rendered animal fat, and salmon oil in the dry diet of Chinook salmon (*Onchorhynchous tshawytscha*) over sixteen weeks [38]. Fish growth was not affected by the substitution, but the flesh fatty acid profiles, particularly PUFA content, varied according to the fatty acid composition of the diet. Several other fish oil replacement studies have reported similar trends; that is, the tissue fatty acid composition largely reflects that of the diet, but vegetable oils generally do not impair growth or performance of the fish [39-41]. After comparing two base diets (100% fish oil or 100% flax oil) substituted with various proportions of sunflower oil, Menoyo et al [40], concluded that flax could be a viable source of dietary lipids for farmed Atlantic salmon. In trout (*Onchorhynchus mykiss*), studies comparing salmon oil, flax oil, soybean oil, poultry fat, pork lard and beef tallow suggested that the EPA and DHA levels in the tissue were regulated and maintained at 10-12% regardless of diet [41].

As a Δ6 desaturated ω-3 PUFA, stearidonic acid (SDA) is also particularly interesting, not only for aquaculture, but for human and livestock nutrition as well. In mammals, the conversion rate of SDA to EPA is around 17-33% [42-44], while the conversion of ALA to EPA is much lower at 0.2-7% [45]. The low rate of conversion of ALA to EPA is believed to be due to the inefficient Δ6 desaturation step; thus, consumption of SDA could effectively bypass this step and support more efficient accumulation of downstream VLCPUFA. As fish utilize a similar pathway for converting ALA to EPA, SDA-enriched oils may be useful for maintaining or increasing EPA levels in the tissue.

To date, most studies investigating the efficacy of SDA have used echium oil as a source of SDA. Echium is one of a few higher plants that accumulate appreciable amounts of SDA (up to 14%). In Arctic charr (*Salvelinus alpines*), replacing 80% of the dietary fish oil with echium oil for 16 weeks led to increases in dihomo-gamma-linolenic acid (DGLA, 20:3ω-6) and eicosatetraenoic acid (ETA, 20:3ω-3), the immediate elongation products of GLA and SDA, respectively, in both flesh and liver tissue [46]. The echium diet, however, was not effective at maintaining EPA and DHA levels in charr. In a similar trial with Atlantic cod (*Gadus morhua*), a 100% echium oil diet for 18 weeks resulted in only slight increases in DGLA, with no increase in ETA [18]. Such studies illustrate the complexity of species-specific differences in dietary fatty acid metabolism.

In Atlantic salmon parr, neither echium oil nor a 1:1 blend of echium and canola oil resulted in a decrease in white muscle DHA content after 42 days [47]. It was suggested that reduced availability of VLCPUFA in the diet led to efficient retention and redistribution of DHA [48]. More recent studies with echium oil in salmon [49] and barramundi [50] failed to show any major benefits of SDA-enriched oils in fish feeding trials. It is worthwhile noting, however, that the SDA content of echium (14%) is relatively low compared to what may be obtainable from a transgenic plant source, and the presence of GLA as competing ω-6 byproduct may also affect the conversion efficiencies observed in feeding trials. Thus, a plant source high in SDA and low in GLA and other ω-6 PUFA may be more effective at shifting fish metabolism toward conversion of ω-3 PUFA.

Given that the flesh fatty acid composition typically reflects the lipid composition of the diet, it becomes clear that direct substitution of fish oil with EPA and DHA-enriched oils may be the most effective means of meeting the ω-3 VLCPUFA requirements of farmed fish. As described earlier, efforts to produce a sustainable land-based source of EPA and DHA are advancing rapidly; however, the intensifying pressure on the global fish oil supply requires a multi-faceted approach to address the sustainability of the aquaculture sector. To that end, the use of different feeding phases in fish production has been explored as a way to make the most efficient use of available dietary lipids (reviewed by [35]). These trials use different starter, grow-out, and finishing feeds to address the different dietary needs at various stages of development. With respect to fish development, the need for dietary VLCPUFA (i.e. fish oil) is greatest during the starter phase, in which rapid growth, high survival, and normal development (particularly of neural tissue) are the most important considerations. Finishing feeds, though not yet common in commercial settings, may also call for a greater VLCPUFA content in order to ensure that the desired post-harvest fatty acid composition is obtained.

There seems to be the greatest potential for fish oil substitution during the grow-out phase, during which flesh fatty acid composition is not as critical and growth rate and feed conversion are the main considerations. In this case, a VLCPUFA-rich finishing diet would facilitate "washing-out" of the less desirable flesh FA prior to harvest. Such feeding strategies would allow fish producers to reduce their overall reliance on fish oil through more efficient feeding strategies, thereby reducing feed costs and enhancing the sustainability of the operation.

OILSEED PROCESSING TECHNOLOGIES FOR AQUAFEED

There has been considerable progress toward developing sustainable alternatives to fish meal and fish oil for aquafeed applications [51]. Today, the pork, beef, and poultry industries are the major users of oilseed meals for feed applications, but advancements in processing technology may expand market opportunities into the aquaculture sector. While the development of specialty oils containing high levels of physiologically important PUFA remains a major focus, advancements in oilseed processing technologies are also crucial for the production of high-quality protein concentrates for fish nutrition. For example, plant seeds contain a substantial amount of fibre, which is considered undesirable for fish feed. Novel processes based on fractionation of the oilseed cake or meal may facilitate the removal of anti-nutritional components left behind by traditional processing [52] and yield a product more amenable for use in fish feed.

The interest in processing technologies has been driven in part by the growth of so-called biorefineries, which aim to maximize the value of all components of a particular feedstock. The rapid expansion of oilseed-based biofuels has sparked increased interest in finding alternative markets for the leftover seed meal, which would help offset the relatively high feedstock costs involved in biodiesel production. Technologies that recover high-value components from the meal or otherwise increase the value of the meal for feed applications would substantially broaden the operating margins for biorefining facilities while helping to meet the needs of specialty markets such as the aquaculture sector.

The typical industrial process for production of edible oils involves pressing the seeds, followed by a solvent extraction, usually with hexane. This method results in a good recovery of high-quality, low cost oil, with residual seed meal being used by the animal feed industry. However, environmental considerations, as well as the desire to increase the value of the meal, have led researchers to explore processing technologies that rely less on organic solvents and take advantage of aqueous or enzyme-assisted technologies. Enzymes can be used to facilitate oil extraction by hydrolyzing the cell wall polysaccharides and oilbody membranes, and have the advantage of yielding non-denatured protein, which has higher value as a protein concentrate (containing fibre) or protein isolate (fibre removed). Advancements in industrial enzyme production are making enzyme-assisted processes increasingly cost-efficient, and several modifications of aqueous processes have been described in papers and patents [52,53]. Several companies have already patented technologies for production of canola and flax protein isolates, and industrial pilot plants using aqueous technologies are already in operation.

Several processes for the manufacture of protein concentrates or isolates for aquaculture have also been developed. These technologies yield non-denatured proteins with better digestibility for fish nutrition. Membrane separation techniques are used to separate oilseed components, remove anti-nutritional factors, and fractionate protein for fish feed production.

The VLCPUFA-enriched oilseeds pose additional challenges for processing since the high PUFA content of the oils reduces their thermal and oxidative stability. Aqueous milling processes, followed by separation of the lipid fraction, are well-suited for the preservation of the high-value VLCPUFA, since the harsh refining steps (bleaching, degumming and deodorization) are not needed for fish feed production. The presence of phospholipids is actually considered beneficial for fish nutrition.

Supercritical fluid extraction is another possible strategy for processing of oilseeds for aquafeed applications, but so far the expansion of this technology has been inhibited by the cost of the process and lack of equipment for continuous processing using this technology [54].

15.4   CONCLUSION

Aquaculture production has grown immensely since the mid-1960s, and has now become the world's fastest growing primary food production sector, providing almost half of all the fish consumed by humans today [55]. At the same time, production from capture fisheries has declined due to chronic overfishing – largely to support the expansion of the global aquaculture industry. In 2006, global aquaculture utilized 68% of fish meal and 88% of fish oil produced worldwide [56], with a few carnivorous species (e.g. trout and salmon) consuming 19.5% and 51% of meal and oil, respectively, despite the fact that they represent only 3% of global aquaculture production [56].

Clearly, neither capture fisheries nor aquaculture can sustainably meet the needs of a human population expected to exceed 9 billion by 2050, unless the aquaculture industry can overcome its reliance on capture fisheries as a source of feed ingredients. Future growth of the industry depends on the development of alternative feed ingredients, a large fraction of which are likely to come from terrestrial plants, particularly from modified oilseeds containing EPA and DHA. Advancements in oilseed biotechnology and processing have set the stage for such designer oilseeds to make a major impact on the aquafeed market in the years to come.

KEYWORDS

- **Aquaculture**
- **Eicosapentaenoic acid**
- **Linoleic acid**
- **Supercritical fluid extraction**
- **Very long chain polyunsaturated fatty acid**

REFERENCES

1. Subasinghe, R., Soto, D., and Jia, J. Global aquaculture and its role in sustainable development. Rev Aquaculture, 1 2 9 (2009).

2. de Silva, S., Francis, D., and Tacon, A. Fish oils in aquaculture. In Fish Oil Replacement and Alternative Lipid Sources in Aquaculture Feeds. CRC Press, 1 20 (2010).

3. Simopoulos, A. P. The importance of the omega-6/omega-3 fatty acid ratio in cardiovascular disease and other chronic diseases. Exp Biol Med. (Maywood), 233 674 688 (2008).

4. Venegas-Caleron, M., Sayanova, O. and Napier, J. A. An alternative to fish oils: Metabolic engineering of oil-seed crops to produce omega-3 long chain polyunsaturated fatty acids. Prog Lipid Res. 49 108 119 (2010).

5. Nykiforuk, C. L., Shewmaker, C., Harry, I., Yurchenko, O. P., Zhang, M., and Reed, C. et al. High level accumulation of gamma linolenic acid (C18:3Delta6.9,12 cis) in transgenic safflower (Carthamus tinctorius) seeds. Transgenic Res. (2011).

6. Cheng, B., Wu, G., Vrinten, P., Falk, K., Bauer, J., and Qiu, X. Towards the production of high levels of eicosapentaenoic acid in transgenic plants: the effects of different host species, genes and promoters. Transgenic Res. 19 221 229 (2010).

7. Gatlin, D. M., Barrows, F. T., Brown, P., Dabrowski, K., Gaylord, T. G., and Hardy, R. W. et al. Expanding the utilization of sustainable plant products in aquafeeds: a review. Aquaculture Res. 38 551 579 (2007).

8. Plourde, M., and Cunnane, S. C. Extremely limited synthesis of long chain polyunsaturates in adults: implications for their dietary essentiality and use as supplements. Appl Physiol Nutr Metab. 32 619 634 (2007).

9. Griffiths, G., and Morse, N. Clinical applications of C-18 and C-20 chain length polyunsaturated fatty acids and their biotechnological production in plants. J Am Oil Chem Soc. 73 171 185 (2006).

10. Uttaro, A. D. Biosynthesis of polyunsaturated fatty acids in lower eukaryotes. IUBMB Life, 58 pp. 563 571 (2006).

11. Metz, J.G., Roessler, P., Facciotti, D., Levering, C., Dittrich, F., and Lassner, M. et al. Production of polyunsaturated fatty acids by polyketide synthases in both prokaryotes and eukaryotes. Science, 293 290 293 (2001).

12. Griffiths, G., Stobart, A. K., and Stymne, S. Delta 6- and delta 12-desaturase activities and phosphatidic acid formation in microsomal preparations from the developing cotyledons of common borage (Borago officinalis). Biochem J. 252 641 647 (1988).

13. Tocher, D. R. Fatty acid requirements in ontogeny of marine and freshwater fish. Aquaculture Res 41 717 732 (2010).

14. Hong, H., Datla, N., Reed, D. W., Covello, P. S., MacKenzie, S. L., and Qiu, X. High-level production of gamma-linolenic acid in Brassica juncea using a delta6 desaturase from Pythium irregulare. Plant Physiol. 129 354 362 (2002).

15. Liu, J. W., De Michele, S., Bergana, M., Bobik, E., Hastilow, C., and Chuang, L. T. et al. Characterization of oil exhibiting high gamma-linolenic acid from a genetically transformed canola strain. J Am Oil Chem Soc. 78 489 493 (2001).

16. Eckert, H., La, V. B., Schweiger, B. J., Kinney, A. J., Cahoon, E. B., and Clemente, T. Co-expression of the borage Delta 6 desaturase and the Arabidopsis Delta 15 desaturase results in high accumulation of stearidonic acid in the seeds of transgenic soybean. Planta, 224 1050 1057 (2006).

17. Ruiz-Lopez, N., Haslam, R. P., Venegas-Caleron, M., Larson, T. R., Graham, I. A., and Napier, J. A. et al. The synthesis and accumulation of stearidonic acid in transgenic plants: a novel source of 'heart-healthy' omega-3 fatty acids. Plant Biotechnol J. 7 704 716 (2009).

18. Bell, J. G., Strachan, F., Good, J. E., and Tocher, D. R. Effect of dietary echium oil on growth, fatty acid composition and metabolism, gill prostaglandin production and macrophage activity in Atlantic cod (Gadus morhua L.). Aquaculture Res. 37 606 617 (2006).

19. Abbadi, A., Domergue, F., Bauer, J., Napier, J. A., Welti, R., and Zahringer, U. et al. Biosynthesis of very-long-chain polyunsaturated fatty acids in transgenic oilseeds: constraints on their accumulation. Plant Cell 16 2734 2748 (2004).

20. Domergue, F., Abbadi, A., Ott, C., Zank, T. K., Zahringer, U., and Heinz, E. Acyl carriers used as substrates by the desaturases and elongases involved in very long-chain polyunsaturated fatty acids biosynthesis reconstituted in yeast. J Biol Chem. 278 35115 35126 (2003).

21. Qi, B., Fraser, T., Mugford, S., Dobson, G., Sayanova, O., and Butler, J. et al. Production of very long chain polyunsaturated omega-3 and omega-6 fatty acids in plants. Nat Biotechnol. 22 739 745 (2004).

22. Wu, G., Truksa, M., Datla, N., Vrinten, P., Bauer, J., and Zank, T. et al. Stepwise engineering to produce high yields of very long-chain polyunsaturated fatty acids in plants. Nat Biotechnol. 23 1013 1017 (2005).

23. Kajikawa, M., Matsui, K., Ochiai, M., Tanaka, Y., Kita, Y., and Ishimoto, M. et al. Production of arachidonic and eicosapentaenoic acids in plants using bryophyte fatty acid Delta6-desaturase, Delta6-elongase, and Delta5-desaturase genes. Biosci Biotechnol Biochem. 72 435 444 (2008).

24. Hoffmann, M., Wagner, M., Abbadi, A., Fulda, M., and Feussner, I. Metabolic engineering of omega-3-very long chain polyunsaturated fatty acid production by an exclusively acyl-CoA-dependent pathway. J Biol Chem. 283: 22352 22362 (2008).

25. Dahlqvist A, Stahl U, Lenman M, Banas A, Lee M, Sandager L et al.: Phospholipid:diacylglycerol acyltransferase: an enzyme that catalyzes the acyl-CoA-independent formation of triacylglycerol in yeast and plants. Proc Natl Acad Sci USA, 97 6487 6492 (2000).

26. Lu, C. F., Xin, Z. G., Ren, Z. H., Miquel, M., and Browse, J. An enzyme regulating triacylglycerol composition is encoded by the ROD1 gene of Arabidopsis. Proc Natl Acad Sci USA, 106 18837 18842 (2009).

27. Bates, P. D., Durrett, T. P., Ohlrogge, J. B., and Pollard, M. Analysis of acyl fluxes through multiple pathways of triacylglycerol synthesis in developing soybean embryos. Plant Physiol. 150 55 72 (2009).

28. Bates, P. D., Browse, J. The pathway of triacylglycerol synthesis through phosphatidylcholine in Arabidopsis produces a bottleneck for the accumulation of unusual fatty acids in transgenic seeds. Plant J. (2011).

29. Wood, C. C., Petrie, J. R., Shrestha, P., Mansour, M. P., Nichols, P. D., and Green, A. G. et al. A leaf-based assay using interchangeable design principles to rapidly assemble multi-step recombinant pathways. Plant Biotechnol J. 7 914 924 (2009).

30. Petrie, J. R., Shrestha, P., Liu, Q., Mansour, M. P., Wood, C. C., and Zhou, X. R. ct al. Rapid expression of transgenes driven by seed-specific constructs in leaf tissue: DHA production. Plant Methods, 6, 8 (2010).

31. Barker, B. Brain, eye, and heart-healthy canola oil in the works. Top Crop Manager. 5-15-2011. Ref Type: Electronic Citation, (2008) (http://www.topcropmanager.com/content/view/4128/38/).

32. Nuseed. Australian scientific collaboration set to break world's reliance on fish for long chain omega-3. Nuseed media release. 5-15-2011. Ref Type: Electronic Citation, http://www.nuseed.com/Assets/448/1/110412_Omega3inCanola_PioneeringAustralianresearchallianceannouncement_Embargo-Tuesday12April2011.pdf. (2011).

33.   Watkins C. Oilseeds of the future: Part 1. INFORM. 5-15-2011. Ref Type: Electronic Cita-
      tion, (http://www.aocs.org/Membership/FreeCover.cfm?itemnumber=1086). (2009)
34.   Monsanto. World's first SDA omega-3 soybean oil achieves major mileston that advances
      the development of foods with enhanced nutritional benefits. 5-15-2011. Ref Type: Elec-
      tronic Citation, http://monsanto.mediaroom.com/index.php?s=43&item=761&printable.
      (2009).
35.   Glencross, B. and Turchini, G. Fish oil replacement in starter, grow-out, and finishing
      feeds for farmed aquatic animals. In Fish Oil Replacement and Alternative Lipid Sources
      in Aquaculture Feeds. 373 404 (2010). CRC Press
36.   Turchini, G. M., Torstensen, B. E., and Ng, W. K. Fish oil replacement in finfish nutrition.
      Rev Aquaculture, 1 10 57 (2009).
37.   Das UN: Essential fatty acids - a review. Curr Pharm Biotechnol. 7 467 482 (2006).
38.   Mugrditchian, D. S., Hardy, R. W., and Iwaoka, W. T. Linseed oil and animal fat as alterna-
      tive lipid sources in dry diets for Chinook Salmon (Oncorhynchus tshawytscha). Aquacul-
      ture 25 161 172 (1981).
39.   Stubhaug, I., Lie, O., and Torstensen, B. E. Fatty acid productive value and beta-oxidation
      capacity in Atlantic salmon (Salmo salar L.) fed on different lipid sources along the whole
      growth period. Aquaculture Nutr., 13 145 155 (2007).
40.   Menoyo, D., Lopez-Bote, C. J., Diez, A., Obach, A., and Bautista, J. M. Impact of n-3 fatty
      acid chain length and n-3/n-6 ratio in Atlantic salmon (Salmo salar) diets. Aquaculture,
      267 248 259 (2007).
41.   Greene, D. H. S. and Selivonchick, D. P. Effects of dietary vegetable, animal and marine
      lipids on muscle lipid and hematology of rainbow-trout (Oncorhynchus mykiss). Aquacul-
      ture, 89 165 182 (1990).
42.   Yamazaki, K., Fujikawa, M., Hamazaki, T., Yano, S., and Shono, T. Comparison of the
      conversion rates of alpha-linolenic acid (18-3(N-3)) and stearidonic Acid (18-4(N-3)) to
      longer polyunsaturated fatty acids in rats. Biochimica et Biophysica Acta, 1123 18 26
      (1992).
43.   James, M. J., Ursin, V. M., and Cleland, L. G. Metabolism of stearidonic acid in human
      subjects: comparison with the metabolism of other n-3 fatty acids. Am J Clin Nutr., 77
      1140 1145 (2003).
44.   Harris, W. S., Lemke, S. L., Hansen, S. N., Goldstein, D. A., DiRienzo, M. A., and Su,
      H. et al. Stearidonic acid-enriched soybean oil increased the omega-3 index, an emerging
      cardiovascular risk marker. Lipids, 43 805 811 (2008).
45.   Brenna, J. T., Salem, N., Sinclair, A. J., and Cunnane, S. C. Alpha-Linolenic acid supple-
      mentation and conversion to n-3 long-chain polyunsaturated fatty acids in humans. Pros-
      tag Leukotr Ess., 80 85 91 (2009).
46.   Tocher, D. R., Dick, J. R., MacGlaughlin, P., and Bell, J. G. Effect of diets enriched in
      Delta 6 desaturated fatty acids (18:3n-6 and 18:4n-3), on growth, fatty acid composition
      and highly unsaturated fatty acid synthesis in two populations of Arctic charr (Salvelinus
      alpinus L.). Comp Biochem Phys B., 144 245 253 (2006).
47.   Miller, M. R., Nichols, P. D., and Carter, C. G. Replacement of dietary fish oil for Atlan-
      tic salmon parr (Salmo salar L.) with a stearidonic acid containing oil has no effect on
      omega-3 long-chain polyunsaturated fatty acid concentrations. Comp Biochem Phys. B,
      146 197 206 (2007).
48.   Tocher, D., Francis, D., and Coupland, K. n-3 Polyunsaturated fatty acid-rich vegetable
      oils and blends. In Fish Oil Replacement and Alternative Lipid Sources in Aquaculture
      Feeds. 209 244 (2010) CRC Press.

49. Miller, M. R., Bridle, A. R., Nichols, P. D., and Carter, C. G. Increased elongase and de-saturase gene expression with stearidonic acid enriched diet does not enhance long-chain (n-3) content of seawater Atlantic Salmon (Salmo salar L.). J Nutr., 138 2179 2185 (2008).
50. Alhazzaa, R., Bridle, A. R., Nichols, P. D., and Carter, C. G. Replacing dietary fish oil with Echium oil enriched barramundi with C-18 PUFA rather than long-chain PUFA. Aquaculture, 312 162 171 (2011).
51. de Silva, S. and Soto, D. Climate change and aquaculture: potential impacts, adaptation and mitigation. FAO Fisheries Technical Paper, 530 137 215 (2009).
52. Mueller, K., Eisner, P., Yoshie-Stark, Y., Nakada, R., and Kirchhoff, E. Functional properties and chemical composition of fractionated brown and yellow linseed meal (Linum usitatissimum L.). Journal of Food Engineering, 98 453 460 (2010).
53. Nennerfelt, L., Evans, D., and Ten Haaf, W. Process for extracting flax protein concentrate from flax meal. Nutrex Wellness, Inc. [US 6998466 B2]. Ref Type: Patent (2006)
54. Brunner, G. Supercritical fluids: technology and application to food processing. Journal of Food Engineering, 67 21 33 (2005).
55. Food and Agriculture Organization. The status of world fisheries and aquaculture. Rome, Food and Agriculture Organization. Ref Type: Report (2009)
56. Tacon, A. G. J. and Metian, M. Global overview on the use of fish meal and fish oil in industrially compounded aquafeeds: Trends and future prospects. Aquaculture, 285 146 158 (2008).

CHAPTER 16

ELECTROCHEMICAL METHODS FOR ESTIMATION OF ANTIOXIDANT ACTIVITY OF VARIOUS BIOLOGICAL OBJECTS

N. N. SAZHINA, E. I. KOROTKOVA, and V. M. MISIN

CONTENTS

16.1 INTRODUCTION

A comparison of the total content of antioxidants and their activity with respect to oxygen and its radicals in juice and extracts of herbs, extracts of a tea and also in human blood plasma was carried out in the present work by use of two operative electrochemical methods: ammetric and voltammetric. Efficiency of methods has allowed studying dynamics of antioxidants content and activity change in same objects during time. Good correlation between the total phenol antioxidant content in the studied samples and values of the kinetic criterion defining activity with respect to oxygen and its radicals is observed.

For the last quarter of the century there was a considerable quantity of works devoted to research of the activity of antioxidants (AO) in herbs, foodstuff, drinks, biological liquids and other objects. It is known that the increase in activity of free-radical oxidation processes in a human organism leads to destruction of structure and properties of lipid membranes. There is a direct communication between the superfluous content of free radicals in an organism and occurrence of dangerous diseases [1,2]. Antioxidants are class of biologically active substances which remove excessive free radicals, decreasing the lipid oxidation. Therefore a detailed research of the total antioxidant activity of various biological objects represents doubtless interest.

At present, there are a large number of various methods for determining the total antioxidants content and also their activity with respect to free radicals in foodstuffs, biologically active additives, herbs and preparations, biological liquids and other objects [3]. However, it is impossible to compare the results obtained by different methods, since they are based on different principles of measurements, different modeling systems, and have different dimensions of the antioxidant activity index. In such cases, it is unreasonable to compare numerical values, but it is possible to establish a correlation between results obtained by different methods.

Ones from the simplest methods for study of antioxidant activity of various biological objects are electrochemical methods, in particular, ammetry and voltammetry. A comparative analysis of the antioxidants content and their activity in juice and extracts of herbs, extracts of a tea and vegetative additives, and also in plasma of human blood is carried out in present work. Operability of methods has allowed studying also dynamics of change of the antioxidants content and their activity in same objects during time.

16.2 EXPERIMENTAL DETAIL

16.2.1 AMMETRIC METHOD FOR DETERMINING THE TOTAL CONTENT OF ANTIOXIDANTS

The essence of the given method consists in measurement of the electric current arising at oxidation of investigated substance on a surface of a working electrode at certain potential. An oxidation of only OH - groups of natural phenol type antioxidants (R-OH) there is at this potential. The electrochemical oxidation proceeding under scheme $R\text{-}OH \rightarrow R\text{-}O^{\cdot} + e^{-} + H^{+}$ can be used under the assumption of authors [4], as model for measurement of free radical absorption activity which is carried out according to equation R-OH → R-O˙ + H˙. Both reactions include the rupture of the same bond O-H. In this case, the ability of same phenol type antioxidants to capture free radicals can be measured by value of the oxidizability of these compounds on a working electrode of the ammetric detector [4]. Ammetric device "TsvetJauza-01-AA" in which this method is used, represents an electrochemical cell with a glassy-carbon anode and a stainless steel cathode to which a potential 1,3 V is applied [5]. The analyte is introduced into eluent by a special valve. As the analyte pass through the cell the electrochemical AO oxidation current is recorded and displayed on the computer monitor. The integral signal is compared to the signal received in same conditions for the comparison sample with known concentration. Quercetin and gallic acid were used in work as the comparison sample. The root-mean-square deviation for several identical instrument readings makes no more than 5% [5].The error in determination of the AO content including the error by reproducibility of results was within 10%. The method involves no model chemical reaction, and measurement time makes 10–15 min.

16.2.2 VOLTAMMETRIC METHOD FOR DETERMINING THE TOTAL ACTIVITY OF ANTIOXIDANTS WITH RESPECT TO OXYGEN AND ITS RADICALS

The voltammetric method uses the process of oxygen electroreduction (ER O_2) as modeling reaction. This process is similar to oxygen reduction in tissues and plant extracts. It proceeds at the working mercury film electrode (MFE) in several stages with formation of the reactive oxygen species (ROS), such as $O_2^{\cdot-}$ and $HO_2^{\cdot}$ [6]:

$$O_2 + e^- \rightleftharpoons O_2^- \tag{1}$$

$$O_2^- + H^+ \rightleftharpoons HO_2^{\cdot} \tag{2}$$

$$HO_2^{\cdot} + H^+ + e^- \rightleftharpoons H_2O_2 \tag{3}$$

$$H_2O_2 + 2H^+ + 2e^- \rightleftharpoons 2H_2O \tag{4}$$

For determination of total antioxidant activity it is used the ER O_2. It should be noted that antioxidants of various natures were divided into 4 groups according to their mechanisms of interaction with oxygen and its radicals (Table 1) [7].

TABLE 1 Groups of biological active substances (BAS) divided according mechanisms of interaction with oxygen and its radicals.

N Group	1 group	2 group	3 group	4 group
Substance names	Catalyze, phtalocyanines of metals, humic acids.	Phenol nature substances, vitamins A, E, C, B, flavonoids.	N, S-containing substances, amines, amino acids.	Superoxide dismutase(SOD), porphyry metals, cytochrome C
Influence on ER O_2 process	Increase of ER O_2 current, potential shift in negative area	Decrease of ER O_2 current, potential shift in positive area	Decrease of ER O_2 current, potential shift in negative area	Increase in ER O_2 current, potential shift in positive area
The prospective electrode mechanism.	EC* mechanism with the following reaction of hydrogen peroxide disproportion and partial regeneration of molecular oxygen.	EC mechanism with the following chemical reaction of interaction of AO with active oxygen radicals	CEC mechanism with chemical reactions of interaction of AO with oxygen and its active radicals	EC* mechanism with catalytic oxygen reduction via formation of intermediate complex.

*The note: E – electrode stage of process, C – chemical reaction.

The first group of substances increased ER O_2 current according mechanism (5) – (7):

$$O_2 + e^- + H^+ \rightleftharpoons HO_2^•$$

$$HO_2^• + e^- + H^+ \rightleftharpoons H_2O_2$$

$$2H_2O_2 \xrightarrow{\text{catalyst}} 2H_2O + O_2$$

For the second group of the antioxidants we suppose following mechanism of interaction of antioxidants with the reactive oxygen species (ROS) (8):

$$O_2 + e^- \underset{}{\overset{k_0}{\rightleftharpoons}} O_2^- + R\text{-}OH + H^+ \underset{}{\overset{k_1^*}{\rightleftharpoons}} H_2O_2 + R\text{-}O^• \qquad (8)$$

The third group of the BAS decreased ER O_2 current via the following mechanism (9) - (11):

$$O_2 \longrightarrow O_2^S \xrightarrow[k_1]{+RSH} HO_2 + \bar{e} \xleftrightarrow{k_{R1}} HO_2^{\bullet-} + RS^{\bullet} \qquad (9)$$

$$HO_2^{\bullet-} + RSH \rightleftarrows H_2O_2 + RS^{\bullet} \qquad (10)$$

$$RS^{\bullet} + RS^{\bullet} \rightleftarrows RS - SR \qquad (11)$$

For the fourth group of substance we could suggest mechanism with catalytic oxygen reduction via formation of intermediate complex similar by SOD (12).

$$(12)$$

For voltammetric study of the total antioxidant activity of the samples automated voltammetric analyzer "Analyzer of TAA" (Ltd. "Polyant" Tomsk, Russia) was used. As supporting electrolyte the 10 ml of phosphate buffer (pH = 6.76) with known initial concentration of molecular oxygen was used [7]. The electrochemical cell (V = 20 ml) was connected to the analyzer and consisted of a working mercury film electrode (MFE), a silver- silver chloride reference electrode with KCl saturated (Ag|AgCl|KCl$_{sat}$) and a silver- silver chloride auxiliary electrode. The investigated samples (10 – 500 µl) were added in cell.

Criterion K is used as an antioxidant activity criterion of the investigated substances:

For the second and third groups of antioxidants: $K = \dfrac{C_0}{t}(1 - \dfrac{I}{I_0})$, µmol/l×min, (13)

For the firth and four groups of antioxidants: $K = \dfrac{C_0}{t}(1 - \dfrac{I_0}{I})$, µmol/l×min, (14)

where I, I_0 —limited values of the ER O_2 current, accordingly, at presence and at absence of AO in the supporting electrolyte, C_0 — initial concentration of oxygen

(μmol/l), i.e. solubility of oxygen in supporting electrolyte under normal conditions, t — time of interaction of AOs with oxygen and its radicals, min.

This method of research has good sensitivity. It is simple and cheap. However, as in any electrochemical method of this type, the scatter of instrument readings at given measurement conditions is rather high (up to a factor of 1.5–2.0). Therefore each sample was tested 3–5 times, and results were averaged. The maximum standard deviation of kinetic criterion K for all investigated samples was within 30%.

16.3   RESULTS AND DISCUSSION

16.3.1   JUICES OF MEDICINAL PLANTS

In the present work juices pressed out from different parts of various medicinal plants, such as basket plant or golden tendril (Callisia fragrans), Moses-in-the-cradle (Rhoeo spatacea), Dichorisandra fragrants (Dichorisandra fragrans), Blossfelda kalanchoe (Kalanchoe blossfeldiana), air plant (Kalanchoe pinnatum) and devil's backbone (Kalanchoe daigremontiana) were investigated [8]. To preserve the properties of the juices, they were stored in refrigerator at $-$ 12°C and unfrozen to room temperature immediately before the experiment. In both methods before being poured into the measuring cell, the test juice was diluted 100 - fold. A quercetin was used as the sample of comparison. Results of measurement of the total content of antioxidants and their activity with respect to oxygen and its radicals in juice of investigated plants, received by above described methods, are presented in Table 2. For samples 1, 4, 7, and 12 where the small content of phenol type antioxidants is observed, voltamperograms (VA-grams) look like, characteristic for substances of the third group of Table 1 (Figure 1). Other samples show the classical phenolic mechanism as substances of the second group (Figure 2), and have high values of the AO content and kinetic criterion K, especially the sample 10.

TABLE 2   The total content of AOs in the juice samples and their kinetic criterion K

Sample no.	Plant name	Content of antioxidantsC, mg/l	K, μmol/(l min)
1	Juice from *Callisia fragrans* (golden tendril) leaves	63,6	0,72
2	Juice from *Callisia fragrans* (golden tendril) lateral sprouts (4–5 mm in diameter)	279,6	1,81
3	Juice from *Rhoeo spathacea* (Moses_in_the_cradle) leaves	461,2	2,45
4	Juice from *Callisia fragrans* leaves	73,2	0,83

TABLE 2 *(Continued)*

5	Juice from *Callisia fragrans* stalks (1–2 mm in diameter)	119,3	1,23
6	Juice from kalanchoe (*Kalanchoe blossfeldiana*) bulblets	251,3	1,78
7	Juice from *Dichorisandra fragrans* stalks (6 mm in diameter)	38,41	0,52
8	Juice from *Dichorisandra fragrans* leaves	179,1	0,77
9	Juice from *Kalanchoe pinnata* (air plant) leaves	201,5	1,99
10	Juice from *Kalanchoe daigremontiana* (devil's backbone) leaves with bulblets	742,4	4,12
11	Juice from *Kalanchoe daigremontiana* stalks (5 mm in diameter)	142,5	1,47
12	Juice from *Callisia fragrans* herb + 20% ethanol	70,2	0,88

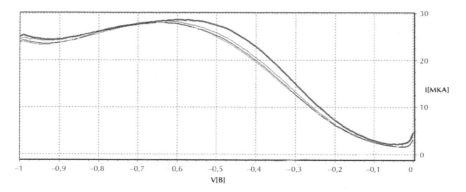

FIGURE 1 Typical VA-grams of the ER O_2 current for compounds from the third group in Table 1. The upper curve is the background current in AOs absence; left curves were recorded in AOs presence.

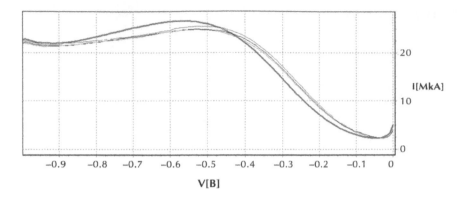

FIGURE 2 Typical VA-grams for AOs of the second substances group in Table 1. The upper curve is the background current.

Correlation dependence between the kinetic criterion K and the total AOs content is presented on Figure 3. The results of measurements spent for juices of medicinal plants, show high correlation (r = 0,96) between these methods. The explanation of received results is resulted in [8].

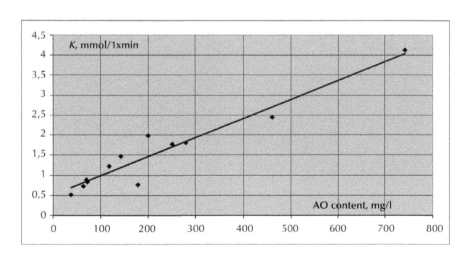

FIGURE 3 Correlation dependence between the kinetic criterion K and the total AOs content (r = 0,96).

16.3.2   WATER EXTRACT OF MINT

The purpose of the present work is to measure the total antioxidant activity of a water mint extract by voltammetric method and to study the mechanisms of influence of mint components on the process of ER O_2. Concurrently measurements of the total phenol AO content were carried out by an ammetric method [9]. The object of the research was water extract of the mint peppery (Mentha piperita). Dry herb was grinded in a mortar till particles of the size 1–2 mm. Further this herb (0.5 g) was immersed into 50 ml of distilled water with T = 95°C and held during 10 min without thermostating. Then the extract was carefully filtered through a paper filter and if necessary diluted before measurements.

VA-grams of the ER O_2 current have been received in various times after mint extraction. It has appeared that the fresh extract (time after extraction t = 5 min) "works" on the mechanism of classical AO, reducing a current maximum and shifting its potential in positive area (Figure 2). However, approximately through t = 60 min, character of interaction of mint components with oxygen and its radicals changes, following the mechanism, characteristic for substances of the 4th group in Table 1 (Figure 4). Transition from one mechanism to another occurs approximately during t = 30 min after extraction and the kinetic criterion K becomes thus close to zero. At the further storage of an extract in vitro character of VA-grams essentially does not change, and the kinetic criterion caused by other mechanism, grows to values 2.0 µmol/l·min during 3 hr after extraction.

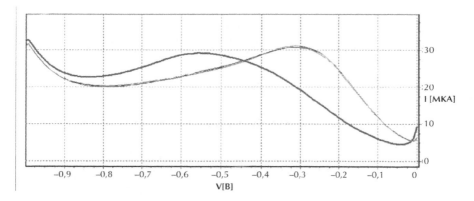

FIGURE 4 VA-grams of the ER O_2 current in the absence (left curve) and in the presence (right curves) of mint extract at t = 60 min after mint extraction.

In parallel for the same mint extract the registration of the total phenol AO content C in mg of gallic acid (GA) per 1 g of dry mint has been made during extract storage time t by ammetric method. Dependence C on t testifies to notable falling of C after extraction of a mint extract (approximately 20% for 2 hr of storage). It is possibly explained by destruction of unstable phenol substances contained in an extract. Therefore, apparently, VA-grams character and K values during the first moment after extraction could be established as influence of classical phenol AOs on the ER O_2 process. The general character of ER O_2 is defined already by the mint substances entering into 4th group of the Table 1. Probably, various metal complexes as mint components are dominated causing increase of ER O2 current and catalytic mechanism of ER O_2. More detailed statement of experimental materials given in [9].

16.3.3 EXTRACTS OF TEA, VEGETATIVE ADDITIVES, AND THEIR MIXES

In the given section results of measurements of the total AO content and activity by two methods in water extracts of some kinds of tea and vegetative additives are presented [10]. These parameters were measured also in extracts of their binary mixes to study possible interference of mixes components into each other. Objects of research were water extracts of three kinds of tea (Chinese green tea «Eyelashes of the Beauty», gray tea with bergamot «Earl gray tea» and black Ceylon tea «Real»), mint peppery (*Mentha piperita*) and the dry lemon crusts. 10 extracts of binary mixes of the listed samples with a different weight parity of components have been investigated also. Preparation of samples and extraction spent in the same conditions, as for mint (the previous section). Gallic acid was used as the comparison sample in an ammetric method. Efficiency of this method has allowed tracking the dynamics of AO content change in investigated samples directly after extraction. On Figure 5 dynamics of total phenol AO content change for 5 samples is shown. The most considerable content C decrease is observed in tea extracts (20–25%) that, possibly, as well as for mint, is explained by destruction of the unstable phenol substances in extracts (katehins, teaflavins, tearubigins, and so on.). For extract of lemon crusts the total AO content is much less and practically does not change during first minutes after extraction.

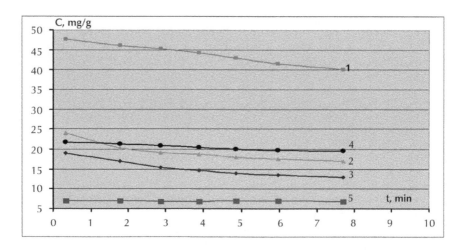

FIGURE 5 Dynamics of the total AO content *C* change (in units of gallic acid) during extract storage time t: 1 – green Chinese tea, 2 – gray tea with bergamot, 3 – black Ceylon tea, 4 – mint, 5 – lemon crusts (for lemon crusts C was increased in 5 times).

Measurement results of the total AO content for extracts of tea and additives (Figure 6 (a)) and extracts of their mixes in a different parity (Figure 6 (b)) are presented. The AO content in mixes (c) is calculated under the additive AO content contribution of mix components, taken from Figure 6 (a) according to their parity.

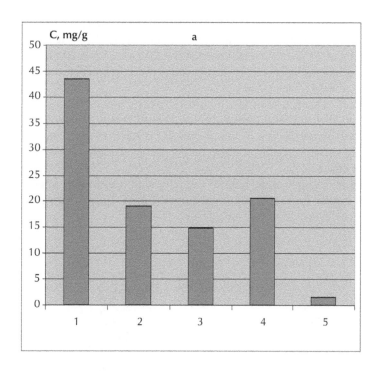

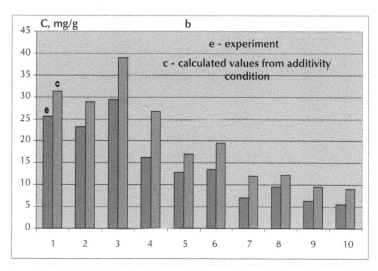

FIGURE 6   The total AO content C (in units of gallic acid):

a - in extracts of: 1 – green Chinese tea, 2 – gray tea with bergamot, 3 – black Ceylon tea, 4 – mint, 5 – lemon crusts.

b——in extracts of tea and additives mixes: 1 – tea 1 + tea 2 (1:1), 2 – tea 1 + tea 3 (1:1), 3 – tea 1 + mint (4:1), 4 – tea 1 + lemon crusts (3:2), 5 - tea 2 + tea 3 (1:1), 6 - tea 2 + mint (4:1), 7 - tea 2 + lemon crusts (3:2), 8 - tea 3 + mint (4:1), 9 - – tea 3 + lemon crusts (3:2), 10 - mint (4:1) + lemon crusts (2:3). In brackets the parity between components of mixes is specified.

As to mixes of tea and additives (Figure 6 (b)) the measured values of the AO content in extracts of investigated mixes (e) have considerable reduction in comparison with the additive contribution of the phenol AO content of mix components, i.e. is observed their strong antagonism. Especially it is considerable for extracts of tea with lemon crusts mixes.

The measured values of the total AO activity with respect to oxygen and its radicals *K* are presented for extracts of tea, additives (Figure 7(a)) and their mixes (Figure 7 (b)).

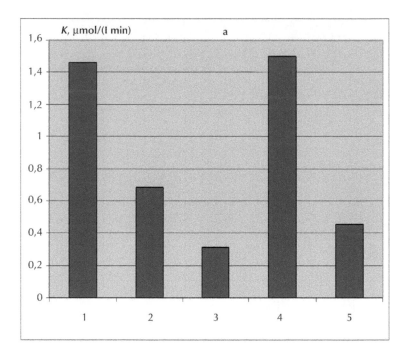

FIGURE 7   *(Continued)*

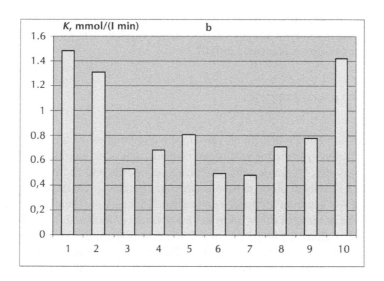

FIGURE 7 The total AO activity with respect to oxygen and its radicals K:

a - in extracts of tea and additives: 1 – green Chinese tea, 2 – gray tea with bergamot, 3 – black Ceylon tea, 4 – mint, 5 – lemon crusts.

b - in extracts of tea and additives mixes: 1 – tea 1 + tea 2 (1:1), 2 – tea 1 + tea 3 (1:1), 3 – tea 1 + mint (4:1), 4 – tea 1 + lemon crusts (3:2), 5 - tea 2 + tea 3 (1:1), 6 - tea 2 + mint (4:1), 7 - tea 2 + lemon crusts (3:2), 8 - tea 3 + mint (4:1), 9 - – tea 3 + lemon crusts (3:2), 10 - mint (4:1) + lemon crusts (2:3). In brackets the parity between components of mixes is specified.

For all samples of extracts, except an extract of the lemon crusts, dominating character of interaction of extract components with oxygen and its radicals has not phenolic, but the catalytic nature. This interaction proceeds on the mechanism (12), characteristic for substances of 4th group in Table 1. Lemon crusts extract "works" on the mechanism of classical AO (8). Unlike values of the phenolic AO content measured by a ammetric method, the kinetic criterion has appeared maximum not only for extract of green tea, but also for mint extract, minimum – for extract of black tea. In spite of the fact that the phenol AO content in extracts of tea and mint has appeared much more, than in lemon crusts, final total activity of tea and mint extracts is defined not by phenol type substances, but, apparently, various metal complexes, present in them, and catalyze of proceeding chemical processes. Considerable shift of the ER O_2 current maximum potential in positive area attests in favor of enough high content of phenol substances in teas and mint extracts.

For the activity of mixes extracts deviations of the measured values K from the values calculated on additivity (here are not presented) are big enough and are observed both towards reduction, and towards increase. For this method, apparently, the additively principle does not "work" since activity of mix components has the different nature and the mechanism of interaction with oxygen and its radicals. Activity of mixes is defined not only chemical interactions between substances, but diffusion factors of these substances to an electrode and so on. Activity of tea mixes extracts changes weaker in comparison with activity of separate tea extracts. The possible explanation of the received results is presented in [10].

PLASMA OF HUMAN BLOOD

A human blood plasma is a difficult substance for researches. Its antioxidant activity is defined, mainly, by presence in it of amino acids, uric acid, vitamins E, C, glucose, hormones, enzymes, inorganic salts, and also intermediate and end metabolism products. The total activity of blood plasma is integrated parameter characterizing potential possibility of AO action of all plasma components, considering their interactions with each other. The purpose of this work was measurement of total activity with respect to oxygen and its radicals of blood plasma of 30 persons simultaneously with measurement of the total AO content in plasma. Blood plasma has been received by centrifuging at 1500 r/min of blood of 30 patients from usual polyclinic with different age, sex and pathology. It is necessary to notice that for the majority of plasma samples VA-grams were stable for 3–4 identical measurements and resulted on 2nd group of Table1 (Figure 2). ER O_2 current potential shift was small (00.03 V) that testifies to presence in plasma of small phenol substances content. For some blood plasma samples (4 samples from 30) VA-gram character was corresponded to substances of 4th group in Table1. For studying of correlation of the results received by two methods, values of K, measured during 30 min after plasma defrosting have been selected for the samples having VA-grams of 2nd group. On Figure 9 these values are presented together with corresponding measured values of the total phenol AO content C, spent also during 30 min after plasma defrosting. Correlation of received results with factor r = 0.81 is observed. It means that in blood plasma of many patients there are phenol substances which define dominant processes of plasma components interaction with oxygen and its radicals. Results of this work were published in [11].

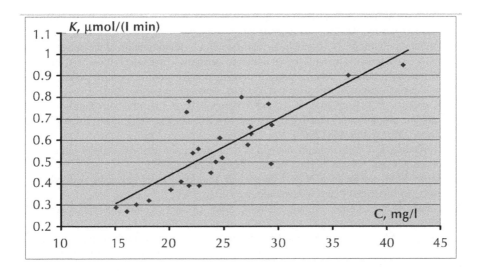

FIGURE 8 Correlation dependence between the kinetic criterion K and the total AO content C (in units of gallic acid) for samples of blood plasma, having VA-grams of 2nd type. r=0.81.

16.4 CONCLUSION

Use of two operative electrochemical methods realized in devices "TsvetJauza-01-AA" and c "Analyzer of TAA" allows quickly and cheaply to define the total content of antioxidants and their activity with respect to oxygen and its radicals in various biological objects. Results of the present work show good correlation of these methods, and the specified devices can be applied widely in various areas.

KEYWORDS

- **Ammetric method**
- **Antioxidants**
- **Human blood plasma**
- **Integral signal**
- **Voltammetric method**

REFERENCES

1. *Study of Synthetic and Natural Antioxidant in vivo and in vitro.* E. B Burlakova (Ed.), Nauka, Moscow (1992).
2. Vladimirov, Yu. A. and Archakov, A. I. *Lipid peroxidation in biologicall membranes*, Nauka, Moscow (1972).

3.   Roginsky V. and Lissy, E. Review of methods of food antioxidant activity determination. *Food Chemistry*, **92**, 235–254 (2005).

4.   Peyrat_Maillard, M. N., Bonnely, S., and Berset, C. Determination of the antioxidant activity of phenolic compounds by coulometric detection. Talanta, 51, 709–714 (2000).

5.   Yashin, A. Ya. Inject-flowing system with ammetric detector for selective definition of antioxidants in foodstuff and drinks. *Russian chemical magazine*, **52**(2), 130–135 (2008).

6.   Korotkova, E. I., Karbainov, Y. A., and Avramchik, O. A. Investigation of antioxidant and catalytic properties of some biological-active substances by voltammetry. Anal. Bioanal. Chem., 375(3) 465–468 (2003).

7.   Korotkova, E. I. Voltammetric method of determining the total AO activity in the objects of artificial and natural origin. Doctoral thesis, Tomsk (2009).

8.   Misin, V. M. and Sazhina, N. N. Content and Activity of Low_Molecular Antioxidants in Juices of Medicinal Plants. *Khimicheskaya Fizika*, **29**(9) 1–5 (2010).

9.   Natalia, S., Vyacheslav, M., and Elena, K. Study of mint extracts antioxidant activity by electrochemical methods. *Chemistry&Chemical Technology*, **5**(1) 13–18 (2011).

10. Misin, V. M., Sazhina N. N., and Korotkova, E. I. Measurement of tea mixes extracts antioxidant activity by electrochemical methods. *Khim. Rastit. Syr'ya*, **2**, 137–143 (2011).

11. Sazhina, N. N., Misin, V. M, and Korotkova E. I. The comparative analysis of the total content of antioxidants and their activity in the human blood plasma. *Theses of reports of 8th International conference "Bioantioxidant"*, Moscow, 301–303 (2010).

CHAPTER 17

OZONOLYSIS OF CHEMICAL AND BIOCHEMICAL COMPOUNDS

S. RAKOVSKY, M. ANACHKOV, and G. E. ZAIKOV

CONTENTS

17.1 INTRODUCTION

The ozone application is based on its powerful oxidative action. The ozonolysis of chemical and biochemical compounds, as a rule, takes place with high rates at low temperatures and activation energies [1]. This provides a basis for the development of novel and improved technologies, which find wide application in ecology [2], chemical, pharmaceutical and perfume industries, cosmetics [3,4], cellulose, paper and sugar industries, flotation, microelectronics, veterinary and human medicine, agriculture, foodstuff industry, and many others [5-8]. At present it is very difficult to imagine a number of high technologies without using ozone. Here should be mentioned the technologies for purification of waste gases, water and soils, manufacture of organic, polymer and inorganic materials, disinfecting and cleaning of drinking and process water, disinfecting of plant and animal products, sterilization of medical rooms and instruments, resolving cosmetics problems, sterilization and therapy in veterinary and human medicine, surface cleaning and functionalization in the manufacture of polymer and microelectronics articles, flotation, deodorization and decolorization of gases, liquids and solid substances, oxidation in the intermediate stages of various technologies, and so on.

17.2 ECOLOGY

17.2.1 WASTE GASES

More than 100 contaminants of the atmospheric air have been identified. Among them SO_2, CO_2, nitrogen oxides, various hydrocarbons, and dust constitute 85%.

The main sources of harmful substances emissions, that is dust, SO_2 and CO_2 in the air, are the processing plants for coke, briquettes, coals, the thermal power stations, air, water, and road transport [9].The exhaust gases contain also, CO, organic and inorganic compounds, and so on.[10]. Dust, sulfur dioxide, carbon dioxide, nitrogen oxides and organic compounds, and so on are the main pollutants released in the environment from metallurgy, the manufacture of fertilizers, and petrochemistry. The most typical air pollutants emitted by some chemical productions are listed in Table 1.

TABLE 1 Main air contaminants emitted by some chemical productions

No.	Production	Pollutants	No.	Production	Pollutants
1.	Nitric acid	NO, NO_x, NH_3	12.	Ammonium nitrate	CO NH_3, HNO_2, NH_4NO_3 - dust
2.	Sulfuric acid: a) Nitroso b) contact	NO, NO_x, SO_x, H_2SO_4, Fe_2O_3, dust	13	Superphosphate	H_2SO_4, HF, dust

TABLE 1   *(Continued)*

3.	Hydrochloric acid	HCl, Cl$_2$	14	Ammonia	NH$_3$, CO
4.	Ocsalic acid	NO, NO$_x$, C$_2$H$_2$O$_2$, dust	15	Calcium chloride	HCl, H$_2$SO$_4$, dust
5.	Sulfamidic acid	NH$_3$, H$_2$SO$_4$, NH(SO$_3$NH$_4$)$_2$,	16	Chlor	HCl, Cl$_2$, Hg
6.	Phosphor	P$_2$O$_3$, H$_3$PO$_4$, HF, Ca$_5$F$_4$(PO$_4$)$_2$, dust	17	Caprolactam	NO, NO$_2$, SO$_2$, H$_2$S, CO
7.	Phosphoric acid	P$_2$O$_3$, H$_3$PO$_4$, HF, Ca$_5$F$_4$(PO$_4$)$_2$, dust	18	PVC	Hg, HgCl$_2$, NH$_3$
8.	Acetic acid	CH$_3$CHO, CH$_3$OH	19	Artificial fibers	H$_2$S, CS$_2$
9.	Nitrogen fertilizers	NO$_2$, NO, NH$_3$, HF, H$_2$SO$_4$, HNO$_3$	20	Mineral pigments	Fe$_2$O$_3$, FeSO$_4$
10.	Carbamide	NH$_3$, CO, (NH$_2$)$_2$CO, dust	21	Electrolysis of NaCl	Cl$_2$, NaOH

The technologies with ozone participation are very promising for SO$_2$ utilization as sulfate, CO as carbonate, nitrogen oxides as nitrates.

Scrubbers used previously for removing only acids, for example HCl and HF, and so on, now can be used for SO$_2$ separation *via* the addition of an oxidizing agent into the water intended for gas treatment.

Ozone as compared with the conventional oxidizers such as hydroperoxide, chlorine, sodium hypochloride, and perchlorate, shows appreciably higher oxidizing efficiency [11].

The rate constant of ozone reaction with NO is about 1010 cm^3/(mol.sec), and with NO$_2$ it is ~107 cm^3/(mol.sec) whereby the oxidation of NO by ozone in the liquid phase is characterized by a higher absorption rate and rise in the concentration of the obtained HNO$_3$. Moreover, the oxidation of NO can be accomplished in the exhaust gases or in the course of absorption [12].

The CO oxidation by ozone is carried out in the presence of Fe, Ni, Co, and Mn oxides. In most cases ozonation is more appropriate than the conventional methods and sometimes it appears the only possible way for its preparation [13].

The purification of exhaust gases from burners working on liquid and gas fuel is accomplished by using oxidation catalysts of ABO3 perovskite-type oxides combined with ozonation. Ozone is injected prior the waste gases flow. A may be La, Pr, or other alkali earth element; B may be Co, Mn, or other transition metal. A may be also partially substituted by Sr, Ca, or any other alkali earth element. A catalyst deposited on ceramic support of honeycomb type may contain $SrCo_{0.3}Mn_{0.7}O_3$. It is oxidized 24% CO in the absence of ozone and 80% in ozone atmosphere [14].

The removal of nitrogen oxides (NO_x) from waste gases is carried out through ozone oxidation in charged vertical column. The lower part of the column is loaded by 5–15% $KMnO_4$, and ozone is blown through the lower and upper part of the column and the waste gases are fed to the center of the column charge. The inlet concentration of 1000 ppm NO_x in the waste gas is reduced to 50 ppm in the outlet gas flow. Another method for cleaning of exhaust gases from NO_x involves the application of plasma generator and ozonator [15] which practically leads to the complete removal of nitrogen oxides.

Sulfur containing compounds are removed from gas mixtures by silicon oil scrubbing and ozone oxidation of absorbed pollutants. A model system providing that the gas mixture contains methandiol and deimethylsulphide is fed at a rate of 9–120 m^3/hr into silicon oil charged scrubber followed by ozone oxidation of the absorbate. The oxidates are extracted with water and the silicon oil is regenerated. Thus the purified gas mixtures practically do not contain any sulfur [16].

A special apparatus is designed for decomposing acyclic halogenated hydrocarbons in gases. It includes a chamber for mixing of the waste gases with ozone coupled with UV-radiation, ozonator, and inlet and outlet units. This method is very appropriate for application in semiconducting industry whereby acyclic and halogenated hydrocarbons are used as cleaning agents [17].

The mechanisms of ozonolysis of volatile organic compounds such as alkenes and dienes are discussed and the products output is determined by matrix isolation FTIR spectroscopy [18].

The synthesis of a material from zeolite via pulverization, granulation and drying at 500–700°C, cooling to 100–200°C, electromagnetic radiation or ozone treatment is described in [19]. The material thus obtained is suitable for air deodorization, drying, and sterilization.

The purification of gases containing condensable organic pollutants can be carried out by gas treatment with finely dispersed carbon, TiO_2, Al_2O_3, Fe_2O_3, SiO_2, and H_2O_2 and ozone as oxidizers [20].

A wet scrubbing process for removing total reduced sulfur compounds such as H_2S and mercaptans, as well as the accompanying paramagnetic particles from industrial waste gases is proposed [21]. For this purpose a water-absorbing clay containing MnO_2 is used. The collected clay is regenerated by ozone oxidation.

The removal of mercury from waste gases is carried out by catalytic ozonation [22]. The used gases mixed with ozone are blown over a zeolite supported Ni/NiO catalyst. This procedure results in 87% conversion of mercury into mercuric oxide which is isolated by filtration. Pt and CuO_x/HgO system is proposed as another catalyst for this process [22].

17.2.2   WASTE WATER

In contrast to the cleaning of natural water by ozonation which is experimentally confirmed as the most appropriate, the purification of waste water by ozone is still an area of intensive future research. This could be explained by the great diversity of pollutants in the used water and the necessity of specific approach for each definite case.

The recycling of water from cyanide waste water (copper cyanide contaminant) is accomplished by ozone oxidation combined with UV radiation and ion exchange method [23]. Further, the oxidate is passed through two consecutive columns charged with cationite and anionite resins. The ozonolysis priority over the conventional chlorinating method is demonstrated by the fact that ozonation fails to yield chlorides whose removal requires additional treatment.

The ozonolysis of CN-ions leading to CNO^--ion formation combined with UV-radiation proceeds at 3 fold higher rate as compared with that without radiation employment. Upon exchange in NaOCN, the Na^+ is substituted by H^+ followed by its decomposition to CO_2 and NH^{4+} via hydrolysis in acidmedium. The ammonium ion is absorbed by the cation-exchange resin. The solution electroconductivity on the anion-exchange resin becomes lower that 10 muS/cm and cyanides, cyanates or copper ions have not been monitored. On the basis of these experimental results, a recycling method and apparatus for detoxification of solutions containing cyanides is proposed. The recycled water from the cyanide waste water may be re-used as deionization water in gold plating. The process ion-exchange resins are regenerated after conventional methods. Actually, this method is not accompanied by the formation of any solid pollutants [23].

The waste water from the electrostatic and galvanic coatings containing CN^--ions and heavy metals is treated consecutively by ozone and CO_2. The cyanides and carbonates being accumulated in the precipitate are already biodegradable [24].

Waste water containing Cr(III) is oxidized by ozone combined with UV-radiation [25]. The radiation reduces the oxidation time about three-fold. The oxidate is consecutively passed through columns charged with cationite and anionite resulting in the obtaining of waste water with electroconductivity below 20 mS/cm. This deionized water can be successfully used for washing of coated articles. The Cr(IV) concentrated solution from the anionite regeneration can be further treated by appropriate ion-exchange method yielding a high purity Cr(IV) solution [25].

Purification of waste water containing $ClCH_2COOH$ and phenol is carried out by ozone treatment and UV-radiation in the presence of immobilized photocatalyst. Thus the rate of decomposition is higher than the rate obtained only in the presence of ozone and UV-radiation [26].

Used water containing small amounts of propyleneglycol nitrate or nitrotoluenes is a subject to combined oxidation with ozone and H_2O_2, under pressure and heating to the supercritical point. Thus a nominal conversion of the contaminants higher than 96.7% is achived [27].

In laboratory experiments for reduction in residual COD in biologically treated paper mill effluents it is subject to ozonation or combined ozonation and UV radiation at various temperatures and pressures [28]. At a ratio of the absorbed ozone to COD less that 2.5:1 g/g, the elimination level with respect to COD and DOC is up to 82% and 64%, respectively. The ozone consumption is essentially higher in the case of the UV combined ozonation at pH >9 and elevated temperature to 40°C.

The decolorization and destruction of waste water containing surface active substances is performed by ozone treatment at ozone concentration of about 80 mg/L. It has been found that for waste water which has undergone partial biological treatment by fluent filters or anaerobe stages up to COD concentration >500 mg/L and ratio 30D5/COD >0.2 ozonolysis with ozone concentrations up to 1.8 g/L does not change CBR, while after the complete biological treatment at COD< 500 mg/L ozonation results in rise of the ratio 30D5/COD from <0.05 to 0.37. This evidences the increase of CBR [28].

The purification of waste water from paper manufacture characterized by enhanced biodegradability, as well as the removal of COD and halogenated compounds is carried out by a method based on ozonation and biofiltration [29].

Waste water from the paper pulp production is clean up by treatment with ozone and activated acid tar. The high values of pH favor the lignin decomposition and carboxylic acids formation since in this case the $^{\square}$OH - radicals and not ozone are the oxidizing agent. The application of activated acid tar has a positive effect on the dynamics of the microbiological growth, substrate consumption, and CBR of organic acids. For example, maleic and oxalic acids are decomposed completely at ozonation of waste water in the presence of activated acid tar. For immobilization of biological culture the use of polyurethane foams appear to be very suitable [30].

The used water in collectors of a plant for paper manufacture is ozonized in a foamy barbotating contactor. The results indicate that ozone is very efficient in the oxidation of coloring and halogenated compounds. Its activity is proportional to the amount of absorbed ozone regardless of the variations in the gas flow rate, the inlet ozone concentration and contactor design. The amount of consumed ozone depends on the operating conditions and waste water characteristics. The absorption rate rises in the presence of stages including ozone decomposition, particularly at higher rates [31].

In biological granular activated carbon (GAC) columns the effect of pretreated ozonation on the biodegradability of atrazine is investigated. The metabolism of isopropyl-^{14}C-atrazine gives higher amount of $^{14}CO_2$ than the ring UL-[14]C - atrazine which shows higher rate of dealkylation that the process of ring cleavage. The pretreatment with ozone increases the mineralization of the ring UL-[14]C - atrazine and consequently raises the GAC-columns capacity. 62% of the inlet atrazine is transformed into $^{14}CO_2$ in columns charged by ozonized atrazine and water. However, in columns with nonozonized atrazine and water only 50% of the inlet atrazine is converted into $^{14}CO_2$ and in columns supplied only with nonozonized atrazine only 38% of it are converted to $^{14}CO_2$ [32].

The waste water from petrochemical industry contains substantial amounts of phenol. The phenols mixtures appear to be more toxic that phenol itself and possess a synergetic effect. Their decomposition to nontoxic products is important and modern problem. The oxidation of phenols such as p-cresol, pyrocatechin, rezorzine, and hydroquinone by O_3 in aqueous medium appears to be a promising method for their degradation. At concentrations of 100 mg/L and pH = 11.5 their complete oxidation is performed for 30 min.

The phenol content in used water is reduced from 145–706 mg/L to 2.5 mg/L at ozone consumption of 1.1–2.6 g/L. In real conditions the ozone consumption varies from 5 to 10 g/L per g phenol (Table 2)

TABLE 2   Results from the ozonation of waste water from coke processing $[O_3]$ = 5 mg/L, flow rate 6 L/min and treatment time 4 hr

Pollutant, mg/l	Before ozonation	After ozonation
Monophenols	710	0.8
Polyphenols	380	198
Cyanides	3	29
Thiocyanides	384	0
Thiosulfates	538	0
Sulfides	43	2
H_2O_2	0	12
Bases	114	27

The treatment of concentrated aqueous solutions of phenol (1.0 g/L) with ozone causes the appearance of a yellowish coloring after absorption of 1 mol O_3 per L mol phenol which gradually fades away. In this case the higher phenols undergo slight oxidation and H_2O_2 is identified in the water after the treatment.

The tests demonstrate that upon varying content of phenols in the used water the ozonation is one of the most promising method for their removal. The ozonation was found to be applicable for the decomposition of rhodanide in neutral and weak acid medium in the temperature range of 9–25°C. The ozone consumption in this case is 2 mg/mg. The ozone uptake for the oxidation of the cyanide ion (CN^-) is 1.8 mg/mg.

The complex Zn cyanides are oxidized similarly to the simple soluble cyanides. Regardless of the higher resistance of Cu cyanide complexes, the ozone treatment reduces the CN^- content substantially (traces). Copper carbonate is precipitated as a reaction product.

The ozonation can be successfully employed as a method for cleaning the waste water from the manufacture of ammonium nitrate containing CN^-, S^{-2}, CSN^- ions. The purification level is 83, 98 and 95% respectively at O_3 consumption of 0.7 g/L/g water for 10 min contact.

The ozonation of waste water from sulfate-cellulose production results in 60% purified used water which is returned in the process cycle.

The cleaning of waste water from the petrochemical plants by ozone is efficient under wide range of pH (5.8–8.70 and temperatures (5–50°C). Thus the level of petrochemicals in the waste water at pH 5.8, ozone consumption of 0.52 mg/mg and 10 min contact is reduced from 19–33 mg/L to 2 mg/L. In Table 3 is shown the efficiency of various treatment procedures for deodorization of petrochemicals contaminated water.

TABLE 3   Result from deodorization of petrochemicals contaminated water (grades/dilution)

Petroleum products	Before treatment	Coagulation and filtration	Charcoal and filtration	Chlorination	Ozonoation	Coagulation, filtration and ozonation
Petrol	5/80	5/60	5/50	5/80	2 ÷ 3/10	1/0
Kerosin	5/100	5/80	5/70	5/100	4 ÷ 5/25	1/0
Petroleum	5/100	5/60	5/60	5/100	5/20	2/5
Oil	5/10	1/0	1/0	5/10	1 ÷ 2/2	1 ÷ 0/0

Ozonation is very economical and highly efficient method for decomposing carcinogenic substances such as 3,4-benzopyrene, and so on, particularly after the biological purification of waste water.

The application of ozone is also advisable for mercury oxidation in chlorine producing, for decontamination of the used water from the poisonous tetraethyllead, for destroying pesticides traces, and so on.

The highest efficiency of cleaning of petrochemicals contaminated water is achieved at the following sequence: primary cleaning (mechanical), secondary (biological), and complete purification (ozonation).

Ozone is also successfully applied for the complete purification of waste water form dyes manufacture. Their color decreases about 10,000 times at exposure to 0.22 g/L ozone and the biological activity of the reaction products is reduced almost to zero. The semi-industrial experiments in chemical plant "Kostenetz" (Bulgaria) were carried out on the dyes manufactured in the plant (Table 4):

TABLE 4   Ozonated days in the plant Kostenetz

Order number	Day's name	Empiric formula	Molecular weight	Structural formula
1.	Direct Congo red	$C_{32}H_{26}O_6N_6S_2Na_2$	698	

TABLE 4 *(Continued)*

2.	Direct blue KM	$C_{34}H_{30}O_5N_5S_1Na_1$	643
3.	Acid blue ZK	$C_{26}H_{20}O_{10}N_3S_3Na_3$	699
4.	Acid chromium green	$C_{16}H_{11}O_{10}N_4S_2Na_2$	529
5.	Acid black ATT	$C_{22}H_{16}O_9N_6S_2Na_2$	618
6.	Acid sulfonic blue	$C_{32}H_{25}O_6N_5S_2Na_2$	685
7.	Acid chromium black C	$C_{22}H_{19}O_5N_6S_1Na_1 +$ $C_{16}H_{13}O_7N_4S_1Na_1$,	502 428
8.	Direct black Z	$C_{34}H_{26}O_7N_8S_2Na_2$	768
9.	Acid chromium yellow R	$C_{17}H_{14}O_3N_3S_1Na_1$	367

In the process of developing a technology for water purification from various organic contaminants, we have proposed a method and apparatus for quantitative determination of chemical compounds, separated by means of liquid chromatography. Special attention is paid to those compounds susceptible to ozone attack such as olefin hydrocarbons and their derivatives, phenols, amines, thiocarbamates, inorganic compounds, and so on. The characterization of the separated compounds is carried out by spectral, mass spectral, refractometer, electrochemical analyses, but the detectors in most of these equipments are rather expensive and complex units without sufficient identification capabilities, particularly for the compounds pointed above. In this sense the apparatus designed by us allows to overcome the disadvantages of the known methods and equipments. The scheme of the ozone detector applicable for identification and liquid chromatography separation of chemical compounds is presented in Figure 1.

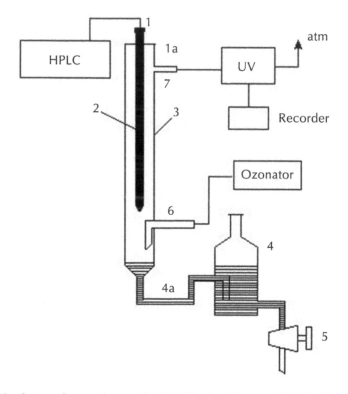

FIGURE 1   Set-up of ozone detector for identification of separated by liquid chromatography chemical compounds. 1—eluent access, 1a—entry for pouring out the elate as thin film on, 2—solid glass stick, 3—outer glass jacket, 4—vessel for eluent accumulation, 4a—hydraulic détente, 5—tap for pouring out the eluate, 6—ozone inlet, and 7—ozone outlet.

The ocean power stations operate on the basis of the temperature difference between the upper and lower water layers. Taking advantage of the one-step solution of ozone in depth, the COD is decreased which prevents the befouling in the piping systems found in the upper layers. The ozone-enriched water from the lower layers is supplied to a cyclic filter and is directly injected into upper seawater for control of the sea microorganisms' growth [33].

The removal of benzofurenes from ash, clays, soils, water, and oils by combustion, ozonolysis of supercritical water, cracking-processes of petroleum, as well as thermal, photochemical and biological decomposition, has been the subject of many investigations

[34]. The direct blowing of ozone through water pipers results in slimes removal from the inner pipers walls [35].

The principal mechanism of advanced oxidation processes (AOPs) function is the generation of highly reactive hydroxy-radicals. Consequently, combination of two or more AOPs expectedly enhances free radical generation, which leads to higher oxidation rates.[36] Among various wastewater treatment options ozone-based AOPs such as ozone/H_2O_2, ozone/UV, and ozone/H_2O_2/UV are key technologies for degrading and detoxifying of all major groups of pesticides, namely carbamates, chlorophenoxy compounds, organochlorines, organophosphates, aniline-based compounds, pyridines and pyrimidines, triazines, and substituted ureas [37, 38]. The AOPs and ozonation are also promising for efficient degradation of pharmaceuticals in wastewater [39].

Photocatalytic oxidation and ozonation appear to be among the most popular treatment technologies for the wastewater compared with other (AOPs) as shown by the large amount of information available in the literature [40]. Photocatalytic oxidation and ozonation is a promising way to perform the mineralization of the wastewater substances like organohalogens, nitrocompounds, organic pesticides and insecticides, surfactants, and coloring matters [41-43].

The ozone blowing at a rate of 2.87 mg/min through aqueous solutions of 15 pesticides with 10 µg/L concentration leads to their transformation and their concentrations are reduced by about 20% for 60 min. The identified intermediates from the rearrangement of EPN, phenitrothione, malathion, diazinon, izoxathion, and chloropiriphos are the corresponding oxones, while for phenthione they are the corresponding sulfoxide, sulfone and sulfooxide oxone analogues, and for disulfothione the intermediate is its sulfo analogue. 2-Amino-4-ethylamino-6-chloro-1,3,5-thriazine is identified as symazin and its 4-isopropylammonium derivative is atrazine intermediate. The isopropylthiolane results in obtaining of its 1-oxo-derivative. Upon replacement of ozone by air changes are actually not registered [44].

At the purification of process water in a soil decontamination plant containing pesticides the removal of organophosphorous compounds such as Thiometon and Disulphoton is of particular importance. In this connection four methods have been applied:

1) Ozonolysis at pH = 2.5,
2) Ozonolysis at pH = 8,
3) UV/H_2O_2 treatment, and
4) Oxidation by Fenton reagent [45].

The laboratory experiments were carried out with pure compounds soluted in buffered deoionized water and in process water which is extracted by solid-liquid extraction of the contaminated soil. The use of ozone in an acid medium turns to be ineffective since the Thimeton reaction practically stops after the formation of PO-derivatives. In all cases when the oxidation is carried out by HO - radicals, a sufficient removal of the pesticides and their metabolites is achieved. Oxadixyl, a cyclic nitrogen containing compound which is present in high concentrations in the process water is the most stable one, thus being the main soil contaminant. The investigated experimental conditions and the results obtained show that these methods are quite acceptable for universal application [45].

The water-soluble agrochemicals ASULAM and MECOPROP are decomposed for 30 min at ozone blowing (flow rate—5.1 mg/min and concentration of 10 ppm) through their aqueous solution (200 ml). The decrease in the flow rate of ozone retards the decomposition process. Similar treatments for 5 hr give in ACEPHATE and DI-CAMBA decomposition to 20%, the rate of decomposition being promoted with H_2O_2 addition [46].

The primary ozonation products of organophosphorous pesticides in water such as diazinone, phenthione (MPP) and ediphenphos (EDDP) [47], are identified by means of GC-MS analysis. The massspectra of the ozonation products of 17 organophos-phorous pesticides point oxones as the primary reaction products. This fact is also confirmed by the SO_4^{2-} generation resulting from the ozonation of their thiophospho-ryl bonds. Oxones are relatively stable towards ozone attack but they are further hy-drolyzed to trialkylphosphate and other hydrolysis products. However, with MPP the thiomethyl radicals are first oxidized to thiophosphoryl bonds giving MPP-sulfoxide, MPP-sulfone, MPP-sulfoxide-oxone are also obtained. Two main oxidation products have been identified at the oxidation of bis-dithio-type ethiones. Phosphate type EDDP is stable towards ozonolysis but its oxidation products are identified after hydrolysis [47].

Pesticides and their degradation products which are not mineralized to CO_2, NH_3, H_2O, and inorganic salts can damage and contaminate the water piping. The treat-ment by photolysis or ozonation may substantially increase the mineralization rate. The photodestructive products of S-triazoles, chloroacetanylide, and parakaute are declorinated and/or oxidized. The ozonation of these herbicides results in formation of products whose side alkyl chains are oxidized or removed, and the aromatic ring is oxidized or soluted but dechlorination does not occur. In some cases the medium may cause microbiological effect [48].

A deodorization method for treatment of ill- smelling air containing ammonia, hydrogen sulfide or amines from refrigerator chambers, lavatories, cattle-sheds, and so on. involves the ozone contact with air in the presence of porous catalyst of 120 m^2/g specific surface. The used gas passes through a layer of carbon black for adsorption of the residual ozone and the nondegradated smelling substances. The active component of the catalyst may be: transition metal or its oxides supported on porous honeycomb type supports [49].

The contaminated smelling gases in drain and fecal water are dried to 60–30% humidity, followed by ozone treatment on an oxidizing catalyst. For example, con-taminated gas containing 30 ppm methylsulfide is dried to 30% humidity and then is treated with ozone in the presence of honeycomb type catalyst containing 83:12.5 = $TiO_2:SiO_2:MnO_2$ with a volume rate of 50,000 h^{-1}. In this case the deodorization ef-ficiency is 99% while at 100% humidity it drops to 80% [50].

17.2.3   NATURAL (DRINKING) WATER

The basic sources of drinking water usually contain various organic admixtures in mg/L: carbon—30, nitrogen—0.8, fatty acids—30, highmolecular organic ac-ids—0.2, naphthenolic acids—1.5, phenols—1.2, luminescent substances: neutral resins——17%, oils, humus——55%, acid resins——19%, hydrocarbons, naphthenolic acids——4–7% [51].

The coloring of natural water varying from pale yellow to brown is due to the presence of humus substances, Their decolorization is carried out by adsorption on co-agulated $Al(OH)_3$ or $Fe(OH)_3$. The increasing demand for fresh water requires the use of more colored water which however is pre-chlorinated and then subject to adsorption up to their complete decolorization [52].

The ozone treatment removes the bad taste and odor of water, the resistant phyto-plankton being oxidized to 20%. The ozone consumption for decolorization of natural water varies in broad limits from 1 to 18 mg/L and depends on the humus composition [51]. The process of humus compounds oxidation by ozone is presented in Scheme 1.

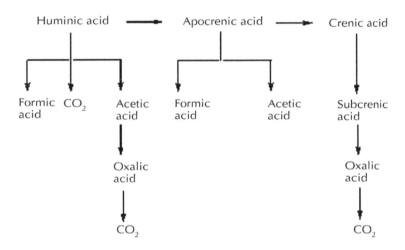

Scheme 1

The various biological processes lead to formation of substances causing the bad taste and smell of water - mercaptanes, sulfides, alcohols, carbonyl compounds, acids, terpenes, amines, and so on. The deodorization effect of ozone is preserved in wide range of pH values, temperatures, and ion composition. Due to its high oxidation po-tential, ozone can also degrade the resistant toxic substances. It is known that during the blooming in the water reservoirs are accumulated biologically active substances among which algotoxine is particularly harmful for the human health. Its complete degradation is achieved by using ozone concentration of 14 mg/L and 50 min contact time [53].

At ozone concentration of 8 mg/L in water, only 6% of *Scendesmus* algae remain alive after 10–15 min. However is to be noted that the ozone action on the various algae is quite different and specific. Thus *Asterionella* type algae are very ozone re-sistant [52].

The oxidative, disinfective, decolorizative, and deodorizative properties of ozone have found wide application for preparation of drinking water [54,55]. Many world corporations such as the French Taileygas, Degramont, Ozonia, the German Fisher, AEG, Siemens, the Swiss Braun & Bovery, the American General Electrics, General Motors and so on, are among the biggest producers of drinking water [56]. Twenty of the biggest world ozonator stations for drinking water treatment are listed (Table 5).

TABLE 5 Some of the world biggest ozonation stations for drinking water treatment

City	State	Water, m³/day	Ozone, kg/h
Moscow	Russia	1 200 000	200
Montreal	Canada	1 200 000	150
Ashford Common	GB	-	146
Coppermills	GB	-	144
Neuilly-sur-Marne	France	600 000	120
Choisy-le-Roi	France	900 000	120
Masan	Korea	-	112.5
Kiev	Ukrain	400 000	80
Helsinki	Finland	495 000	63
Manchester	GB	480 000	50
Orly	France	320 000	50
Singapore	Singapore	230 000	48
Minsk	Belaruss	200 000	47
Mery-sur-Oise	France	300 000	45
Lodz	Poland	190 000	40
		290 000	
Chiba	Japan	270 000	38
Nizhni Novgord	Russia	350 000	30
Bodenzee	Germany	480 000	30
Nant	France	380 000	30
Amsterdam	Holland	125 000	25
Brussels	Belgium	250 000	24
Quebec	Canada	220 000	19
Wrotzlaw	Poland	180 000	17

Ozonation stations for drinking water preparation work in Portugal, Greece, Bulgaria, USA, and so on all over the world.

The technology of drinking water ozonolysis leading to its deodorization, disinfecting, and decolorization includes gas units for dry air producing, ozonator, barbators, or other mixing units, apparatus for HO•-radicals generation, promoting the ozone formation and filter-catalyst for decomposition of the residual ozone [57]. The preferable materials for barbators are zeolites or ceramics on the base of aluminum and/or silicon with pores diameter of 5–100Å, and the catalysts should contain one or more of the following components—Pt, MnO_2, CuO, Ni_2O_3, Zn, YiO_2, and SiO_2 [58].

For disinfection of drinking water, the ozone-containing air is mixed with water in the water pipe, after which the bubbled water is supplied into a special designed chamber filled by porous ceramics charge for intensification of ozone dissolution into water. Some problems concerning the optimization of the UV stimulated oxidation of organic contaminants in soil water with ozone and hydroperoxide is discussed [59].

The disinfection of drinking water by ozonation can be also carried out at the outlet of various water supplying units before and after consummation, such as taps, fountains, water heaters, dental units, and so on [60]. In this connection are used small dimensional and low capacity personal ozonators coupled or installed additionally to these equipments.

The attempts for treatment of drinking water with ionization radiation appear to be ineffective mainly because of nitrates and H_2O_2 formation. These problems, however, do not occur when the radiation is performed in the presence of sufficient ozone amount [61]. The addition of ozone before and in the course of radiation converts the radiation into a purely oxidation process, a modification of the so called - Advanced Oxidation Process (AOP). The combined ozone-ionization radiation method is unique since two processes for HO-radicals generation are simultaneously induced. This method yields higher concentrations of HO-radicals as compared with other AOP. This in its turn results in precipitation of smaller amounts at equal ozone concentrations or reduces the ozone consumption at one and the same amount of precipitates. This makes the process particularly suitable for remediation of weakly contaminated soil water. The ozone concentrations at this process are higher than those used for drinking water disinfection. Thus, the drinking water subject to this treatment is disinfected more rapidly. The purification of waste water using this method is more efficient than the UV-radiation treatment only. This combination reduces COD without causing the rise of BOD. In this sense this method appears to be more attractive than the conventional two-stage processes of ozonation and biological treatment. The injection of gaseous ozone into the radiation chamber improves the water flow turbulence and substantially increases the efficiency of contaminants degradation.

The European organization "European regulation" has allowed the use of ozone for iron and manganese removal from mineral water, these initiating studies on secondary products formation, amount and composition, particularly during ozonation of bromine water. This water, usually contain BrO_3^- ions and various halogenated organic compounds (HOC). Some projects regulate a standard for admissible concentration limit (ACL) of HOC of 25 µg/l BrO_3^--ions, while after other it is from 10 to 0.5 µg/l. The present standard for LD of HOC is 100 µg/l [62]. Upon ozonation of bicarbonate mineral water - one sodium-rich water, and the other, calcium rich, it has been established that: 1) the process depends on the initial Fe and Mn concentrations and the Na and Ca concentrations; 2) HOC and bromate ions are secondary products and 3) the specific components characterizing a definite mineral water affect the ozonolysis. This requires definition of the conditions for Fe and Mn removal for each concrete type of mineral water. The removal of iron and manganese is efficient when the amount of ozone is three times greater than the stoichiometric one (0.43 mg O_3/mg Fe and 0.87 mg O_3/mg mn).

The disinfection of swimming pool water is carried out by controlled addition of compounds bromine and iodine donors (NaBr or NaI) and water oxidation by hypochloride, ozone or potassium peroxymonosulfate [63]. We have designed a compact module system (stationary and mobile) which includes module for air preparation, ozonator, absorption column and unit for the complete decomposition of the residual ozone. Its capacity is 12–15 g/hr and it is very suitable for sanitation of drinking water and used water from swimming pool (3.5 L/sec capacity).

17.2.4   SOILS

The decontamination of dredging wastes containing spilled petroleum is carried out by using a system for the direct injection of ozone (100–1000 ppm) containing air into pipes at the bottom; means for pumping out the spilled petroleum to a tank for its storage; units for its mixing with ozone for decomposition of organic matter in the tank; units for returning of the processed petroleum for the complete degradation of the residual organic matter by means of aerobic bacteria. The efficiency of organic matter decomposition is by about 20% higher than that by using the conventional methods including only treatment by ozone [64].

Another technology for treatment of contaminated soils [65] involves the ozonolysis of the water slurry of the contaminated soil layer, in the presence of H_2O_2 or in its absence and consecutive biological decomposition. The priority of the proposed technology over the conventional ones is that the removal of organic contaminants is by *on-site* treatment of the soils.

The *on site* recyclation of petroleum contaminated soils by ozonation requires the good knowledge of the geological setting and the recycling system. Usually the recycling stages include:

System for air evacuation from the soil,
Means for injection of ozone combined with air evacuation from the soil,
Development of means for ozone generation,
Screening of soils contaminants,
Use of systems for safety control,
Organization of the cleaning units - pumping and injecting drills,
Decision for air injection into the soil,
Methodology for measurement and regulation of ozone concentration, and
Hydraulic purification of the highly contaminated soil areas, and so on [66].

The treatment by ozone can be successfully applied for decomposing mineral oils, polycyclic aromatic compounds (PACs), phenols and some pesticides to biodegradable, and nontoxic compounds. It should be noted that although ozone decomposes a great number of microorganisms, the soil microflora can be easily remediated after this treatment [67]. The latter can be carried out on- and out of the contaminated area. A number of problems related to the ozonolysis of organic compounds, the PACs degradation products, the effect of ozone on the soil microflora, the laboratory and pilot results, and so on are discussed.

A method and apparatus for decontamination of wastes such as soil, porous granulated sludge, are suggested. The polluted materials are placed in a special vessel and ozone is bubbled through. The waste gases obtained from the ozone oxidation are further treated with ozone [68].

A method for ozone treatment both *in situ* and *on site* has appeared to be applicable for the degradation of organic substances adsorbed on the solid surface [69].

Wet-scrubber method (based on MnO_2) for removing total reduced S-containing compounds such as H_2S, mercaptanes, and non-magnetic particles from industrial waste gases using ozone, is suggested [21].

17.3   INDUSTRY

17.3.1   MANUFACTURE OF ORGANIC COMPOUNDS

We will focus our attention to some of the numerous ozone applications in organic polymer and inorganic productions, passivation, cleanup and preparation of surfaces for electronics, superconductors, and so on.

The reaction of ozone with olefins is the main reaction for ozonides preparation (1,2,4-trioxalanes) which can be converted into a mixture of carbonyl compounds, the exact composition of which depends on the olefin structure. This reaction is used for synthetic purposes as with small exceptions the C=C bond is cleaved quantitatively under very mild conditions. Usually the ozonolysis is carried out at low temperatures (as a rule at $-70 \div -30°C$) as demonstrated in [3] Chapters 2 and 3. Mostly, ozonolysis of olefins, except in solid phase, is carried out in various solvents such as paraffins, halogenated hydrocarbons, oxygen, sulfur and nitrogen containing hydrocarbons like ahcohols, glycols, aldehydes, ketones, acids, ethers, esters, amines, amides, nitriles, sulfoxides, mineral acids, bases and mixtures from them, and so on. In most cases the reaction products are not separated from the reaction mixture except for the purposes of ozonide obtaining, and the oxidates are subject to hydrolysis, reduction or oxidation [1,3].

The hydrolysis is applied for preparation of aldehydes [70] and ketones [71]. The reductive decomposition is carried out in the presence of Zn/AcOH, SO_3^{-2}, HSO_3^{-1}, $LiAlH_4$, $SnCl_2$, ArP_3, Ph_3P, SO_2, $(CH_3)_2S$, and H_2/catalyst. The reduction by metal hydrides leads to alcohol formation in the decomposition products mixture while the use of other reductors yields carbonyl compounds. Actually, there are no clear requirements for selection of the most appropriate reductor, but it turned out that dimethylsulfoxide is probably [72] the most preferable reductor in methanol solution. The ozonolysis in the presence of tetracyanoethylene yields directly carbonyl moieties omitting the reduction step [73]. A novel modification of the conventional methods of ozonolysis is exemplified by the ozonolysis of silica gel adsorbed alkenes (acetylenes) [74] and selective ozonolysis of polyalkenes, controlled with the help of dyes introduced in the reaction mixture [73]. Some examples of ozonolysis using various reductors for ozonides decomposition are given in Table 6 [75-80].

TABLE 6   Preparation of aldehydes, ketones and alcohols during alkenes ozonolyis with consecutive reduction

No.	Substrate	Conditions	Product	Yield, %
1.		EtOAc, –20–30°C 2. Pd/CaCO$_3$/H$_2$		61
2.		n-C$_6$H$_{14}$ Zn/HOAc		59
3.		1. MeOH/–40°C 2. Na$_2$SO$_3$/H$_2$O		65
4.	Me(CH$_2$)$_5$CH=CH$_2$	1. MeOH/–30 -60°C 2. Me$_2$S	Me(CH$_2$)$_5$CHO	75
5.	n-Bu- OOCCH=CHCOO- Bu-n	1. MeOH/-65 -70°C 2. (MeO)$_3$P	n-Bu-OOCCHO	78
6.		1. MeOH/-35° -75°C 2. NaI/MeOH/HOAc		75
7.		EtOAc/ (CN)$_2$C=C(CN)$_2$, -70°C	MeCO(CH$_2$)$_3$COMe + (CN)$_2$C—C(CN)$_2$	61
8.	Me(CH$_2$)$_5$CH=CH$_2$	1. n-Pentane/-38° - -42°C 2. LiAlH$_4$/Et$_2$O	Me(CH$_2$)$_5$CH$_2$OH	93
9.	H$_2$C=CH(CH$_2$)$_8$ COOH	1. MeOH/0°C 2. NaBH$_4$/NaOH/ EtOH/H$_2$O	HOCH$_2$(CH$_2$)$_8$COOH	91

Ozonolysis is a powerful chemical method for economical and ecologically pure preparation of various carbonyl compounds [81].

The ozonolysis of cyclic alkenes in protic solvents followed by reductive decomposition of the hydroperoxides formed is a classical method for dialdehydes synthesis [82]. This method is applied for the preparation of 3-ethoxycarbonylglutaric dialdehyde by ozonolysis of ethyl 3-cyclopentenecarboxylate. The dialdehyde is an intermediate in the synthesis of dolazetrone mesitilate which is an active medical substance in ANZEMET anti-emetik. Zn/AcOH, phosphines, amines, and sulfides such as 3, 3'-thiodipropionic acid and its salts are the most suitable reductors for this system. Their efficiency is comparable to that of methyl sulphide but without its shortcomings. The polymer immobilized 3,3'-thiodipropionic acid is also very active in this reduction reaction [82].

The efficient conversion of oximes into the corresponding carbonyl compounds can be also carried out through ozone oxidation [83].

The application of ozone is a new and convenient way for the preparation of cyanoacetylaldehyde (3-oxopropylonitrile) (1) and its stable dimethyl acetal (3,3-dimethoxypropylonitrile) (2) used as valuable intermediates for organic synthesis [84]. For this purpose the ozonolysis of (E)-1,4-dicyano-2-butene or 3-butylonitrile is carried out. Then the oxidates are treated by DMS yielding 1. Further, compound 1 can be used either directly in the next reactions or is transformed into 2. The 2 output amounts to from 67 to 71%. The acetal can be again hydrolyzed to 1 by treatment with ion-exchange resin Amberlyst-15 [84].

The easily available allylphenyl ethers are ozonized at –40°C, treated with DMS giving solutions of the corresponding phenoxyacetaldehydes which are purified by column chromatography. Their reaction with 1-methyl-1-phenylhydrazine yields the corresponding hydrazones [85].

Quinolinealdehyde derivatives can be synthesized by ozonolysis of the respective quinolineolefins at –60–72°C in methanol or ethanol solution with consecutive reduction with DMS at the same temperatures. Then the reaction mixture is heated to room temperature for 1 hr. The yield of the products amounts to 29%. [86].

6-Chloro-2-hexanone is prepared from 1-methylcyclopentane *via* three-step scheme: a) ozonolysis of 1-methylcyclopentane in carboyilic acid solution to 1-methylcyclopentanol, b) conversion of 1-methylcyclopentanol into 1-methylcyclopentyl hypochloride using NaClO, c) cyclization of 1-methylcyclopentyl hypochloride in 6-chloro-2-hexanone [87].

Ozonation is a stage in the synthesis of optically pure (R)-(+)-4-methyl-2 cyclohexen-1-one from (R)-(+)-pulegone and hydroxy ketone from (-)-*cis*-pulegol [88].

Polycarbonyl compounds and aldehyde-acids are the ozonolysis products of (+)-4a-[1-(triethylsiloxy)-ethenyl]-2-carene [89].

Carbonyl oxides obtained from vinyl ethers ozonolysis undergo rapid [3+2] cycloaddition with imines giving the corresponding 1,2,4-dioxazolidines in yields of 14–97% [90].

Carbonyl oxides can be also used for the synthesis of 3-vinyl-1,2,4-trioxalanes (α-vinylozonides) by a [3+2] cycloaddition with α,β-unsaturated aldehydes. However, α, β-unsaturated ketones are practically inactive in this reaction. The reaction of 3-vinyl-1, 2, 4-trioxalanes with ozone leads to the formation of the corresponding dizonides [91].

The carbonyl oxides prepared from the ozonolysis of enol ethers (for example, 1-metoxy-4,8-dimethylnone-1,7 diene) undergo stereoselective intramolecular cyclo-addition with inactivated alkenes yielding bicyclic dioxalanes. The latter are easily converted into β-hydroxycarbonyl species and 1, 3 - diols by catalytic hydrogenation thus providing a new approach to the synthesis of 1,3-oxygenated products [92]. The ozonolysis of vinyl ethers is discussed in detail [93].

The synthesis of 1,2-dioxalanes is carried out by ozonolysis of 1,1-disubstituted nonactivated olefins. Thus the ozonolysis of cyclopropyl-1,1-disubsituted olefins does not produce carbonyl oxides, but formaldehyde oxide. The latter can react with the initial olefin leading to the formation of 2-dioxalanes with 10% yield. At the addition of "foreign" olefins or aldehydes other dioxalanes and normal ozonides can be obtained [94].

The cycloalkenes ozonolysis in the presence of methyl pyruvate results in tri-substituted ozonides formation. The latter contain three reaction centers (peroxides, proton on the ozonides cycle and methoxy-carbonyl group) accessible for various functionalization. The cyclohexene ozonolysis in the presence of methylpyrovate gives ozonide whose treatment with PPh_3 or Et_3N yields $CHO(CH_3)_4CHO$ and $CHO(CH_3)_4COOMe$ (after esterification), respectively. This method proved to be very convenient and practical way for the synthesis of linear compounds containing various terminal groups from symmetrical cycloalkenes, in two steps and with good yields [95].

The direct conversion of olefins into esters is accomplished during mono-, di- and thi-substituted olefins ozonolysis in 2.5 M methanol sodium hydroxide-dichloromethane solution. The methyl esters are obtained in high yields. Thus, 3-benzyloxy-1-nonene (5b) is transformed into 2-benzyloxyoctanoate (7b) in 78% yield [96].

The ozonolysis of acyclic alkenes including terpenes is reviewed [97].

The role of ozonolysis as ecological process for the selective and specialized oxidation of petrochemical olefins and cyclic alkenes, for the manufacture of biologically active substances and normal organic compounds is thoroughly discussed [98].

In Ref. 99 are summarized the data on ozonolysis of acyclic and cyclic mono-, di- and trienes in the various stages of the synthesis of insect pheromones and juvenoids [99].

N-acylated esters of (cyclohexa-1,4-dienyl)-L-alanine are ozonized aiming at the synthesis of novel unnatural amino acids. The combined reduction and ozonolysis followed by condensation with a suitable nucleophile results in transformation of the aromatic ring of L-Phe to isooxasolyl, N-phenylpyrazolyl and to bicyclic pyrazolo[1,5-a]pyrimidine groups. The preparation of heterocyclic alanine derivatives is reported [100].

Thymidine diphospho-6-deoxy-a-D-ribo-3-hexulose synthesized by D-glucose ozonolysis of methyl-glucophosphate. is used as a central intermediate in the biosynthesis of di- and tridioxy sugars [101].

The ozonolysis of vinyl halides followed by reductive regio- and stereo-controlled intramolecular cyclocondensation is a key stage in the synthesis of amino sugars [102].

The ozonolysis of pyrols, oxazoles, imidazoles, and isooxazoles demonstrates another application of this reaction to the organic synthesis. Pyrols are efficient protective groups of the amino functions in the synthesis of α-aminoalcohols, α-aminoketones, α-aminoaldehydes and some peptides. The oxazole ring is also known as a protective group in the peptide synthesis [103].

The 1-substituted imidazoles ozonolysis leads to the formation of the corresponding N-acylamides, which are important amine or acyl derivatives [104].

Some novel tetraacetal oxa-cages and complex tetraquinone oxa-cages are synthesized from alkylfuranes by three stage reaction. Oxo rings in the tetraacetals are obtained by the ozonolysis of cis-endo-1,4-diones (norbornene derivatives) in dichloromethane solution at $-78°C$ and consecutive reduction with DMS. The tetraquinone oxa-cages are obtained in the cis-endo-1,4-diones ozonolysis in dichloromethane solution at $-78°C$ and TEA treatment [105].

A method for cleavage and oxidation of C_{8-30} olefins to compounds containing terminal carboxylic acid groups is reported [106]. The process is promoted by oxidation catalysts, such as Cr, Mn, Fe, Co, and so on.

The conversion of ethane to methanol and ethanol by ozone sensitized partial oxidation at near atmospheric pressure has been quantitatively studied. The effect of temperature, oxygen concentration in the inlet gas, contact time in the reactor and ozone concentration in oxygen has been evaluated. The selectivity in regard to ethanol, methanol, as well as the combined selectivity towards formaldehyde-acetaldehyde-methanol-ethanol is also discussed [107].

Upon studying the ozone-induced oxidative conversion of methane to methanol and ethane to ethanol it has been established that these reactions do not occur in the absence of ozone which clearly suggests that the partial oxidation is initiated by the oxygen atoms generated from the ozone decomposition [87,108].

Ortho-selective nitration of acetanilides with nitrogen dioxide in the presence of ozone, at low temperatures results in ortho-nitro derivatives formation in high yields [109].

The ozone mediated reaction of nonactive arenes with nitrogen oxides in the presence of suitable catalysts is reported as a new method for the synthesis of the corresponding nitro derivatives with high yields [110].

The ozone induced reaction of polychloro benzenes and some related halogeno compounds with nitrogen dioxide is a novel non-acid methodology for the selective mono nitration of moderately deactivated aromatic systems [111]. In the presence of ozone and preferably with methanesulfonic acid as a catalyst, the polychloro benzenes undergo mononitration with nitrogen dioxide at low temperatures giving polychloronitrobenzenes in nearly quantitative yields [111].

The ozone mediated reaction of aromatic acetals and acylal with nitrogen dioxide is suggested as a novel methodology for the nuclear nitration of acid sensitive aromatic compounds under neutral conditions [112].

Mineral and acid-free nitro compounds are prepared from CH_2Cl_2 soluted pyridine or its derivatives treated with $NO_2/O_2/O_3$ - mixture for 8 hr [113].

The nitration of aromatic compounds with nitrogen oxides in the presence of ozone is a catalytic process [114].

The stereoselective synthesis of vinyl ethers is accomplished by N - (arylidene (or alkylidene) amino) - 2-azetidinones reaction with ozone and $NaBH_4$ treatment resulting in di- and trisubstituted olefins derivatives [90].

The steroeselective synthesis of (2s, 3s) norstatine derivatives is carried out through aldehydes ozonolysis in the presence of lithium methoxyallene [94].

An interesting method of succinic acid preparation from butadiene rubber ozonolysis is suggested [115].

The preparation of α-phenyl ketone, ω-carboxylate-ended telehelic methyl methacrylate oligomers by the ozonolysis of regioregular methyl methacrylate-phenylacetylene copolymers is described [116]. The oligomers have a number molecular mass varying from 1600 to 4500 (with respect to the initial copolymers) and polydespersity less than 2.

Ozone is also very convenient agent in the manufacture of organic ceramics [117].

The ability of ozone to destroy the double C=C bonds in organic compounds is the reason for its wide application in the preparation of bifunctional compounds. This is the principle that lies in the organization of the manufacture of dodecanedicarboxylic acid (1) and azelaic acid (2). The initial substrate for the preparation of (1) is cyclododecane which is obtained from the trimerization of butadiene and partial hydrogenation and (2) is prepared on the base of oleic acid:

$$CH_3(CH_2)_7CH=CH(CH_2)_7COOH \; + \; O_3 \; \longrightarrow \; CH_3(CH_2)_7CH \underset{O}{\overset{O-O}{\diagdown}} CH(CH_2)_7COOH$$

$$CH_3(CH_2)_7CH \underset{O}{\overset{O-O}{\diagdown}} CH(CH_2)_7COOH \; \overset{[O]}{\longrightarrow} \; CH_3(CH_2)_7COOH \; + \; HOOC(CH_2)_7COOH$$

The main part of dicarboxylic acids is used in the manufacture of polyester fibers, and the azelaic acid esters (n-hexyl, cyclohexyl-, iso-octyl- and 2-ethylhexyl ester) are excellent plastisizers and synthetic lubricants.

The action of ozone is also used in pharmaceuticals in the preparation of valuable hormone products. Thus the C=C bond at C_{17} in the side chain of stigmasterol is destroyed by ozone yielding progesterone - an initial source for many hormones such as cortisone, male and female sex hormones, and so on.

The method of selective decolorization of fabrics containing cellulose materials, such as cotton and oxidizing dyes [118], includes the following steps: application of oxidation blocking agent to the fibers; contact of the fibers with the oxidizing reagent in gas or evaporated state in the presence of moisture till the oxidation and decolorization of the dyer is carried out; interruption of this contact before the beginning of a substantial destruction of the fiber. The oxidizing agents are selected among ozone, chlorine and nitrogen oxides and sulfur dioxide flow. The application of ozone for decolorizing indigo-painted cotton jeans after this procedures results in jeans material which does not turn yellow for 6 months while the untreated goes yellow much more rapidly [118].

The changes in the composition and chemistry of UV/ozone modified wool fiber surfaces [119] are investigated by means of photoelectronic spectroscopy (XPS). The oxidation of the disulfide sulfur to sulfone groups ($^-SO_3H$) containing S^{6+} approaches almost 90% conversion. This is much more that the conversion levels by using oxygen plasma. Ozonolysis results in rise of C-O - groups content, particularly of the carbonyl ones [119].

The process for producing cellulose fibers and moldings such as fibers, filaments, threads (yarn), films, membranes in the form of flat, pipe and empty fibrous membranes, and so on is carried out using ozone [120]. They are produced by extrusion of cellulose solutions in tertiary amine aminooxides and in some cases in water (particularly N-methylformaline N-oxide and water); regenerating bath and water for washing by treatment with hydrogen peroxide, peracetic acid, ozone, or chlorine dioxide for the regeneration of tertiary amines oxides. The introduced compounds can be enzymatically or catalytically destroyed before solvent and water regeneration.

Pitch-based carbon or graphite fibers of high tensile strength are manufactured by primary treatment of the pitch-based fibers with high concentrated ozone for a short period; then the treatment by ozone-free gases follows; in the next step they are subject to carbonization of graphitization to give elliptical fibers of long, wool similar structure. Upon treatment in the absence of ozone the fibers stitched during graphitization [121].

The EPR analysis of thermal decomposition of peroxides in ozonized polypropylene fiber for grafting shows that the decomposition of the peroxide groups begins at about 70°C. The generation of several peroxides radicals is registered; the access of the fiber to the spin sample is enhanced through oxidation. Small amounts of RO_2 - radicals with lifetime of couple of weeks have been identified. The integral intensity of the EPR-signal rises with time and ozonation temperature [122].

Ozonolysis modifies the diffusion pattern of liquid monomers in polypropylene matrix as their addition occurs in the amorphous phase. The appearance of intra-morphological structure and peroxides localization upon ozonolysis of polypropylene (granules and fibers) is monitored using electron spin resonance and transmission electron microscopy [123].

Upon ozonolysis, UV-radiation and plasma treatment polymer peroxides are generated on the surface of films from polypropylene, polyurethanes, and polyester fibers [124]. Their thermal and reductive-oxidative decomposition have been studied by means of functional analysis using peroxidase and iodide. However, their disposition in the polymer specimen is quite various thus impeding their analysis and depends on the treatment agent. For example, the treatment with plasma generates easily accessible peroxides in polyurethane films while the UV-radiation and ozonation leads to formation of peroxides incapable of reacting with aqueous solution of peroxidase. The redox decomposition of the peroxide groups by ferro-ions at 25°C has shown that less than 50% of the peroxides may react with ferro-ions at rate constants similar to those of hydrogen peroxide in aqueous solution. The thermal degradation of peroxides does not follow first order kinetics, most probably because of the generation of various peroxide species, characterized by different rates of decomposition. The lowest rate constant of decomposition observed at 62°C is 3.10^{-3} min^{-1} which does not depend on the polymer nature and the method of peroxides generation [124].

Ozone-induced graft polymerization onto polymer surface is an important and convenient method for polymer modification [125].

Ozone-induced graft copolymerization of polyethylene glycol monomethyl ether methacrylate onto poly(etherurethane) improves the hydrophility and water absorption. The autoaccelerated effect in ozone-induced polymerization has been also discussed [126].

The modification of the surface properties of polypropylene and block copolymer is carried out by ozone treatment. Thus, the wet ability of the polymers is improved - the contact angle of water becomes □ 67°. The break of the polymer chain and carboxyl groups formation are accelerated by using high ozone concentrations. The cut-off fragments resulting from the ozone treatment are removed from the surface by ultrasound and organic solvents [127].

The low temperature nonelectrolytic nickel plating onto three types of polypropylene is carried out by substrate pretreatment with ozone. The latter modifies the polymer surface for galvanization while the combination of polar with anchor effect as a result of the ozone etching enhances the adhesive properties of the polymer surface. The washing of the material after ozonolysis is obligatory for ensuring a good adhesion of the material [128].

The surface properties of polypropylene (I), 100 μm films from (I), ethylene-propylene block polymer (II), or ethylene-propylene polymer prepared by the random method (III) can be improved by ozonation at concentrations of 1.38, 0.64, and 0.41 mol %, respectively. The adhesion of dyes, coloring agents, dye layers, on (I) is substantially improved and depends on the substrates in the following order: (II) >(III) >(I). It has been found that for each sample there are optimum conditions of ozonation. Reactive dyes such as epoxy and acrylo urethane resins impart a better adhesion force that the conventional nonreactive acrylic or vinylchloride resin [129].

Freshly extruded, 50 μm C_2H_4-ethyl acrylate-maleic anhydride copolymer film is treated by 500 mL/m^2 oxygen containing 10 g/m^2 ozone and is calendered with 200 μm monolayer C_2ClF_3 - polymer film at 15°C to produce moisture-proof packaging material with intra-layer adhesion of 800 g/15 mm and thermosealing force of 3.6 kg/15 mm against 250 and 1.3, respectively in the absence of ozone treatment [130].

Methylmethacrylate polymers with good thermal degradation resistance are prepared by ozonolysis [131] whereby the end unsaturated groups are converted into carbonyl or carboxyl ones. Ozone/air mixture is blown through a 10 g Acrypet VH solution in CH_2Cl_2 at −78°C followed by exposure only to air for 60 min and after solvent removal the solid residue is dried in vacuum at 297°C for 8 hr. The polymer thus obtained is characterized by an initial temperature of thermal degradation of 315°C against 297°C for the polymer without ozone treatment [131].

The surfaces of propene polymer moldings are treated with ozone for improving their hydrophility. The ozone treatment of the surfaces for 8 hr reduces the contact angle of water from 100 to 81° [132].

Ozonation of PVC latex is also carried out for removal of vinyl chloride residues. The aqueous dispersions of saturated polymers are treated with ozone and the vinyl monomers are removed [133].

Graft polymerization of vinyl monomers onto Nylon 6 fiber is carried out after ozone oxidation of the fibers or films from Nylon 6 with vinyl monomers such as acrylamide, methylmethacrylate, and vinyl acetate. The molecular mass of Nylon 6 decreases slightly upon ozonation. For acrylamide graft polymerization system the preliminary treatment in air or vacuum by γ-rays radiation prior ozonation results in higher graft percent. For methyl methacrylate the apparent graft percent does not rise with the ozonation time. However the apparent graft percent for vinyl acetate is increased with the ozone time treatment [134].

The adhesion of PVC-, fluropolymer, or polyester-coated steel panels is improved by treatment with 5–50% solution of H_2O_2 or ozone [135].

Synthesis of water-soluble telehelic methyl-ketone-ended oligo- N,N-dimethyl-acrylamides by the ozonolysis of poly(N,N-dimethylacrylamide-stat-2,3-dimethylbu-tadiene)s is reported [136].

The synthesis of telehelic methylmethacrylate and styrene oligomers with fruoro-phenyl ketone end groups is accomplished by the ozonolysis of copolymers containing 4-flurophenyl butadiene units [137].

Graft polymerization of acrylic acid is carried out into preliminary ozonized silox-ane matrixes [138].

The manufacture of base discs for laser recording material is realized by covering of the plastic substrates surface with solid polymer layer possessing directing channels and/or signal holes. The surface of the plastic substrate is preliminary cleaned by UV/ ozone exposure before the formation of the hardened polymer layer [139].

The controlled degradation of polymers containing ozonides in the main chain takes place in the presence of: 1) periodate suported on Amberlyst A26 (SPIR); 2) diphenylphosphine deposited on polystyrene (SPR); boron hydride supported on Am-berlyst A26 (BER) [129]. Poly(butyl)methacrylate-copolymers are prepared by emul-sion polymerization. These materials and the homopolymer poly (butyl) methacrylate are ozonized at various temperatures and treated by any of the reagents described above, thus giving telechelic oligomers in 99% yield. Molecular mass varies when the ozonation temperature is changed. The end aldehyde groups are registered using ^1H and ^{13}C -NMR spectroscopy; the end hydroxyl groups are observed by means of FTIR and ^{13}C-NMR; the presence of hydroxyl groups is confirmed by tosylate formation; the carboxyl groups are identified by FTIR and ^{13}C-NMR and quantitatively determined by titration; in the SPR-generated oligomers the content of aldehyde groups consti-tutes about 80% of the end functional groups. The oligomers obtained in BER and SPIR contain 99% hydroxyl and carboxyl groups [140].

Polyethylene fibers are subject to ozone treatment for modifying their surface [141]. The analysis of the surface is carried out by means of X-ray photoelectron (ESCA) and IR (FTIR) spectroscopy. Carbon (C) and oxygen (O) were the main atoms monitored with ESCA (C-1s, O-1s areas) on the treated fibers. The analysis of C-1s peaks (C_1, C_2 and C_3) reveals that the oxidation level depends of the ozonation time. The components of 1s peak (O_0, O_1, and O_2) are very useful for carrying out the surface

analysis. They demonstrate the presence of carbonyl groups (1740–1700 cm⁻¹) even
onto untreated fibers whose intensity rises with treatment time. Ozonolysis is directed
from the surface to the fiber bulk. This is confirmed by the great enhancement of the
carbonyl bond band after 3 hr ozonation. The thermal analysis suggests structural and
morphological changes of the fiber when ozonation time exceeds 2 hr [141].

The processing of fiber-reinforced plastics is performed by blowing an ozone-oxygen mixture (flow rate of 0.4 L/min) for 5 hr through a CH_2CL_2 solution containing
glass reinforced fiber particles (1 mm diameter) filled with $CaCO_3$ and unsaturated
polyester. The fibers emerge on the surface till a fine powder is precipitated [142].

The manufacture of laminates through heat-sealing method involves the application of electric crown, UV-radiation or ozone exposure [143].

The manufacture of pour point depressants for oils , particularly useful for diesel
oils, is carried out by oxidative degradation of waste polyethylene and/or polypropylene (I) with ozone at 30–150°C. 1000 g waste (I) is exposed to ozone action at 150°C
for 5 hr giving the pour point depressant. The addition of 1% depressant to diesel oil
reduces the solidification of the oil from –20 to –35°C and the temperature of filter
plugging from –9 to –20°C [144].

17.3.2   INORGANIC PRODUCTIONS

Stainless steel parts are treated by ozone for surface passivation. The parts are heated
in oxidative or inert atmosphere at a temperature of condensation lower or equal to
–10°C. The unreacted ozone is re-used [145].

The main passivating agent is oxygen in combination at least with ozone. The system is particularly appropriate for passivation of metal (for example, stainless steel, Ti,
and so on.) equipment, used in chemical plants and exposed to strong corrosion action
at high temperatures and pressures [146].

For increasing the corrosion resistance of metals and alloys they were exposed
from 1 sec to 10 min in cold plasma under pressure 1–103 Pa and 100–5000 V in atmosphere containing O_2, O_3, N_2, H_2, air, CO_2, CO, N-oxides, H_2O (gas), combustible gas,
and/or inert gas. Thus, 17% C0-ferrite stainless steel is subject to plasma treatment for
4 min at 103 Pa, 100 mA, and 250 V in nitrogen-oxygen mixture with 20% oxygen.
The corrosion resistance was evaluated by treatment of the sample with a solution containing 17 ml 28% $FeCl_3$, 2.5 ml HCl, 188.5 ml H_2O, and 5 g NaCl. The sample shows
relatively good resistance as compared with the untreated one [147].

Electrochemical tests reveal the influence of dissolved ozone on the corrosion behavior of Cu-30 Ni, and 304L stainless steel in 0.5N sodium chloride solutions [148].
These experiments include: measurements of the corrosion potential as a function of
the time and ozone concentration, cyclic polarization experiments, isopotential measurements of the current density and study of the film components. The results of these
experiments show that for Cu-30 Ni and 304L -stainless steel the corrosion potential
is shifted to the more noble values (300 mV) at $[O_3]< 0.2–0.3$ mg/L. At higher concentrations it remains unchanged. The dissolved ozone reduces the corrosion level for
Cu-30 Ni - alloys which is evaluated by the substantial decrease in the current density
at constant applied potential. The improvement of the corrosion resistance should be

related to the decrease of the thickness of the corrosion products film and to the higher oxygen fraction as compared with the chloride in the same film. For stainless steel the differences in the passivating films in ozonized and nonozonised solutions are negligible as it is shown by spectroscopy [148].

Laboratory experiments have been carried out to study the ozone application for acid oxidizing leaching of chalcopyrite with 0.5 M H_2SO_4. For evaluating the reaction mechanism the effect of particles size distribution, stirring and acid concentration, the dissolution reaction and reaction kinetics on the leaching have been investigated. The reaction rate is governed by ozone diffusion to the reaction mixture. Ozone is an efficient oxidizer and the process is most effective at 20°C [149].

The rate of acid leaching of chalcopyrite depends on the use of ozone as an oxidizing agent in sulfuric acid solutions. The leaching of chalcolyrite follows a parabolic law [150]. The use of ozone as an oxidizer provides conditions for the formation of elemental sulfur on the leached surfaces. The rate of leaching is reduced with temperature as the ozone solubility decreases with temperature. The results show that ozone is the best oxidizing agent for acid oxidizing leaching of chalcopyrite and may be applied in pilot plants for regular manufacture.

We have studied the possibility of using ozone for improvement Ag extraction form polymers deposits in a flotation plant in the town of Rudozem (Bulgaria). It has been found that upon bubbling of ozone (1% vol. concentration, flow rate –300 L/hr) through flotation machine the degree of Ag extraction is increased by 1–2% [150].

The redox leaching of precious metals from manganese-containing ores carried out by other authors show also positive results [151].

Molybdenite flotation from copper/molybdenum concentrates by ozone conditioning results in relatively pure copper-free molybdenum. The process including multistep ozone flotation proves to be a technical and economical profitable method [152].

The manufacture of potassium permanganate is accomplished by melting of Mn-containing compounds with KOH, dissolution of the melt and the solution oxidation. For reducing the energy consumption $Mn(NO_3)_2$ is used as a Mn-containing compound. It can be easily alloyed with KOH in a 1:5–1:10 ratio at 250–300°C. Then the product is dissolved in 20–25% KOH solution, the solid residues is removed, saluted in 3–5% KOH and the solution is oxidized by ozone-air mixture [153].

The manufacture of silicon carbide ceramics is performed as melted organosilicon polymer is oxidized with 0.001% vol. ozone. Polycarboxylstyrene is the preferred polymer. The ceramics obtained is characterized by high thermal resistance and acceptable physical properties [117].

Mixtures containing In- and Sn-compounds are molded and sintered in a furnace in air atmosphere containing ≥ 1000 ppm ozone. The ITO ceramics thus obtained are characterized by high density at low temperature sintering for a short time [154,155].

A method for oxidizing carbonaceous material and especially for bleaching gray kaolin for subsequent use as coating or filler for paper in the presence of ozone is reported in [156].

The manufacture of mercury (I) chloride includes the reaction of Hg with hydrochloric acid in the presence of water, subsequent removal and drying at 95–105°C.

Ozone (0.1–0.1 g/g product) is bubbled through the reacting mass to increase the product yield and quality [157].

Arsenic acid is prepared from $(As_2Cl_6)^{2-}$ -ions by ozonolysis [158].

The removal of color and organic matter in industrial phosphoric acid by ozone and the influence on activated carbon treatment is described [148]. Industrial phosphonic acid containing 42–45% P_2O_5 and 220–300 mg/l organic matter (OM) is subject to combined treatment with ozone and activated carbon. The independent ozonation results in removal of the initial dark color of the acid and the organic matter. It is only through absorption on activated carbon that the level of OM could be reduced to 80% per 25 g/kg P_2O_5. The ozonation prior adsorption enhances the efficiency of the activated carbon effect and decreases its specific consumption [159].

The fabrication of high-Tc superconductors is carried out using ozone-assisted molecular beam epitaxy (MBE). It includes the simultaneous evaporation of the elementary components and application of ozone as a reactive oxygen source. The ozone priority over the other oxygen forms is that it is rather stable and could be produced and supplied to the substrate in a very pure state using simplified apparatus ensuring a well defined flow of oxidizing gas. In order to prepare films with high temperatures of superconductivity the growth should be carried out at relatively low stresses in the system. In addition, the surface during film growth can be analyzed by reflectance high energy electron diffraction. The most recent improvements in ozone-assisted MBE for family of $YBa_2Cu_3O_7$- delta films are described. The results show that this technique is very appropriate for growth of high quality superconductive films and could be ideal for the manufacture of such structures [160].

The growth of superconductive oxides under vacuum conditions compatible with MBE requires the use of activated oxygen. The atomic oxygen or ozone appears to be such species [150]. The characteristics of a radio-frequency plasma source for molecular beam epitaxial growth of high-Tc superconductor films (200 Å) from $DyBa_2Co_3O_7$ on $SrTiO_3$ are described and discussed [161].

An apparatus for the preparation of pure ozone vapor for use *in situ* growth of superconducting oxide thin films are designed [151]. Pure condensed ozone is produced from distillation of diluted ozone-oxygen mixture at 77K. The condensed ozone is heated until the pressure of its vapors approaches the necessary for an adequate flow of ozone- gas to the chamber of thin films growth. The thin films from $YBa_2Cu_3O_7$ with zero resistance at temperatures of about 85K grow at ozone pressure in the chamber of 2.10^{-3} Torr even without subsequent untempering. It should be noted that in contrast to other highly reactive oxygen species ozone can be prepared and stored in very pure form which makes it very convenient for studies on the kinetics of growth and oxidation of films with a well defined gas [162].

A patent for preparation of oxides superconductors $M-M^1-M^2-M^3$ includes ozone oxidation where: M = elements of III B groups such as: Y, Sc, La, Yb, Er, Ho or Dy; M^1 = elements from II A group, such as: B, Sr, Ba or Ca; M^2 - Cu and one or more elements from I B group such as Ag or Au; M^3 = O and on or more elements belonging to VI A group like S or Se and/or elements from VII A group as F, Cl or Br. The superconductors thus prepared have high critical temperature and high critical current densities [163].

The manufacture of bismuth-, copper- alkaline earth oxide high-temperature superconductors is carried out through calcination and/or sintering in ozone-containing atmosphere with $Bi(OH)_2$, $Ca(NO_3)_2$ and $Sr(NO_3)_2$, and subsequent treatment by $CuCO_3(OH)_2$; the components are dried, calcinated in ozone containing air, molded and dried in air atmosphere. Their temperature of superconductivity is 107K and the critical current density is 405 A/cm^2 [164].

A low temperature method for preparation of superconductive ceramic oxides is described [165]. It includes the treatment of the substrate heated surface by ozone to the complete evaporation of the other components thus forming a superconductive ceramic oxide.

The manufacture of rare earth barium copper oxide high-temperature superconductor ceramics is carried out by sintering at 930–1000°C in ozone atmosphere containing oxygen. Y_2O_3, $BaCO_3$, and CuO are mixed, calcinated, pulverized, pressed, and sintered at 950°C in the presence of ozone-oxygen mixture after which they are gradually cooled to the critical superconductive temperature 94K [166].

A review devoted to the growth of co-evaporated superconducting yttrium barium copper oxide (YBa_2CuO_7) thin films oxidized by pure ozone is presented [167].

The removal of organic pollutants from the surface of supports for microelectronics purposes is conducted by UV/ozone treatment [168].

Low temperature silicon surface cleaning is carried out by fluoric acid (HF) etching, washing, cleaning with deionized water, N--gas blowing, /UV-ozone treatment. This is a treatment preceding the process of silicon epitaxy [169-171].

The method for total room temperature wet cleaning of silicon surface comprises the use of HF, H_2O_2, and ozonized water and is with 5% more economical than that the standard procedure of wet cleaning [172].

The removal of resist films supported on semiconductive substances is accomplished by using ozone [173].

The application of ozone for reducing the temperature and energy in the process of very large scale integration (VLSI) is described [174,175]. The main problems under discussion are as follows: cleaning of Si-support with ozone by two methods: dry process combined with UV and wet one - with ozonized water; stimulation of Si thermal oxidation with ozone; deposition of SiO_2 under atmospheric pressure and low temperature via chemicals evaporation using tetraethylorthosilicate and ozone; 4) deposition of Ta_2O_3 at atmospheric pressure and low temperature applying $TaCl_5$ and ozone; 5) tempering with ozone and UV-radiation of Ta_2O_5 films used for dielectric for memory units , this treatment reduces the current permeability into the film; 6) etching of photoresistant materials by $O_2/O_3/CF_4$ and the effect of ozonator charge and injection of CF_4 [174,175].

The cleaning of synchrotonic radiation optics with photogenerated reactants has a number of priorities over the methods of discharge cleaning [176]. Upon discharge cleaning, the discharge particles should react with the surface contaminants until its sheilding by the rough discharge elements which may pollute or destroy the surface. Contrary, if the particles can be photon generated near the surface, the problem with the protection drops off and in some cases the cleaning can be more efficient. An estimation of the various methods for cleaning was made comparing the rates of polymethylmethacrylate films removal. A number of various light and geometry sources have been tested. The highest rate of cleaning was achieved upon using UV/O$_3$ cleaning method at atmospheric pressure. This method has been widely applied for cleaning semiconductive surfaces from hydrocarbon contaminants. It is noted that it is also effective in removing graphite-like pollutants from synchrotonic radiation optics. It proves to be more simple, economical and selective method as compared with other cleaning method. Although, it requires the drilling of a hole in the vacuum chamber, the cleaning of the optics can be carried out without dissemble which saves much time. This method is successfully applied for cleaning grates and reflectors in several beams [176].

Native oxide growth and organic impurity removal on silicon surface is carried out by ozone-injected ultrapure water [166]. In order to manufacture high-efficient and reliable ULSI-units the further integration and minimization is in progress. The cleaning with H$_2$SO$_4$/H$_2$O$_2$/H$_2$O is accompanied by serious problems: a great amount of chemical wastes is obtained which must be suitably treated. The cleaning technology with ultrapure water includes ozone injection in concentrations of 1–2 ppm. This method is very efficient in removing the organic impurities from the surface for a short time and at room temperature. The process wastes can be treated and recycled. The ozonized water components can be easily controlled [177].

The reaction mechanism of chemical vapor deposition using tetraethylorthosilicate and ozone at atmospheric pressure is reported and discussed [178].

Covering of semiconductor devices with silica films is carried out *via* CVD using Si(OEt)$_4$ and ozone. The deposition process is followed by heating in oxygen atmosphere with simultaneous UV radiation. The SiO$_2$-film thus formed can improve the water resistance of semiconductive devices [179].

The fabrication of nondoped silicate glass film with flat structure by O$_3$-TEOS deposition includes three stages: 1) substrate treatment; 2) formation of Al conducting layer on it using oxygen plasma through heating; 3) formation of SiO$_2$-film on the substrate by plasma using ozone and TEOS [180].

In Ref. 170 are reviewed the future trends for interlayer dielectric films production and their formation technologies in ULSI multilevel interconnections [170]. The properties of the interlayer dielectric films and their preparation technologies should satisfy the following three requirements: 1) available for disposition (setting) on large surface; 2) low dielectric constant; 3) low deposition temperature. Two techniques have been developed for the selective deposition of SiO$_2$-films which is the best way for achieving a complete planirization of the interlayer surface of dielectric films: low temperature liquid phase deposition using a saturated aqueous solution of hydrofruorosilicon acid H$_2$SiF$_6$; 2) half-selective SiO$_2$-films deposition at 390°C with Si(OC$_2$H$_5$)$_3$ and ozone and preliminary treatment with tetraflurocarbonic (CF$_4$) plasma on TiW or TiN surfaces [181].

Lead zirconate and titanate thin films are successfully prepared by reactive evaporation. The elements Pb, Zr, and Ti are evaporated in ozone-oxygen mixture. The films obtained have equal thickness and composition per a large surface (in the range of ±2% from the 4-inch support) [182].

Ceramic coatings on substrates are formed in the presence of ozone [183]; the substrates (silicon checks) are covered (dipped) in a solution of one or more (partially) hydrolyzed pre-ceramic silicon alholates with general forrmula $R_xSi(OR)_{4-x}$ ($R = C_{1-20}$ alkyl, aryl, alkenyl or alkinyl; x = 0-2). Further, the solvent is evaporated to form a coating which is subsequently heated in the presence of ozone up to 40–100°C thus converting into a ceramic coating. It in its turn may be covered also by a protective layer containing Si, or Si and C, or Si and N, or Si, C and N, or SO_2, and oxide. These coatings which are abrasive-, corrosive and thermo-resistant have also a small number of defects and suppress the diffusion of ionic contaminants such as Na and Cl ions and are particularly convenient for electronic units [183].

Method of forming zinc oxide light-shielding film for liquid crystals shields includes the injection of vapors of alkyl zinc compound with ozone or atomic oxygen in the activation chamber after which they pass through a chamber for ZnO film deposition at low temperature heating at about 200°C. The method is applicable for large scale production of these films [184].

An evidence for a new passivating indium rich phosphate prepared by UV/O_3 oxidation of indium phosphide, InP, is provided [185]. The phosphate does not exist as crystal compound and its composition is $InP_{0.5}O_{2.75}$. The passivating ability of the latter with respect to InP surface is discussed.

For improving the light-absorption characteristics of oxide optical crystals they are heated in to ozone-oxygen atmosphere [186,187]. Thus the light absorption from the optical crystal in a wave range different from that with which the crystal affects its own absorption is reduced to the most possible level. Devices using such oxide optical crystals with improved absorption characteristics work with high efficiency as optical amplifiers, optical isolators, optical recording medium, and optical generators.

The manufacture of solar cell modules with transparent conducting film covered by amorphous Si layer and electrode on the backside linked to the transparent isolator layer is described [188]. The preparation includes the application of a laser beam for electrode molding and exposure in oxidizing atmosphere containing 0.5–5% O_3 or $\geq 10\%$ O [188].

A method for strong oxidation in ozone atmosphere is proposed for surface activation of photoconductive PbS films [189].

Adhesion-producing materials for electroless plating and printing circuits contain particles of thermostable material slightly soluble in the oxidizing agent; the particles are dispersed in the thermostable resin which becomes almost insoluble in the oxidizing agent at hardening. These materials find application for printing circuits and are treated by an oxidizing agents (for example, chromic acid, chromate, permanganate, or ozone) to create a concave material surface [190].

KEYWORDS

- Advanced oxidation processes
- Ozone application
- Ozonolysis
- Polycyclic aromatic compounds
- Polyethylene fibers

REFERENCES

1. Rakovsky, S. K. and Zaikov, G. E. Kinetics and Mechanism of Ozone. Reactions with Organic and Polymeric Compounds in Liquid Phase, NOVA Sci. Publ. Inc., Commack, N.Y. (second edition), p. 340 (2007).
2. Da Silva abd, L. M. and Jardim, W. F. Qim. Nova, 29, 310 (2006).
3. Bailey, P.S. Ozonattion in Organic Chemistry, Academic Press, New York, v.I, 1978, v.II, 1982, Adv. Chem. Ser. - A, (Ed. HH Wasserman), 39 -I, 39 - II, 1978, 1982.
4. Rice, R.G. Ozone: Science & Engineering, 24, 1 (2002).
5. Pedroza, F. C., Aguliar, M. S., Luevanos, A. M., and Anaya, J. G. Ozone: Science & Engineering, 29, 307 (2007).
6. De Smedt, F., De Gendt, S., Claes, M., Heyns, M. M., Vankerckhoven, H., and Vinickier, C. Ozone: Science & Engineering, 24, 379 (2002).
7. Khadre, M. A., Yousef, A. E., and Kim, J. G. Journal of Food Science, 9, 1242 (2001).
8. Shatalov, A. A. and Pereia, H. Carbohydrate Polymers, 3, 275 (2007).
9. Pergut, E. A. and Gorelik, D. O. Instrumental Methods to Control Air Pollution. Leningrad, Khimiya (1981).
10. Chanlett, E. T. Environmental Protection, McGraw-Hill (1979).
11. Flaekt, A. B. Res. Discl., 326, 453 (1991).
12. Ruck, W. Vom. Wasser, 80, p. 253 (1993).
13. Mccoustra, M. S. and Horn, A. B. Chemical Society Reviews, 23, 195 (1994).
14. Tabata, K., Matsumoto, I., and Fukuda, Y. Jpn. Kokai Tokkyo Koho, JP 01,270,928 (Oct 30, 1989),
15. . Pollo, I., Jarosszynska- Wolinska, J., Malicki, J., Ozonek, J., and Wojcik, W. Pol. PL 137,426 (Jul 31, 1987).
16. De Guardia, A., Bouzaza, A., Martin, G., and Laplanche, A. Pollut.Atmos., 152, 82 (1996).
17. Takeyama, K. and Nitta, K. Jpn. Kokai Tokkyo Koho, JP 01,236,925 (Sep 21, 1989).
18. Moortgat, G. K., Horie, G., and Zahn, B. C. Pollut.Atmos., 29, 44 (1991).
19. Tomita, K. Jpn.Kokai Tokkyo Koho, JP 02,152,547 (Jun 12, 1990).
20. Vicard, J. F. and Vicard, G. PCT Int. Appl. WO 92 19,364 (Nov 12, 1992).
21. Iannicelli, J. U.S. US 4, 923,688 (May 08, 1990).
22. Buettner, F. and Koch, P. Ger. Offen, DE 9,931,891 (Apr 04, 1991).
23. Wada, H., Naoi, T., Y. Kuroda, Nippon Kagaku Kaishi, 9, 834 (1994).
24. Matt, K. and Guttenberger, H. G. Eur. Pat. Appl, EP 330,028 (Aug. 30, 1989).
25. Waga, H., Naoi, T., and Kuroda, Y. Nippon Kagaku Kaishi, 4, 306 (1995).
26. Tanaka, K. PPM, 27, 10 (1996).
27. Welch, J. F. and Siehwarth, J.D. U.S. US 4,861,497 (Aug. 30 1989).

28. Oeller, H. J., Daniel, I., and Weinberger, G. Water Sci.Technol., Forest Industry Wastewaters V, 35, 269 (1997).
29. C.H. Moebius., and M. Cordes-Tolle, Water Sci.Technol., Forest Industry Wastewaters V, 35, 245 (1997).
30. Nakamura, Y., Sawada, T., and Kobayashi, F. Water Sci.Technol., Forest Industry Wastewaters V, 35, 277 (1997).
31. Zhou, H. and Smith, D. W. Water Sci.Technol., Forest Industry Wastewaters V, 35, 251 (1997).
32. Huang, C. M. and Banks, M. K. J.Environ.Sci.Health, Part B, 31, 1253 (1996).
33. Sumimoto, H. and Yamazaki, T. Jpn.Kokai Tokkyo Koho, JP 03,111,670 (May 13, 1991).
34. Kawamoto, K. Kogaito Taisaku, 27, 617 (1991).
35. Ukita, S. Jpn.Kokai Tokkyo Koho, JP 03 80,996 (Apr 05, 1991).
36. Gunukula, R. B. and Tittlebaum, M. E. Journal of Environmental Science and Health A, 36, 307 (2001).
37. Ikehata, K. and El-Din, M. G. Ozone: Science & Engineering, 27, 1 (2005).
38. Ikehata, K. and El-Din, M. G. Ozone: Science & Engineering, 27, 173 (2005).
39. Ikehata, K., Naghashkar, N. J., and El-Din, M. G. Ozone: Science & Engineering, 28, 353 (2005).
40. Agustina, T. E., Ang, H. M., and Vareek, V. K. Journal of Photochemistry and Photobiology C: Photochemistry Reviews, 6, 264 (2005).
41. Gilbert, E. Ozone: Science & Engineering, 24, 75 (2002).
42. Yanga, Y., Min, J., Qina, Q., and Zhaia, X. Journal of Molecular Catalysis A: Chemical, 267, 41 (2007).
43. Cernigoja, U., Stangara, U. L., and Trebse, P. Applied Catalysis B: Enviromental, 75, 229 (2007).
44. Miriguchi, Y., Hayashi, H., and Umetani, T. Osaka-shi Suidokyoku Suishtsu Shikensho Chosa Kenkyu narabini Shiken Seiseki, 46, 10 (1994), /Pub. 1995/.
45. Munz, C., Galli, R., and Egli, Chem.Oxid., 2, 247, (1992), /Pub. 1994/.
46. Kojima, H., Katsura, E., H.Ogawa, and H.Kaneshima, Hokkaidoritsu Eisei Kenkyushoho, 41, 71 (1991).
47. N. Ohashi, Y. Tsuchiya, H. Sasano, and A. Hamada, Jpn .J. Toxicol. Environ. Health, 40, 185 (1994).
48. C.J. Hapeman-Somish, ASC Symp.Ser., (Pestic. Transform. Prod.: Fate Signif. Environ., 459, 133 (1991).
49. Terui, S., Sano, K., Nishikawa, K., and Inone, A. Jpn.Kokai Tokkyo Koho, JP 02 139,017 (May 29, 1990).
50. Terui, S., Sano, K., Kanazaki, T., Mitsui, K., and Inoue, A. (Jpn.Kokai Tokkyo Koho), JP 01 56,124 (Mar 03, 1989).
51. Golubovskya, E. K. Biological Bases of Water Purification, Visshaya Shkola, Moscow, (1978).
52. Gottschalk, Ch., Libra, J., and Saupe, A. Ozonation of Drinking Water and of Wastewater, Willey-VCH, p. 189 (2000).
53. Beltran, F. J. Ozone Reaction Kinetics for Water and Wastewater Systems, Taylor & Frances, (2007).
54. von Gunten, U., Water Research, 37, 1443 (2003).
55. von Gunten, U., Water Research, 37, 1469 (2003).
56. Geering, F., Ozone: Science & Engineering, 21, 187 (1999).
57. Rakness, K. L., OZONE in Drinking Water Treatment: Process Design, Operation, And Optimization, American Water Works Association, p. 302 (2005).

58. Ogawa, K. and Seki, N. Jpn.Kokai Tokkyo Koho, JP 04,243,597 (Aug 31, 1992).
59. Hoshima, Y. Jpn.Kokai Tokkyo Koho, JP 02 40,289 (Feb 09, 1990).
60. Filippi, A., Tilkes, F., Beck, E.G., and Kirschner, H., Dtsch.Zahnaerztl. Z., 46, 485 (1991).
61. Gehringer, P. Radiation-induced oxidation for water remediation. Oesterr. Forschungszent. Zeitbersdorf, [Ber.] OEFZS, (OEFZS - 4738), pp. 21 (1995).
62. Charles, L., Pepin, D., Ph. Puig, J.Eur.Hydrol., 27, 175 (1996).
63. Thornhill, R. W. UK Pat. GB 2,219,790 (Dec 20, 1989).
64. Nitsuta, Y. and Nitta, Y. Jpn. Kokai Tokkyo Koho, JP 04,367,799 (Dec 21, 1992),
65. Gordon, G. and Pacey, G. E. Chem. Oxid., 2, 230 (1992), Pub. 1994.
66. Bruckner, F., In situ remediation of petroleum contaminated soils by ozone treatment. ECOINFORMA 92, Int.Tag.Ausstellung Umweltinf.Umweltkommun., 2, 211 (2nd 1992), Pub. 1993.
67. Preusser, M., Ruholl, H., Schneidler, H., Wessling, E., and Wortmann, C., Wiss.Umwelt, 3, 135, (1990).
68. Rimpler, M. Ger. Offen., DE 19,512,448 (Oct 10, 1996).
69. Wessling, E., Erzmetall, 44, 196 (1991).
70. Mori, M., Yamakoshi, H., Nojima, M., Kusabayashi, S., Mccullough, K. J., Griesbaum, K., Kriegerbeck, P., and Jung, I. C., Journal of the Chemical Society-Perkin Transactions I, 12, 1335 (1993).
71. Mccullough, K. J., Teshima, K., and Nojima, M., J.Chem.Soc.-Chem Commun., 11, 931 (1993).
72. Sugiyama, T., Yamakoshi, H., and Nojima, M., J.Org.Chem, 58, 4212 (1993).
73. Ishiguro, K., Nojima, T., and Sawaki, Y., J.Phys.Org.Chem, 10, 787 (1997).
74. Nojima, T., Ishiguro, K., and Sawaki, Y., J Org Chem, 62, 6911 (1997).
75. Fukagawa, R. and Nojima, M., Journal of the Chemical Society-Perkin Transactions 1: 17, 2449 (1994).
76. Mccullough, K. J., Teshima, K., and Nojima, M. J. Chem. Soc-Chem.Commun., 11, 931 (1993).
77. Mccullough, K. J., Sugimoto, T., Tanaka, S., Kusabayashi, S., and Nojima, M. Journal of the Chemical Society-Perkin Transactions, 6, 643 (1994).
78. Satake, S., Ushigoe, Y., Nojima, M., and Mccullough, K. J. Journal of the Chemical Society-Chemical Communications, 14, 1469 (1995).
79. Teshima, K., Kawamura, S. I., Ushigoe, Y., Nojima, M., and Mccullough, K. J. Journal of Organic Chemistry, 60, 4755 (1995).
80. Mccullough, K. J., Tanaka, S., Teshima, K., and Nojima, M. Tetrahedron, 50, 7625 (1994).
81. Schobert, B. D., Chim.Oggi, 13, 21 (1995), /Pub.1995/.
82. Appell, R. B., Tomlinson, I. A., and Hill, I., Syn. Comm., 25, 3589 (1995).
83. Yang, Y. T., Li, T. S., and Li, Y. L. Synthetic Comm., 23, 1121 (1993).
84. Jachak, M., Mittelbach, M., Kriessmann, U., and Junek, H. Synthesis-Stuthgart, 3, 275 (1992).
85. Jellen, W., Mittelbach, M., and Junek, H. Monatsh.Chem., 127, 167 (1996).
86. Matsumoto, H., Kanda, H., Y. Obada, H. Ikeda, and Murakami, T. Jpn.Kokai Tokkyo Koho, JP 08 03,138 (Jan 09, 1996).
87. Allen, D. E., Ticker, Ch. E., Hobbs, Ch. C., and Chidambaram, R. U.S. Pat., US 5,491,265 (Feb 13, 1996).
88. Lee, H. W., Ji, S. K. I., Lee, Ch., and Lee, J. H., J.Org.Chem., 61, 2542, (1996).
89. Mikaev, F. Z. and Galin, F. Z. Izv. Akad. Nauk SSSR, ser. khim., 10, 1984 (1995).
90. Mccullough., K. J., Mori, M., Tabuchi, T., Yamakoshi, H., Kusabayashi, S., and Nojima, M. J.Chem.Soc.-Perkin Transactions 1, 14, 41 (1995).

91. Mori, M., Tabuchi, T., Nojima, M., and Kusabayashi, S. J.Org.Chem., 57, 1649 (1992).
92. Casey, M., Ulshaw, A. J., SYNLETT, 3, 214 (1992).
93. Kuczkowski, R. L. Ozonolysis of vinyl ethers, Series: Advances in Oxygenated Process: A Research Annual, 3, (1991).
94. Reiser, R., Seeboth, R., Suling, C., Wagner, G., Wang, J., and Schroder, G. Chem.Ber., 125, 191 (1992).
95. Hon, Y. S. and Yan, J. L. Tetrahedron, 53, 5217 (1997).
96. Marshall, J. A., Garofalo, A. W., and Sedrani, R. C. SYNLETT., 5, 643 (1992).
97. Odinokov, V. N., Bashk. Khim. Zhurn., 1, 29 (1994).
98. Odinokov, V. N. Bashk. Khim. Zhurn., 1, 11 (1994).
99. Yu, G., Ishmuratov, R., Kkharisov, Ya., Odinokov, V. N., and Tolstikov, G. A. Uspehi Khimii, 64, 580 (1995).
100. Zwilichovsky, G., and Gurvvich, V. J.Chem.Soc., Perkin Trans.1., 19, 2509 (1995).
101. Mueller, T. and Schmidt, R. R. Angew.Chem., Int.Ed.Engl., 34, 1328 (1995).
102. Pitzer, K. and Hudlicky, T. Synlet., 8, 803 (1995).
103. Kashima, Ch., Maruyama, T., and Arao, H. Yuki Gosei Kagaku Kyokaishi, 54, 132 (1996).
104. Kashima, C. Harada, K., and Hosomi, A. Heterocycles, 33, 385 (1992).
105. Wu, H. J. and Lin, Ch. Ch. J.Org.Chem., 60, 7558 (1995).
106. McVay, K. R., Gaige, D. G., and Kain, W. S. U.S. Pat. US 5,543,565 (Aug 06, 1996),
107. Gesser, H. D., Zhu, G., and Hunter, N. R. Catalysis Today, 24, 321 (1995).
108. Gesser, H. D., Hunter, N. R., and Das, P. A. Ñatalysis Letters, 16, 217 (1992).
109. Hitomi, S., Ishibashi, T., Takashi, M., and Kenkichi, T. Tetrahedron Letters, 32, 6591 (1991).
110. Suzuki, H., Murashima, T., Shimizu, K., and Tsukamoto, K. Chemistry Letters, p. 817 (1991).
111. Suzuki, H., Mori, T., and Maeda, K. SYNTHESIS- Stutgart, 8, 841 (1994).
112. Suzuki, H., Yonezawa, S., Mori, T., and Maeda, K. J.Chem.Soc.-Perkin Trans l., 11, 1367 (1994).
113. Suzuki, H., Murashima, T., and Tsukamoto, K. Jpn.Kokai Tokkyo Koho, JP 04,282,368 (Oct 07, 1992).
114. Suzuki, H. and Murashima, T. Kagaku to Kagyo (Tokyo), 46, 43 (1993).
115. Lyakumovich, A. G., Samuilov, Y. D., Goldberg, Ya. M., and Kondrashova, M. N. Russ. Pat., RU 2,036,895 (Jun 09, 1995).
116. Ebdon, J. R. and Flint, N. J. Macromolecules, 27, 6704 (1994).
117. Sawaki, T., Watanabe, S., and Shimada, K. Jpn.Kokai Tokkyo Koho, J P 02 26,868 (Jan 29, 1990).
118. Wasinger, E. and Hall, D. U.S. Pat. US 5 261,925 (Nov 16, 1993).
119. Bradley, R. H., Clackson, I. L., and Sykes, D. E. Applied Surface, 72, 143 (1993).
120. Connor, H. G. and Budgell, D. PCT Int. Appl, WO 96 18,761 (Jun 20, 1996).
121. Miura, M., Naito, T., Hino, T., and Kuroda, H. Jpn. Kokai Tokkyo Koho, JP 01,250,415 (Oct 05, 1989).
122. Queslati, R. and Roudesli, S. European Polymer Journal, 27, 1383 (1991).
123. Catoire, B., Verney, V., Hagege, R., and Michel, A. Polymer, 33, 2307 (1992).
124. Kulik, E. A., Ivanchenko, M., Kato, K., Sano, S., and Ikada, Y. J. Polym. Sci., Part A- Polymer Chemistry, 33, 323 (1995).
125. Fujimoto, K., Takebayashi, Y., Inoue, H., and Ikada, Y. J. Polym. Sci., Part A- Polymer Chemistry, 31, 1035 (1993).
126. Wang, Ch., Wang, A., Che, B., Zhou, C., Su, I., Lin, S., and Wang, B. Gaofenzi Xuebao, 1, 114 (1997).

127. Jian, Y. and Shiraishi, Sh. Hyomen Gijutsu, 40, 1251 (1989).

128. Jian, Y. and Shiraishi, S. H. Hyomen Gijutsu, 40, 1256 (1989).

129. Jian, Y. and Shiraishi, S. H. Hyomen Gijutsu, 41, 273 (1990).

130. Yamamoto, H. Jpn. Kokai Tokkyo Koho, JP 02,198,819 (Aug 07, 1990).

131. Yoshito, M. Jpn. Kokai Tokkyo Koho, JP 03,45,604 (Feb 27, 1991).

132. Shiraishi, S. H., Ken, U., Suchiro, K., and Nitsuta, K. Jpn. Kokai Tokkyo Koho, JP 03,103,448 (Apr 30, 1991).

133. Marshall, R. A., Parker, D. K., and Hershberger, J. W. U.S. Pat., US 5,498,693 (Mar 12, 1996).

134. Iwasaki, T. Kobunshi Ronbunshu, 50, 115 (1993).

135. Hosoda, Y., Ikishima, K., and Yanai, A. Jpn. Kokai Tokkyo Koho, JP 04,197,472 (Jul 17, 1992).

136. Ebdon, J. R. and Flint, N. J. J. Polym. Sci., Part A- Polymer Chemistry, 33, 593, (1995).

137. Ebdon, J. R. and Hodge, P. Polymer, 34, 406 (1993).

138. Eftushenko, A. M., Chikhachova, I. P., Timofeeva, G. V., Stavrova, S. T., and Zubov, V. P. Visoko molek. soed., ser. A, 35, 312 (1993).

139. Tashibana, S. H. and Yokoyama, R. Jpn. Kokai Tokkyo Koho, JP 01,248,339 (Oct 03, 1989).

140. Rimmer, S., Ebdon, J. R., and Shepherd, J. M. Reactive and Functional Polymers, 26, 145 (1995).

141. Chtourou, H., Riedl, B., Kokta, B. V., Adnot, A., and Kaliguine, S. J. Appl. Polym. Sci., 49, 361 (1993).

142. Vallet, A., Delmas, M., Fargere, T. H., and Sacher, G. PCT Int.Appl, WO 96 18,484 (Jun 20, 1996).

143. Kawabe, K. and Hatakeyama, Y. Jpn. Kokai Tokkyo Koho, JP 09 01,665 (07 Jan 07, 1997).

144. Czerwiec, W. and Kwaitek, A. Eur. Pat., EP 383,969 (Aug 29, 1990).

145. Shimizu, S. H. and Iwata, N. Jpn. Kokai Tokkyo Koho, JP 08 337,867 (Dec 24, 1996).

146. Lagana, V. EUR Pat. EP 504,621 (Dec 23, 1992).

147. Berneron, R. and de Gelis, P. EUR Pat. EP 340,077 (Nov 02, 1989).

148. Lu, H. H., Diguette, D. J. Effect of dissolved ozone on the corrosion behaviour of Cu-30 Ni and 304L stainless steel in 0.5N sodium chloride solutions. Report 1989, Order No. AD-A213804, p. 37.

149. Havlik, T., and Skrobian, M. Rudy, 37, 295 (1989).

150. Havlik, T. and Skrobian, M. Can. Metall.Q., 29, 133, (1990).

151. Clough, T., Siebert, J. W., and Riese, A. C. Aust. Pat. AU 607, 523 (7 Mar. 07, 1991).

152. Ye, Y., Jang, W. H., Yalamanchili, M. R., and Miller, J. D. Trans.Soc.Min., Metall., Explor., 288, 173 (1990), Pub. 1991.

153. Chkoniya, T. K. U.S.S.R. Pat., SU 1, 541, 190 (Feb 07, 1990).

154. Saski, T., Sasaki, H., and Iida, K. Jpn. Kokai Tokkyo Koho, JP 08 169, 769 (Jul 02, 1996).

155. Saski, T. and Sasaki, H. Jpn. Kokai Tokkyo Koho, JP 07 277, 821 (Oct 24, 1995).

156. Caropreso, F. E., Castrantas, H. M., and Byne, J. M. U.S. Pat. US 4, 935, 391 (Jun 19, 1990).

157. U.S.S.R. Pat. SU 1,567, 520 (May 30, 1990), O.V. Emel'yanova, and B.V. Emel'yanov.

158. Ruhlandtsende, A. D., Bacher, A. D., and Muller, U. Zeitschrift fur Naturforschung, Section B, 47, 1677 (1993).

159. Belfadhel, H., Ratel, A., Ouederni, A., Bes, R. S., and Mora, J. C., Ozone: Sci.Eng., 17, 637, (1995).

160. Achutharaman, V. S., Beauchamp, K. M., Chandrasekhar, N., Spalding, G. C., Johnson, B. R., and Goldman, A. M. Thin Solid Films, 216, 14 (1992).

161. Locquet, J. P., J. Vacuum Sci. and Tech., A. Vacuum surfaces and films, 10, 3100 (1992).
162. Berkley, D. D., Goldman, A. M., Johnson, B. R., Morton, J., and Wang, T., Rev. Sci. Instrum., 60, 3769 (1989).
163. Usui, T. and Osanai, Y. Jpn. Kokai Tokkyo Koho, JP 01 122,921 (May 16, 1989).
164. Odan, K., Miura, H., and Bando, Y. Jpn. Kokai Tokkyo Koho, JP 01 224,258 (Sep 07, 1989).
165. Goldman, A. M., Berkley, D. D., and Johnson, B. R. PCT Int. Appl, WO 90 02,215 (Mar 08, 1990).
166. T. Nakamori, Jpn. Kokai Tokkyo Koho, JP 01 179,752 (Jul 17, 1989).
167. Johnson, B. R., Beauchamp, K. M., Berkley, D. D., Liu, J. X., Wang, T., and Goldman, A. M. Proc. SPIE-Int.Soc.Opt.Eng., (Process.Films High Tc Supercond.Electron), 1187, 27 (1990).
168. Vig, J. R., Proc. Electrochem. Soc (Semicond.Clean.Tachnol.), .90, 105 (1990).
169. Suemitsu, M., Kaneko, T., and Nobuo, N. Jpn. J .Appl. Phys., Part 1, 28, 2421 (1989).
170. Kaneko, T., Suemitsu, M., and Miyamoto, N., Jpn. J .Appl. Phys., Part 1, 28, 2425 (1989).
171. Inada, A. and Kawasumi, K. Jpn. Kokai Tokkyo Koho, JP 01 135,023 (May 26, 1989),
172. Ohmi,. T. Semicond. Int., 19, 323 (1996).
173. Kamikawa, Y. and Matsumura, K. Jpn. Kokai Tokkyo Koho, JP 02 106,040 (Apr 18, 1990),
174. Maeda, K. Kurin Tekunoroji, 4, 62 (1994).
175. Choi, H., Kyoo, K., Jung, S. D., and Jeon, H. Han'guk Chaehyo Hakhoechi, 6, 395 (1996).
176. Hansen, R. W. C., Welske, J., Wallace, D., and Bissen, M. Nucl.Instrum.Methods Phys. Res., Sect.A, 347, 249 (1994).
177. Ohmi, T., Sagawa, T., Kogure, M., and Imaoka, T. J. Electrochem. Soc., 140, 804 (1993).
178. Takaaki, K., Akimasa, Y., and Yasuji, M. Jpn. J. Appl. Phys., 31, 2925 (1992).
179. Hishamune, Y. Jpn. Kokai Tokkyo Koho, JP 01 82,634 (Mar 28, 1989).
180. Kurihara, H., Saito, M., and Kozuki, T. Jpn. Kokai Tokkyo Koho, JP 09 08,026 (Jan 10, 1997).
181. Homma, T., Mater. Chem. Phys., 41, 234 (1995).
182. Toris, K., Saiton, S., and Ohji, Y. Jap. J. Appl. Physics, Part, 133, 5287 (1994).
183. Haluska, I. A. Fr. Demande, FR 2,697,852 (May 13, 1994).
184. Nishida, S. H. Ger. offen, DE 3,916,983 (Nov 30, 1989).
185. Hollinger, G., Gallet, D., Gendry, M., Besland, M. P., and Joseph, J., Appl. Phys .Lett., 59, 1617 (1991).
186. Tawada, H., Saitoh, M., and Isobe, C. H. Eur Pat. EP 407, 945 (Jan 16, 1991).
187. Tamada, H., Yamada, A., and Saitoh, M. J. Appl. Phys., 70, 2536 (1991).
188. Sadamoto, M., Ashida, Y., and Fukuda, N. Jpn. Kokai Tokkyo Koho, JP 08 148,707 (Jun 07, 1996).
189. Bianchetti, M. F., Canepa, H. R., Larrondo, S., and Walsel de Reca, N. E. An Asoc.Quim. Argent., 84, 73 (1996).
190. Enomoto, R. and Asai, M. Ger. Offen, DE 3,913,966 (Nov 09, 1989).

CHAPTER 18

ANTIOXIDANT ACTIVITY OF MINT

N. N. SAZHINA, V. M. MISIN, and E. I. KOROTKOVA

CONTENTS

18.1   INTRODUCTION

In the present work the total content of antioxidants and their activity relatively oxygen and its radicals are measured in water extract of mint peppery (*Mentha piperita L.*) by two electrochemical methods: ammetry and voltammetry. Mechanism of oxygen reduction and interaction with antioxidants are changed during the 1st hr after extraction. The correlation between antioxidant activity and composition of mint extracts is established.

Now the great attention is given to research of the quantity and activity of antioxidants (AO) in herbs, foodstuff, drinks, and other objects. It is known that the increase in activity of oxidation processes in an organism leads to destruction of structure and properties of lipid membranes. There is a direct dependent between the superfluous quantity of free radicals in organism and occurrence of dangerous diseases [1].

Antioxidants are biologically active substances, which remove excessive free radicals, decreasing the lipid oxidation. Therefore, a detailed research of total antioxidant activity (TAA) and mechanisms of interaction with oxygen and its radicals are highly prospective. [2].

Mint peppery (*Mentha piperita L.*) is one of the most widespread types of herbs. It is widely used in medicine, cookery, household, pharmaceutical and cosmetic industry. Mint is applied as an outside and inside medical pharmaceutical in the form of infusions and teas for treatment of gastro enteric diseases, as a demulcent at palpitation, depression and sleeplessness, as sedative means, as anesthesia at burns and stings of insects, removes stress and a headache. Antioxidant properties of a mint allow to prevent occurrence of a cataract and other illnesses connected with ageing of an organism. Mint peppery contains essence (2–3%) in which basic component is the menthol defining taste and anesthezing properties of mint, and also other substances – a ethers, phellandrene, pinene, jasmole, piperitone, menthofuran etc. There are also tannic and resinous substances, carotin (0.01%), ascorbic acid (0.01%), routines (0.015%) and other polyphenolic combinations [3].

In the literature there are publications devoted to measurement of the AO quantity and activity in extracts of mint [4-7]. However it, basically, difficult chemical methods, and the results received by these methods are badly comparable among themselves, including the absence of uniform measure units. Use of two operative electrochemical methods has allowed studying the antioxidant activity dynamic of mint after extraction.

18.2   EXPERIMENTAL DETAIL

The purpose of the present work was to measure the antioxidant activity (AOA) of a water mint extract by a voltammetric method and to study the mechanism of influence of mint components on process of electrochemical oxygen reduction (ER O_2). In parallel in mint extract measurements of the total phenol type antioxidant content have been spent by an ammetric method.

Object of research was the water extract of the mint peppery (*Mentha piperita L.*), collected and dried in 2008 year in the Tver region. For extract reception the dry grass was crushed in a mortar till the particles size 1–2 mm. Further this grass (0.5 g) was filled by the distilled water with T = 95°C and volume 50 ml and maintained within 10 min without thermo stating. Then an extract was carefully filtered through the paper filter «a dark blue tape» and if necessary diluted before measurements [7]. Acidity of an extract (pH) has monitored as 6.6.

In the present work two electrochemical methods were used. The ammetric method [8] allows defining the total phenol type antioxidant content in investigated samples. Measurements were spent on the device «Zwet-Jauza-01-AA» [9]. The essence of the given method consists in measurement of the electric current arising at oxidation of investigated substance (or mixes of substances) on a surface of a working electrode at constant potential +1,3V. At this potential there are oxidation only OH-groups natural phenolic type antioxidants [7-9]. The received signal from the tested sample is compared to a signal of an individual antioxidant – gallic acid, measured under the same conditions. The measurement error is no more than 10%, measurement time - 10–15 min [9].

The voltammetric method uses an electroreduction of oxygen (ER O_2) as modeling reaction. This process is similar to oxygen reduction in tissues and in stuffs of plants. It proceeds at the working electrode in several stages with formation of the reactive oxygen species (ROS), such as $O_2^{\cdot-}$ and $HO_2^{\cdot-}$ [10]:

$$O_2 + e^- \rightleftarrows O_2^{\cdot-} \quad E^\circ = +0.012 \text{ V} \tag{1}$$

$$O_2^{\cdot-} + H^+ \rightleftarrows HO_2^{\cdot-} \tag{2}$$

$$HO_2^{\cdot-} + H^+ + e^- \rightleftarrows H_2O_2 \qquad E^\circ = +0.682 \text{ V} \tag{3}$$

$$H_2O_2 + 2H^+ + 2e^- \rightleftarrows 2H_2O \qquad E^\circ = +1.770 \text{ V} \tag{4}$$

For determination of TAA in work [10] it is offered to use the first wave of ER O_2 corresponding to stages (1) – (3) when on a surface of a mercury film electrode active oxygen radicals and hydrogen peroxide as an end product are formed. It takes attention that antioxidants of varied nature were divided on 4 groups according their interaction with oxygen and its radicals (table) [11].

TABLE   Groups of biologically active substances divided according mechanisms of influence on oxygen and its radicals.

Group number	1-group	2-group	3-group	4-group
Substance names	Catalasa, phtalocyanines of metals, humin acids.	Phenol nature substances, vitamins A, E, C, B, flavanoids, ubichinon, glucose.	N, S, Se-containing substances, amines, amino acids, active aldehydes	Superocsi-dismutase (SOD), porphyry metals, cyto-chrome C
Influence on ER O_2 process	Increase in current of ER O_2, potential shift in negative area	Decrease of current of ER O_2, potential shift in positive area.	Decrease of current of ER O_2, potential shift in negative area.	Increase in current of ER O_2, potential shift in positive area
The prospective electrode mechanism.	EC* mechanism with the following reaction of hydrogen peroxide disproportion and partial regeneration of molecular oxygen.	EC mechanism with the following chemical reaction of interaction of AO with active oxygen radicals	CEC mechanism with the precede and following chemical reactions of interaction of AO with oxygen and its active radicals	EC mechanism with catalytic oxygen restoration through formation of intermediate complexes.

*The note: E – an electrode stage of process, C – chemical reaction.

As antioxidant activity criterion of investigated substances the kinetic criterion K is used which reflects quantity of oxygen and the active oxygen radicals which have

reacted with AO (or their mixes) in a minute of time: $K = \dfrac{C_0}{t}(1 - \dfrac{I}{I_0})$, mk mol/l×min,

where I, I_0 — currents of electroreduction of oxygen, accordingly, at presence and at absence of AO in a solution of supporting electrolyte, C_0 — initial concentration of oxygen in a solution (mkmol/l), i.e. solubility of oxygen in supporting electrolyte under normal conditions, t — time of an exposition of a working electrode at constant potential under the limiting current of the oxygen, characterizing course of reaction of interaction of AO with active oxygen radicals.

For voltammetric study of TAA of samples automated voltammetric analyzer "Analyzer of TAA" ("SPE Polyant" Tomsk, Russia) was used. As supporting electrolyte the phosphate buffer solution in volume of 10 mL with known initial concentration of molecular oxygen was used, where was added dosed out (δ = 10–500 mkL) quantity of the investigated sample [10]. The electrochemical cell (V = 10 ml) was connected to the analyzer and consisted of a working mercury film electrode (MFE), a silver- silver chloride reference electrode with KCl saturated (Ag|AgCl|KCl$_{sat}$) and a silver- silver chloride auxiliary electrode. An open type cell can be used in this investigation. The reference electrode and the working electrode were held in the electrochemical cell. The working electrode potential was initially set at 0 V for about 120s. During this step the solution was stirred by a magnetic stirrer. After stirring the potential was scanned negatively, causing oxygen reduction, which gives a current first wave at E = –0.1 □ -0.3V. Its value is proportional to the amount of oxygen in the solution bulk. Potential rate scan was 0.1 V/s, potential range was E = 0 to –0.9 V. The voltammetric method of research has good sensitivity. It is simple and cheap. However, as in any similar electrochemical method, the error of some experiments was about 30%.

18.3   DISCUSSIONS AND RESULTS

On Figure 1–3 voltammogramm of the samples, received at various times after mint extraction are resulted: after extraction t = 5 min (Figure 1), t = 30 min (Figure 2), and t = 60 min (Figure 3). It is visible that the fresh mint extract (Figure 1) "works" on the mechanism of phenol nature AO, reducing a current and shifting potential in positive area (2nd group in the table). However, approximately in an hr after extraction (Figure 3), character of interaction of components of mint with oxygen and its radicals varies, following the mechanism, characteristic for substances of 4th group in the table. Transition from one mechanism to another occurs approximately through 30 min (Figure 2) and the kinetic criterion K becomes thus close to zero. Shift of potential of a current half wave of oxygen restoration (Δ) remains approximately identical (0.20 – 0.25 V). At the further storage of an extract in vitro voltammogramm character essentially does not vary, and the module of kinetic criterion □K□ grows to values 2.0 ± 0.5 mkmol/l×min in 3 hr after extraction.

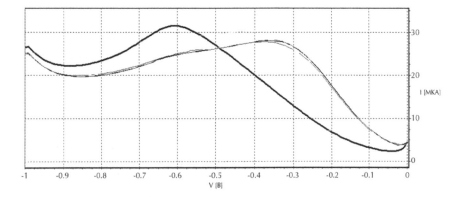

FIGURE 1   Voltammogramms of current of oxygen electroreduction at absence (left curve) and at presence(right curve) of mint extract in cell at t = 5 min after mint extraction. Extract dose δ = 100 mkl in 10 ml of a buffer solution.

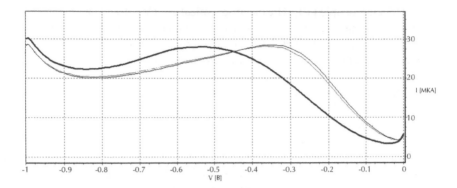

FIGURE 2 Voltammogramms of current of oxygen electroreduction at absence (left curve) and at presence(right curve) of mint extract in cell at t = 30 min after mint extraction. Extract dose δ = 100 mkl in 10 ml of a buffer solution.

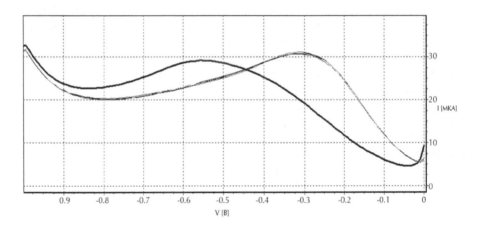

FIGURE 3 Voltammogramms of current of oxygen electroreduction at absence (left curve) and at presence(right curve) of mint extract in cell at t = 60 min after mint extraction. Extract dose δ = 100 mkl in 10 ml of a buffer solution.

On Figure 3–5 the example of dependence of kinetic criterion K and potential shift Δ on concentration of a mint extract in the supporting electrolyte is shown at storage times of an extract over 1 hr. During experiment the influence of extract dose δ in supporting electrolyte on the $|K|$ and Δ was investigated. As results $|K|$ and Δ decrease from $|K|$ = 1.5 ± 0.5 mkmol/l×min, Δ = 0.25 V at δ = 100 mkl (Figure 3), to $|K|$=1.3 ± 0.3 mkmol/l×min and Δ = 0.10 V for δ = 25 mkl (Figure 4) and $|K|$=0.8 ± 0.2 mkmol/l×min, Δ = 0.05 V for δ = 10 mkl (Figure 5). At increase δ above 100 mkl the voltammogramm character small varies, and at δ>300 mkl decrease $|K|$ is observed.

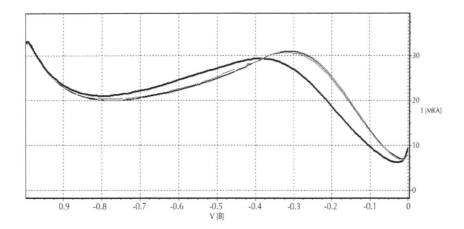

FIGURE 4 Voltammogramms of current of oxygen electroreduction at absence (left curve) and at presence (right curve) of mint extract in cell at t >60 min after mint extraction. Extract dose δ = 25 mkl in 10 ml of a buffer solution.

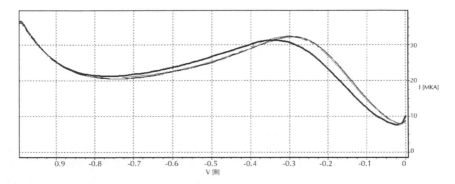

FIGURE 5 Voltammogramms of current of oxygen electroreduction at absence (left curve) and at presence (right curve) of mint extract in cell at t >60 min after mint extraction. Extract dose δ = 10 mkl in 10 ml of a buffer solution.

In parallel with voltammogramm registration for the same mint extract the total phenolic antioxidant content C (in units of gallic acid) has been measured from an extract storage time t by ammetric method (Figure 6). Dependence C on t testifies to notable enough falling C after extraction and filtering of a mint extract (approximately 20% for 2 hr of storage). It, possibly, is connected with destruction of the unstable phenol substances containing in an extract. Therefore, apparently, voltammogramm character and values of kinetic criterion (Figure 1–3) during the first moment after extraction could establish as influence of classical phenolic antioxidant on the process of electroreduction of oxygen and its radicals (Figure 6).

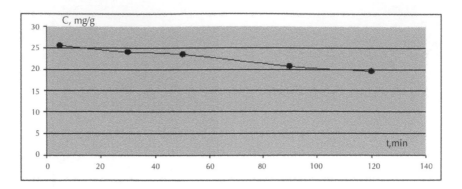

FIGURE 6   The total phenol antioxidant content C in a mint extract depending on extract storage time t.

As it has been mentioned in the introduction, one of the basic components of mint essence is menthol $C_{10}H_{20}O$ [12], voltammogramm character therefore has been studied at introduction in a buffer solution in cell a water menthol solution. Menthol (60 mg) has been dissolved in 100 ml of hot water and doses of this solution 200 – 500 mkl were added in 10 ml of a buffer solution. Voltammogramms have been registrated under the same conditions, as a mint extract. On Figure 7 the typical example of such voltammogramm is presented from which it is visible that menthol "works" as classical AO, decreasing the current of oxygen electroreduction. For a dose of a menthol solution δ = 500 mkl the kinetic criterion K has made 1.5 ± 0.4 mkmol/l×min, slowly decreasing at reduction of menthol concentration. Such behavior has also a cyclohexanol (similar menthol substance) which activity at the same concentration is included into the same interval of values K, as for menthol in a similar way operates. For such components of mint as routines and ascorbic acid, their voltammogramms are a classical example of 2nd groups of antioxidants from table [13]. It is possible to conclude that the total mechanism of interaction of mint components with oxygen and its radicals is defined by the substances not listed above, and other components, for example some substances of the phenolic nature giving shift of potential in a positive side and dominating in the chemical mechanism for a fresh extract. At the further storage of an extract, these unstable substances destruct and the general character of oxygen reduction is defined already by the substances containing in mint and entering into 4th group of the table. Probably, they are any metal complexes which operate as catalysts of passing chemical reactions.

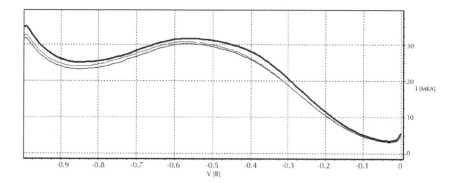

FIGURE 7 Voltammogramms of current of oxygen electroreduction at absence (top curve) and at presence (bottom curve) of water menthol solution. Menthol solution dose δ = 500 mkl in 10 ml of a buffer supporting electrolyte.

18.4   CONCLUSION

The total content of antioxidants and their activity relatively oxygen and its radicals are measured in extract of mint peppery (*Mentha piperita L.*) by ammetric and voltammetric methods. Mechanism of oxygen reduction and interaction with antioxidants are changed during the first hours after extraction. The correlation between antioxidant activity and composition of mint extracts is established.

KEYWORDS

- **Ammetry and voltammetry**
- **Antioxidants**
- **Electrochemical methods**
- **Total antioxidant activity**

REFERENCES

1. Burlakova, E. B. In Bioantioxidant: yesterday, today, tomorrow. Biological Kinetics. Reviews. M. Chemistry, 2, 10–45 (2005).
2. Dillard, C. J. and German J. B. Phytochemicals: nutraceuticals and human health. J. Sci. Food Agric, 80, 1744–1756 (2000).
3. Dudchenko, L. G., Kozjakov, A. S., and Krivenko V. V. Gingery-aromatic and gingery-flavour plants //K: Naukova dumka. 304 (1989).
4. Proestos, C., Chorianopoulos, N., Nychas G. J. E., and Komaitis, M. RP-HPLC analysis of the phenolic compounds of plant extracts. Investigation of their antioxidant capacity and antimicrobial activity. J. Agric. Food Chem, 53(4) 1190–1195 (2005).
5. Autoui, A. K., Mansouri, A., Boskou, G. and Refalas, P. Tea and herbal infusions: Their antioxidant activity and phenolic profile. Food Chem, 89(1) 27–36 (2005).

6. Katalinic, V., Vilos, M., Kulisic, T. and Jukic, M. Screening of 70 medicinal plant extracts for antioxidant capacity and total phenols. Food Chem, 94(4) 550–557 (2006).

7. Misin, V. M, Sazhina, N. N., Zavjalov, A. J. and Yashin JA. I. Measurement of the phenol content in extracts of medicinal grasses and their mixes by ammetric method. Chemistry of vegetative raw materials, 4, 127–132 (2009).

8. Jashin, A. J., Jashin, J. I., Chernousova, N. I., and Pahomov, V. P. Express electrochemical method of antioxidant activity definition in foodstuff. Beer and waters, 6, 44–46 (2004).

9. Yashin, A.Ya. Inject-flowing system with ammetric detector for selective definition of antioxidants in foodstuff and drinks. Russian chemical magazine, 52(2) 130–135 (2008).

10. Korotkova, E. I., Karbainov, Y. A., and Avramchik, O. A. Investigation of antioxidant and catalytic properties of some biological-active substances by voltammetry. Anal. and Bio-anal. Chem., 375(1–3), 465–468 (2003).

11. Korotkova, E. I., Lukina, A. N., et al. In Estimation methods of antioxidant activity of biologically active substances in medical and preventive appointment. The collection of reports. Moscow 182–194 (2004).

12. Handbook of Chemistry and Physics. 71st edition. CRC Press. Ann Arbon. Michigan. (1990).

13. Korotkova, E. I. The voltammetric method for definition of total antioxidant activity in objects of an artificial and natural origin. Dissertation of the Doctor of Chemistry. Tomsk., (2009).

CHAPTER 19

WILD ORCHIDS OF COLCHIS FORESTS AND SAVE THEM AS OBJECTS OF ECOEDUCATION, AND PRODUCERS OF MEDICINAL SUBSTANCES

E. A. AVERJANOVA, L. G.KHARUTA, A. E. RYBALKO, and K. P. SKIPINA

CONTENTS

19.1 INTRODUCTION

Orchids (Orchidaceae Zinn.) Colchian forests - rare and endangered species. They are of interest for ecotourism because of its decorative properties. At the same time they can serve as a source of strong means of restoring health. Propagation and cultivation of Orchids *in vitro* allows us to solve three problems at once. First - the protection and restoration of natural populations suffering under the influence of civilization. The second is environmental education. And the third - to provide effective medicinal substances that restore health by means of biotechnological.

The requirements set priorities of our research activities. And the current state of the biosphere could be better. The living conditions of people in urban areas are particularly grim. The result is a dramatic and widespread decline in public health. Meantime, ways of maintenance and restoration of vital forces of person exist of yore. By this methods are forgotten in modern time, how unnecessary rejected. Because natural sources of curative substances are already too not enough. And biotechnology comes here to the aid.

It is, in particular, the so-called "salep", known for centuries. The source of it is the underground storage organs of plants of orchid family, terrestrial species distributed in temperate climate zone. On the properties of salep we will later, and now just note that the world human population grows and natural populations of terrestrial orchids are thinning, melting away before our eyes, of course, is not in any way providing the needs of people. It is time to replace the usual gathering of funds in the healing nature to the intensive cultivation of the active substance through biotechnology.

Orchidaceae is one of the largest families among angiosperms. According to one estimate the family includes 800 genera and 25,000 species [1]. Orchids - a family of evolutionarily the most perfect of plants pollinated by insects. Orchids are distributed in all climatic zones, on all continents, not only grow in Antarctica. Most rich family in the tropics, subtropics and then go, but also in the temperate zone, it presents much (about 900 species in the temperate latitudes of the northern hemisphere, about 120 in Europe). List of Flora Orchids Sochi Black Sea region has 46 species [2] distributed from the coast to the upper boundary of the alpine meadows. Species of most large size, long-term healing in the sense of obtaining raw materials, and simultaneously the most ornamental found mostly in a narrow strip of sea water's edge on the slopes of the foothills up to 250–400 m above sea level. This is primarily *Orchis mascula*, O. purpurea, O. provincialis, *Platanthera bifolia*, Ophris oestrifera, and So on (Figure 1–4).

FIGURE 1   *Orchis mascula.*

FIGURE 2   *Orchis purpurea.*

FIGURE 3   *Orchis provincialis.*

FIGURE 4   *Ophris oestrifera.*

FIGURE 5   Tuberoid of orchidea.

The abundance of species, unfortunately, is not accompanied by a high number of plants in natural populations. Orchids are usually rare and endangered species throughout all areas [3]. That is why the work on cultivation of this species group is so relevant. This is possible without causing harm to populations only by using methods of cultivation *in vitro*.

19.2   FEATURES OF THE BIOLOGY AND ECOLOGY

Representatives of the orchid family Colchis flora - terrestrial perennial herbaceous plants in height from 20 to 100 cm, with underground storage organs, often very decorative.

Propagation by seeds, often vegetative. The cycle of development is not easy and spread over time from seed to flowering new plants can take 5–7 (15) years [4]. The seeds have no nutrient reserves, due to this are very light and can travel through the air for long distances. But the effectiveness of seed reproduction is low, because seedling is formed only in suitable soil conditions and using a special fungus-symbiont.

Flowering of many species (respectively, and fruit set) does not happen every year. In unfavorable conditions, plants are able to fall into a state of secondary dormancy. They are not making the leaves and stalk, are not showing up on the soil surface during the year, and even a few years, subsisting by the fungus-symbiont. Pollination of flowers is only by insects of specific group of species. The structure of the flower are usually very closely matches the size and morphology of insect pollinators [5]. Frequent situations where seed production populations of orchids is very low due to insufficient number of suitable insect in the vicinities.

Orchids are not capable of strong competition, they just die, if their habitat is overgrown with tall grasses and shrubs. However, many species, which we call the forest, withstand a very weak light in the tall woods with a solid crown canopy. Other orchids do not occur in shaded areas, they can be found in open meadows and forest edges, among the bushes.

Requirements for soils are different for different species, for orchids of Colchis botanical province are usually neutral or alkaline reaction, good aeration. Withstands poor, dry soil, but most species reach their maximum development on rich, moist soils.

In the Figure 5 we can see what it looks like the main source of the healing qualities of Orchids - tuberoid. It is an underground organ for storage of nutrients. In gathering tuberoids as medicinal raw plant is completely destroyed. Here is a mature plant Steveniella satyrioides in a phase of beginning of the growing season, the snapshot taken in February. It is clearly visible size of the tuberoid, it does not exceed 2.5 cm in length. Does it make sense to dig tuberoids from natural populations, destroying them completely? After all, medicinal raw materials we receive very little.

The features of biology and environmental preferences cause the low number of Orchids everywhere. In addition, Colchis forest area of Sochi is a powerful influence of civilization, growing from year to year.

The rapid development of the resort, particularly heightened in connection with the Olympic construction site leads to a tremendous reduction in habitat Orchids. It is now impossible to determine exactly how many tens and hundreds of plants irretrievably lost, where now new roads and other infrastructure in Sochi are. But we can confidently say that this process is not stopped. Construction continues, the city is growing strongly, more low-mountain belt of land occupied by the cottage, chalet and other settlements. The band of suitable habitat and so narrow in many species (from the coast to the heights of 200–300 m above sea level) is rapidly decreasing. At the moment we are witnessing, for example, the loss of populations of Orchis picta and Serapias vomeracea around the holiday village Navalishino. Such examples could be cited.

Local populations disappear, lost some part of the gene pool and it is irreparable. Viability of the species as a whole is reduced. Leaves the planet's ecosystem biodiversity.

Prevention of such events - a long and laborious work, for which already has all the scientific, technological and organizational conditions.

Sochi branch of the Russian Geographical Society and the Department of Physiology of the Sochi Institute of People's Friendship University in the SEC "Biotechnology" developed a project to restore the Sochi Orchids. For solution to this problem is planned emergency assistance - Transfer orchids from the destroyable habitats to the special mini-reserves with further development of technologies for conservation by modern methods of biotechnology. Such an approach to make an impact on natural populations of orchids, in most cases, almost imperceptible.

We have started a cycle of research which include breeding in aseptic culture, adapting the plants in greenhouses, then in the open field and repatriation, in other words transport into specially organized reserves in order to restore extinguished local populations and creating new ones.

In the process of reproduction of native forms will inevitably arise promising clones to create new ornamental crops. A lot of attention to the implementation of the project will receive callus cultures, as a source of medical funds. Based on extensive scientific base and technological development of domestic and foreign scholars and practitioners, we have already started work on the introduction of *in vitro* culture of several species of terrestrial orchids Colchis flora. The main work will be deployed as early as 2012 under a contract with the Olympic Committee Sochi-2014 for the conservation of biodiversity.

19.3   ORCHIDS AS AN OBJECT OF ECOEDUCATION

We aim to preserve orchids Colchian forests and restoration of natural populations is not only in the environmental aspect, but also plan to use them as objects of environmental education, and ecoeducation and upbringing of the youth people. "Wild Orchid" - it sounds quite fresh and new for travelers and Sochi residents. On the basis of laboratory, greenhouse, hothouse complex, the pilot sites open land, as well as protected areas, which are grown under natural conditions different types of Orchids, planned to create a scientific and practical tour route in a single complex with the environmental path. Visitors will be able to directively learn modern scientific and technological methods of conservation of flora, in the best of their ability and skills to participate in the processes of the laboratory for further breeding and cultivation of orchids, the study of biological singularities and, if desired, to create media products that call preserve native nature (photo albums, posters, videos, and so on.). Visitors can also contribute to the arrangement of ecological trails and nursery for plants. This route has a fascination for most of the year, as determined by differences in the timing of flowering of different species. Green leaves of orchids can be seen in Sochi, all year round, the earliest flowering observed in March (Steveniella satyrioides (Figure 6)), a mass flowering begins in April, which continues with the change of mid-July, then in early September, autumn orchid blossoms (Spiranthes spiralis (Figure 7)), and blooms until the end of October. In terms of greenhouse blooms will be more.

FIGURE 6   *Steveniella satyrioides.*

FIGURE 7   Spiranthes spiralis.

With the participation of other interesting plants we will create a guided tour, an unusual and attractive in all seasons.

19.4   MEDICINAL PROPERTIES

Medicinal properties of Orchids are not well understood. Scattered literature data [6-20] summarized in the table (Table 1).

Table 1   Medicinal properties of Orchids

№	Genus	Phytochemistry	Actions, therapeutics
1	Anacamptis Rich.	Starch, mannan, dextrin, sucrose, albumins	Enveloping, restorative. Cystitis, gastritis, colitis, impotence
2	Corallorhiza Chatel.	Alkaloids, slimy substance	Hypotensive, antipyretic, sedative. Rheumatism
3	Cypripedium L.	Alkaloids, saponins, sugars, binder, resinous substances, essential oils, phenolic compound, tsipripedin, glucoside of o-hydroxycinnamic acid, tannins, anthocyanins hrizantemin, lipids, coumarins, luteolin, apigenin, chrysin, ascorbic acid, calcium oxalate, hydroquinone, arbutin	Hypnotic, anticonvulsant, sedative, vasodilator, laxative, antipyretic, sedative, diuretic. Gastritis, gynecological bleeding, headaches, mental disorders, skin diseases.

TABLE 1   *(Continued)*

4	Dactylo-rhiza Neck. ex Nevski	Mucilage, starch, glucoside: loro-glossin, albumen, volatile oil, ash, dactylorhins,	Aphrodisiac, expectorant and nervine tonic
		coumarin lo-roglossin, flavonoids, anthocyanins: cyanine, seranin, ofrisanin, serapianin, orhitsianin	Diabetes, diarrhea, dysentery, paralysis, convalescence, impotence and malnutrition.
			coating, anti-inflammatory, emollient, ditoksikatsionnoe tool
			antitumor, restorative and tonic, sedative.
			Gastric disorders, flatulence,
			hemorrhoids, diarrhea, emaciation
5	Epipactis Zinn.	Alkaloids, coumarins	Laxative. The noise in the head
6	Epipogium J. S. Gmel. ex Borkh.	carotenoids: neoksantin, lutein, violaxanthin, α-carotene, β-carotene	Restorative, tonic, pain reliever
7	Goodyera R.Br.	Alkaloids, loroglossin. rutin, kaempferol-3-0-rutinozid, izoramnetin-3-0-rutinozid, gudayerin	Emollient, detoxification. In scurvy in diseases of the kidneys, eyes, scrofula, to improve the appetite, as a remedy for snake bites, diseases of the stomach, bladder, female diseases and diseases of the eye
8	Gymnadenia R.Br.	Flavonoids, coumarin loroglossin, mannan, glucosides, urea, glucoside of o-hydroxycinnamic acids, flavonoids: quercetin, kaempferol, astragalin, izokvertsitrin, 3-β-glucoside-7-β- glucoside, kaempferol 3-β-glucoside-7-β-glucoside, quercetin, anthocyanins: hrizante-min, cyanine, seranin, ofrisanin, orhitsianin I, orhitsianin II, serapianin	Enveloping, emollient, restorative, toning, nourishing for debilitated patients, an expectorant, wound healing, increases the potency, emollient, expectorant, abortifacient, an aphrodisiac. Epilepsy, neuro-psychiatric disorders, toothache, abscesses, neuroses, dysentery, infertility, gastritis, enteritis, enterocolitis
9	Herminium Guett.	Alkaloids, slimy substance	Painkiller for toothache, enveloping

TABLE 1   *(Continued)*

10	Orhis L.	Blennogenic matter - Mann, pentosans, starch, metilpentozany, proteins, dextrin, sucrose, bitter, volatile oil, carotene, kvertsitrin, pectins, coumarin glycoside loroglossin.	Nutritious, enveloping, restorative, stimulant, contraceptive. Enterocolitis, metiorizm, hemorrhoids, atonic constipation, cystitis, impotence, cancer, colds, snake bites, externally for fungal infections of nails
11	Platanthera Rich.	Starch, saponins, mannan	Enveloping, restorative, febrifuge, diuretic, antispasmodic, hypotensive, and uterine contraceptive. Impotence, menstrual disorders, cystitis, enterocolitis, atonic constipation
12	Spiranthes Rich.	Flavonoids, alkaloids, phenolic compounds: n-hydroxybenzaldehyde, n-gidroksibenzilovy alcohol salt of ferulic acid, polynuclear aromatic compounds: spirantol A, spirantol B orhinol, spirasineol A, spirantezol	Restorative. Kidney disease, generalized weakness, tuberculosis, exhaustion

In generalizing it should be noted that the main active principle tuberoids almost all kinds of substances are blennogenic providing enveloping effect, in combination with other medicinal qualities most often used as a tonic, nourishing way to restore people's health, weakened by a variety of ailments of infectious and traumatic nature. Such a wide range of applications suggests that there is a strong positive effect on the immune system. The universality of the properties allows the use of drugs from Orchids to restore the health of people regardless of their age and condition, without fear of side effects.

19.5   CONCLUSION

The relevance of research and practical measures for introduction to culture of this group of plants is evident. The aseptic culture will give the necessary high multiplication factor, which will, in addition to receiving a sufficient amount of material to isolate the curative substance, and realize research of the reintroduction and repatriation, with the goal of biodiversity conservation of the unique ecosystem of the region in many ways.

KEYWORDS

- Ecoeducation
- Orchidaceae
- Orchids
- *Orchis mascula*

REFERENCES

1. Stewart, J. and Griffiths, M. Manual of Orchids. Timber Press, Portland, Oregon (1995).
2. Solodko, A. S. and Makarova, E. L. Orchids Sochi Black Sea, Sochi (2011).
3. The Red Data Book of the Russian Federation. Plants and fungi, Moscow (2008).
4. Vakhrameyeva, M. G., Denisova, L. V., Nikitina, S. V., and Samsonov, S. K. Orchids of our country. Nauka, Moscow (1991).
5. Ivanov, S. P., Kholodov, V. V., and Fateryga, A. V. Orchids of the Crimea: the composition of pollinator diversity of pollination systems and methods and their effectiveness. Proceedings of Tauric National University. A series of Biology and Chemistry. 22(61) (2009). Number 1. S. 24-34.
6. Luning, B. Alkaloid content of Orchidaceae. C. L. Withner (Ed). The Orchids Scientific Studies, John Wiley and Sons, New York, London (1974).
7. Slaytor, M. B. The Distribution and Chemistry of Alkaloids in the Orchidaceae. Orchid Biology Reviews and Perspectives 1. J. Arditti (Ed.) Cornell University Press. Ithaca/London, pp. 96–115 (1977).
8. Hew, C. S., Arditti, J., and Lin, W. S. Three orchids used as herbal medicines in China: an attempt to reconcile Chinese and Western Pharmacology. Orchid Biology: Reviews and perspectives VII. J Arditti et al. (Eds.), Kluwer Academic Publishers, Dordrecht/Boston/London, pp. 213–83 (1997).
9. Hory, N. Materia Medica of India and their Therapeutics. Neeraj Publishing House (1982).
10. Kizu, H., Kaneko, E. I., and Tomimori, T. Studies on Nepalese crude drugs. XXVI. Chemical constituents of Panch aunle, the roots of Dactylorhiza hatagirea D. Don. Chem. Pharm. Bull. 47 (11), 1618–1625 (1999).
11. Lachen, Z. I., et al. Glycosidic components of tubers Gymnadenia conopsea. J. Nat. Prod., 71, 799–805 (2008).
12. Schroeter, A. I. Medicinal Flora of the Soviet Far East. Meditsina, Moscow, pp. 328 (1975).
13. Hegnauer, R. Chemotaxonomie der Pflanzen. Basel und Stuttgart: Birkhauser Verlag., pp.540 (1963).
14. Gorovoy, P. G., Salokhin, A. V., Doudkin, R. V., and Gavrilenko, I. G. Orchids (Orchidaceae) Far East: Taxonomy, Chemical Compositions and use of the Possibility of Protection (Review), Turczaninowia, 13(4), 32–44 (2010).
15. A. L. Budantsev, et al. (Ed.) Wild useful plants of Russia. St. Univ. Sphfa, p. 663 (2001).
16. Fruentov, N. Medicinal Plants of the Far East. Khabarovsk, p. 352 (1987.).
17. G. P. Yakovlev, et al. (Ed.) Dictionary of medicinal plants and animal products: Textbook. manual. St. Petersburg. Special Literature, p. 407 (1999).

18. Plant Resources of Russia and adjacent states: Flowering plants, their chemical composition and use. The family Orchidaceae - Yatryshnikovye. St. Petersburg.: Science, T. 8, 84–99 (1994).
19. Singh, A. and Duggal, S. Medicinal Orchids: An Overview. Ethnobotanical Leaflets, 13, 351–63 (2009).
20. Seredin, R. M. and Sokolov, S. D. Medicinal plants and their uses. Stavropol Publishing House, (1969).

CHAPTER 20

FIXATION OF PROTEINS ON MNPs

A. V. BYCHKOVA, M. A. ROSENFELD, V. B. LEONOVA,
O. N. SOROKINA, and A. L. KOVARSKI

CONTENTS

20.1   INTRODUCTION

The problem of creation of magnetically targeted nanosystems for a smart delivery of drugs to target cells has been solved with the application of a fundamentally novel method of obtaining of stable protein coatings. The method is based on the ability of proteins to form interchain covalent bonds under the action of free radicals which are generated locally on nanoparticles surface. By the set of physical and biochemical methods it has been proved that the free radical cross-linking of proteins allows obtaining stable single layer protein coatings of several nanometers in thickness. The cross-linked coatings were formed on the surface of individual magnetic nanoparticles (MNPs) (d ~17 nm) by free radical processes taking place strictly in the adsorption layer. Spin labels technique has been applied for studying of macromolecules adsorption on nanoparticles. The free radical cross-linking of proteins on the surface of nanoparticles has been shown to keep native properties of the protein molecules (was demonstrated on thrombin coating). The method may be used to reach various biomedical goals concerning a smart delivery of drugs and biologically active substances revealing new possibilities of design of single layer multiprotein polyfunctional coatings on all the surfaces containing metals of variable valence (for example, Fe, Cu, and Cr).

The MNPs have many applications in different areas of biology and medicine. The MNPs are used for hyperthermia, magnetic resonance imaging, immunoassay, purification of biologic fluids, cell and molecular separation, and tissue engineering [1-6]. The design of magnetically targeted nanosystems (MNSs) for a smart delivery of drugs to target cells is a promising direction of nanobiotechnology. They traditionally consist on one or more magnetic cores and biological or synthetic molecules which serve as a basis for polyfunctional coatings on MNPs surface. The coatings of MNSs should meet several important requirements [7]. They should be biocompatible, protect magnetic cores from influence of biological liquids, prevent MNSs agglomeration in dispersion, provide MNSs localization in biological targets and homogenity of MNSs sizes. The coatings must be fixed on MNPs surface and contain therapeutic products (drugs or genes) and biovectors for recognition by biological systems. The model which is often used when MNSs are developed is presented in Figure 1.

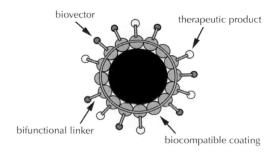

FIGURE 1   The classical scheme of magnetically targeted nanosystem for a smart delivery of therapeutic products.

Proteins are promising materials for creation of coatings on MNPs for biology and medicine. When proteins are used as components of coatings it is of the first importance that they keep their functional activity [8]. Protein binding on MNPs surface is a difficult scientific task. Traditionally bifunctional linkers (glutaraldehyde [9-10], carbodiimide [11-12]) are used for protein cross-linking on the surface of MNPs and modification of coatings by therapeutic products and biovectors. Authors of the study [9] modified MNPs surface with aminosilanes and performed protein molecules attachment using glutaraldehyde. In the issue [10] bovine serum albumin (BSA) was adsorbed on MNPs surface in the presence of carbodiimide. These works revealed several disadvantages of this way of protein fixing which make it unpromising. Some of them are clusters formation as a result of linking of protein molecules adsorbed on different MNPs, desorption of proteins from MNSs surface as a result of incomplete linking, uncontrollable linking of proteins in solution (Figure 2). The creation of stable protein coatings with retention of native properties of molecules still is an important biomedical problem.

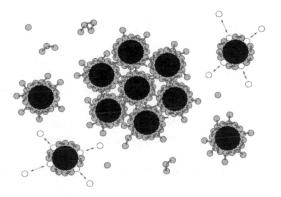

FIGURE 2 Nonselective linking of proteins on MNPs surface by bifunctional linkers leading to clusters formation and desorption of proteins from nanoparticles surface.

It is known that proteins can be chemically modified in the presence of free radicals with formation of cross-links [13]. The goals of the work were to create stable protein coating on the surface of individual MNPs using a fundamentally novel approach based on the ability of proteins to form interchain covalent bonds under the action of free radicals and estimate activity of proteins in the coating.

20.2 MATERIALS AND METHODS

20.2.1 MAGNETIC SORBENT SYNTHESIS

Nanoparticles of magnetite Fe_3O_4 were synthesized by co-precipitation of ferrous and ferric salts in water solution at 4°C and *in the alkaline medium*:

$$Fe^{2+} + 2Fe^{3+} + 8OH^- \rightarrow Fe_3O_4\downarrow + 4H_2O$$

1.4 g of $FeSO_4 \cdot 7H_2O$ and 2.4 g of $FeCl_3 \cdot 6H_2O$ ("Vekton", Russia) were dissolved in 50 mL of distilled water so that molar ratio of Fe^{2+}/Fe^{3+} was equal to 1:2. After filtration of the solution 10 mL of 25 mass % NH_4OH ("Chimmed", Russia) was added to it on a magnetic stirrer. 2.4 g of PEG 2 kDa ("Ferak Berlin GmbH", Germany) was added previously in order to reduce the growth of nanoparticles during the reaction. After the precipitate was formed the solution (with 150 mL of water) was placed on a magnet. Magnetic particles precipitated on it and supernatant liquid was deleted. The procedure of particles washing was repeated for 15 times until neutral pH was obtained. MNPs were stabilized by double electric layer with the use of US-disperser ("MELFIZ", Russia). To create the double electric layer 30 mL of 0.1 M phosphate-citric buffer solution (0.05 M NaCl) with pH value of 4 was introduced. MNPs concentration in hydrosol was equal to 37 mg/mL.

20.2.2 PROTEIN COATINGS FORMATION

The BSA ("Sigma-Aldrich", USA) and thrombin with activity of 92 units per 1 mg ("Sigma-Aldrich", USA) were used for protein coating formation. Several types of reaction mixtures were created: "A1-MNP-0", "A2-MNP-0", "A1-MNP-1", "A2-MNP-1", "A2-MNP-1-acid", "T1-MNP-0", "T1-MNP-0", and "T1-0-0". All of them contained:
1. 2.80 mL of protein solution ("A1" or "A2" means that there is BSA solution with concentration of 1 mg/mL or 2 mg/mL in 0.05 M phosphate buffer with pH 6.5 (0.15 M NaCl) in the reaction mixture; "T1" means that there is thrombin solution with concentration of 1 mg/mL in 0.15M NaCl with pH 7.3),
2. 0.35 mL of 0.1 M phosphate-citric buffer solution (0.05 M NaCl) or MNPs hydrosol ("MNP" in the name of reaction mixture means that it contains MNPs),
3. 0.05 mL of distilled water or 3 mass % H_2O_2 solution ("0" or "1" in the reaction mixture names correspondingly).

Hydrogen peroxide interacts with ferrous ion on MNPs surface with formation of hydroxyl-radicals by Fenton reaction:

$$Fe^{2+} + H_2O_2 \rightarrow Fe^{3+} + OH^{\cdot} + OH^-$$

"A2-MNP-1-acid" is a reaction mixture, containing 10 µl of ascorbic acid with concentration of 152 mg/mL. Ascorbic acid is known to form free radicals in reaction with H_2O_2 and generate free radicals in solution but not only on MNPs surface.

The sizes of MNPs, proteins and MNPs in adsorption layer were analyzed using dynamic light scattering (Zetasizer Nano S "Malvern", England) with detection angle of 173° at temperature 25°C.

20.2.3 STUDY OF PROTEINS ADSORPTION ON MNPS

The study of proteins adsorption on MNPs was performed using ESR-spectroscopy of spin labels. The stable nitroxide radical used as spin label is presented in Figure 3. Spin labels technique allows studying adsorption of macromolecules on nano-sized magnetic particles in dispersion without complicated separation processes of solution components [14]. The principle of quantitative evaluation of adsorption is the following. Influence of local fields of MNPs on spectra of radicals in solution depends on the distance between MNPs and radicals [14-16]. If this distance is lower than 40 nm for magnetite nanoparticles with the average size of 17 nm [17] ESR spectra lines of the radicals broaden strongly and their intensity decreases to zero. The decreasing of the spectrum intensity is proportional to the part of radicals which are located inside the layer of 40 nm in thickness around MNP. The same happens with spin labels covalently bound to protein macromolecules. An intensity of spin labels spectra decreases as a result of adsorption of macromolecules on MNPs (Figure 4). We have shown that spin labels technique can be used for the study of adsorption value, adsorption kinetics, calculation of average number of molecules in adsorption layer and adsorption layer thickness, concurrent adsorption of macromolecules [18-20].

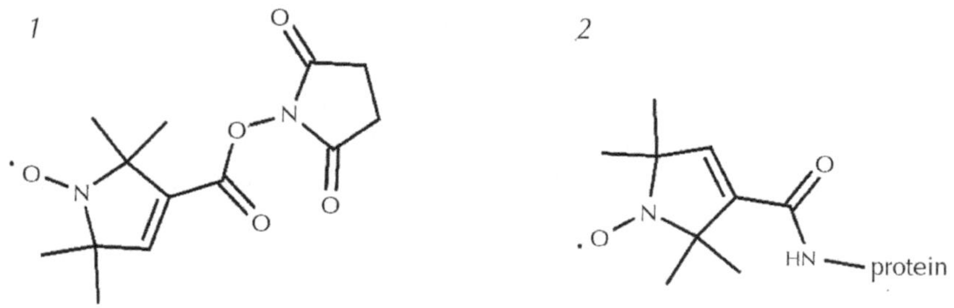

FIGURE 3 The stable nitroxide radical used for labeling of macromolecules containing amino groups (*1*) and spin label attached to protein macromolecule (*2*).

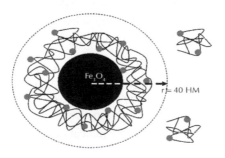

FIGURE 4 The MNPs and spin-labeled macromolecules in solution.

The reaction between the radical and protein macromolecules was conducted at room temperature 25 µl of radical solution in 96% ethanol with concentration of 2.57 mg/mL was added to 1 mL of protein solution. The solution was incubated for 6 hr and dialyzed. The portion of adsorbed protein was calculated from intensity of the low-field line of nitroxide radical triplet I_{+1}.

The method of ferromagnetic resonance was also used to study adsorption layer formation.

The spectra of the radicals and MNPs were recorded at room temperature using Bruker EMX 8/2.7 X-band spectrometer at a microwave power of 5 mW, modulation frequency 100 kHz and amplitude 1 G. The first derivative of the resonance absorption curve was detected. The samples were placed into the cavity of the spectrometer in a quartz flat cell. Magnesium oxide powder containing Mn^{2+} ions was used as an external standard in ESR experiments. Average amount of spin labels on protein macromolecules reached 1 per 4–5 albumin macromolecules and 1 per 2–3 thrombin macromolecules. Rotational correlation times of labels were evaluated as well as a fraction of labels with slow motion ($\tau > 1$ ns).

20.2.4   COATING STABILITY ANALYSIS AND ANALYSIS OF SELECTIVITY OF FREE RADICAL PROCESS

In previous works it was shown that fibrinogen (FG) adsorbed on MNPs surface forms thick coating and micron-sized structures [18]. Also FG demonstrates an ability to re-place BSA previously adsorbed on MNPs surface. This was proved by complex study of systems containing MNPs, spiedlabeled BSA and FG with spin labels technique and ferromagnetic resonance [20]. The property of FG to replace BSA from MNPs surface was used in this work for estimating BSA coating stability. 0.2 L of FG ("Sigma-Aldrich", USA) solution with concentration of 4 mg/Ll in 0.0   phosphate buffer with pH 6.5 was added to 1 Ll of the samples "A1-MNP-0", "A2-MNP-0", "A1-MNP-1", "A2-MNP-1". The clusters formation was observed by dynamic light scattering.

The samples "A2-MNP-0", "A2-MNP-1", "T1-MNP-0", "T1-MNP-1" were centrifuged at 120000 g during 1our on «Beckman Coulter» (Austria). On these conditions MNPs precipitate, but macromolecules physically adsorbed on MNPs remain in supernatant liquid. The precipitates containing MNPs and protein fixed on MNPs surface were dissolved in buffer solution with subsequent evaluation of the amount of protein by Bradford colorimetric method [21]. Spectrophotometer CF-2000 (OKB "Spectr", Russia) was used.

Free radical modification of proteins in supernatant liquids of "A2-MNP-0", "A2-MNP-1" and the additional sample "A2-MNP-1-acid" were analyzed by IR-spectroscopy using FTIR-spectrometer Tenzor 27 ("Bruker", Germany) with DTGS-detector with 2 c^{-1} resolution. Comparison of "A2-MNP-0", "A2-MNP-1" and "A2-MNP-1-acid" helps to reveal the selectivity of free radical process in "A2-MNP-1".

20.2.5   ENZYME ACTIVITY ESTIMATION

Estimation of enzyme activity of protein fixed on MNPs surface was performed on the example of thrombin. This protein is a key enzyme of blood clotting system which catalyzes the process of conversion on FG to fibrin. Thrombin may lose its activity as a result of free radical modification and the rate of the enzyme reaction may decrease. So estimation of enzyme activity of thrombin cross-linked on MNPs surface during free radical modification was performed by comparison of the rates of conversion on FG to fibrin under the influence of thrombin contained in reaction mixtures. 0.15 Ll of the samples "T1-MNP-0", "T1-MNP-1" and "T1-0-0" was added to 1.4 Ll of FG solution with concentration of 4 mg/ml. Kinetics of fibrin formation was studied by Rayleigh light scattering on spectrometer 4400 ("Malvern", England) with multibit 64-channel correlator

20.3 DISCUSSION AND RESULTS

The ESR spectra of spin labels covalently bound to BSA and thrombin macromolecules (Figue 5) allow obtaining information about their microenvironment. The spectrum of spin labels bound to BSA is a superposition of narrow and wide lines characterized by rotational correlation times of 1^{-9} sec and $2 \cdot \times 0^{-8}$ s respectively. This is an evidence of existence of two main regions of spin labelonlocalization on BSA macromolecules [22]. The portion of labels with slow motion is about 70%. So a considerable part of labels are situated in internal areas of macromolecules with high microviscosity. The labels covalently bound to thrombin macromolecules are characterized by one rotational correlation time of 0.2 ns. These labels are situated in areas with equal microviscosity.

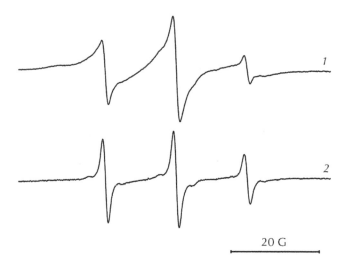

FIGURE 5   The ESR spectra of spin labels on BSA (*1*) and thrombin (*2*) at 25°C.

The signal intensity of spiedlabeled macromolecules decreased after introduction oesMNPs into the solution *that testifies to the* protein adsorption on MNPs (Figur. 6). Spectra of the samples "A1-MNP-0" and "T1-MNP-0" consist of nitroxide radical triplet, the third line of sextet of Mn^{2+} (the external standart) and ferromagnetic resonance spectrum of MNPs. Rotational correlation time of spin labels does not change after MNPs addition. The dependences of spectra lines intensity for spiedlabeled BSA and thrombin in the presence of MNPs on incubation time are shown in Table 1. Signal intensity of spiedlabeled BSA changes insignificantly. These changes correspond to adsorption of approximately 12% of BSA after the sample incubation for 100 min. The study of adsorption kinetics allows establishing that adsorption equilibrium in "T1-MNP-0" takes place when the incubation time equals to 80 min and ~41% of thrombireis adsorbed. The value of adsorption A may be estimated using the data on the portion of macromolecules adsorbed and specific surface area calculated from MNPs density (5200 mg/m³), concentration and size. Hence BSA adsorption equals to 0.3 mg/m² after 100 mes incubation. The dependence of thrombin adsorption value on incubation time is shown in Figur. 7. Thrombin adsorption equals to 1.2 mg/m² after 80 mes incubation.

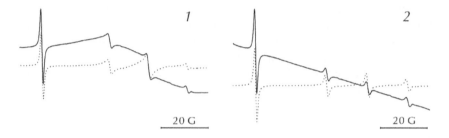

FIGURE 6   The ESR spectra of spin labels on BSA (*1*) and thrombin (2) macromolecules before (dotted line) and 75 mes after (solid line) MNPs addition to protein solution at 25°C. External standart – MgO powder containing Mn^{2+}.

TABLE 1   The dependence of relative intensity of low-field line of triplet I+1 of nitroxide radical covalently bound to BSA and thrombin macromolecules, and the portion N of the protein adsorbed on incubation time t of the samples "A1-MNP-0" and "T1-MNP-".

	Spin-labelled BSA		Spin-labelled thrombin	
t, min.	I_{+1}, rel. units	N, %	I_{+1}, rel. units	N, %
0	0.230 ± 0.012	0 ± 5	0.25 ± 0.01	0 ± 4
15	-	-	0.17 ± 0.01	32 ± 4
35	0.205 ± 0.012	9 ± 5	0.16 ± 0.01	36 ± 4
75	0.207 ± 0.012	10 ± 5	0.15 ± 0.01	40 ± 4
95	-	-	0.15 ± 0.01	40 ± 4
120	0.200 ± 0.012	13 ± 5	0.14 ± 0.01	44 ± 4

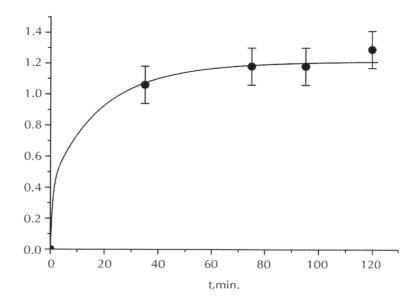

FIGURE 7   Kinetics of thrombin adsorption on magnetite nanoparticles at 25°C. Concentration of thrombin in the sample is 0.9 mg/mL, MNPs—4.0 mg/ml.

The FMR spectra of the samples "A1-MNP-0", "T1-MNP-0" and MNPs are characterized by different position in magnetic field (Figure 8). The centre of the spectrum of MNPs is 3254 G, while the centre of "A1-MNP-0" and "T1-MNP-0" spectra is 3253 G and 3449 G respectively. Resonance conditions for MNPs in magnetic field of spectrometer include a parameter of the shift of FMR spectrum $|M_1| = \frac{3}{2}|H_1|$, where H_1 is a local field created by MNPs in linear aggregates which form in spectrometer field. $H_1 = 2\sum_1^\infty \frac{2\mu}{(nD)^3}$, where D is a distance between MNPs in linear aggregates, μ is MNPs magnetic moment, n is a number of MNPs in aggregate [23]. Coating formation and the thickness of adsorption layer influence on the distance between nanoparticles decrease dipole interactions and particles ability to aggregate. As a result the centre of FMR spectrum moves to higher fields. This phenomenon of FMR spectrum centre shift we observed in the system "A1-MNP-0" after FG addition [20]. The spectrum of MNPs with thick coating becomes similar to FMR spectra of isolated MNPs. So the similar centre positions of FMR spectra of MNPs without coating and MNPs in BSA coating point to a very thin coating and low adsorption of protein in this case. In contrast according to FMR centre position the thrombin coating on MNPs is thicker than albumin coating. *This result is consistent with the data* obtained by ESR spectroscopy.

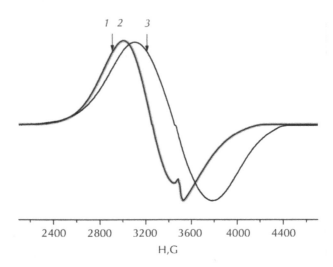

FIGURE 8 The FMR spectra of MNPs (*1*), MNPs in the mixture with BSA (the sample "A1-MNP-0") after incubation time of 120 min (*2*) and MNPs in the mixture with thrombin (the sample "T1-MNP-0") after incubation time of 120 min (*3*).

The FG ability to replace BSA in adsorption layer on MNPs surface is demonstrated in Figure 9. Initially there is bimodal volume distribution of particles over sizes in the sample "A2-MNP-0" that can be explained by existence of free (unabsorbed) BSA and MNPs in BSA coating. After FG addition the distribution changes. Micron-sized clusters form in the sample that proves FG adsorption on MNPs [18]. In the case of "A2-MNP-1" volume distribution is also bimodal. The peak of MNPs in BSA coating is characterized by particle size of maximal contribution to the distribution of ~23 nm. This size is identical to MNPs in BSA coating in the sample "A2-MNP-0". It proves that H_2O_2 addition does not lead to uncontrollable linking of protein macromolecules in solution or cluster formation. Since MNPs size is 17 nm, the thickness of adsorption layer on MNPs is approximately 3 nm.

After FG addition to "A2-MNP-1" micron-sized clusters do not form. So adsorption BSA layer formed in the presence of H_2O_2 keeps stability. This stability can be explained by formation of covalent bonds between protein macromolecules [13] in adsorption layer as a result of free radicals generation on MNPs surface. Stability of BSA coating on MNPs was demonstrated for the samples "A1-MNP-1" and "A2-MNP-1" incubated for more than 100 min before FG addition. Clusters are shown to appear if the incubation time is insufficient.

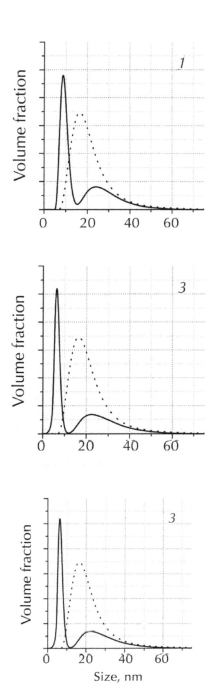

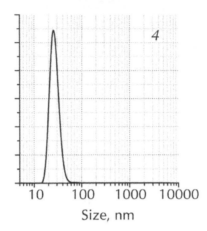

FIGURE 9 Volume distributions of particles in sizes in systems without (*1, 2*) and with (*3, 4*) H_2O_2 ("A2-MNP-0", "A2-MNP-1") incubated for 2 hr before (*1, 3*) and 20 min after (*2, 4*) FG addition. Dotted line is the volume distribution of nanoparticles in sizes in dispersion.

The precipitates obtained by ultracentrifugation of "A2-MNP-0", "A2-MNP-1", "T1-MNP-0" and "T1-MNP-1" were dissolved in buffer solution. The amount of protein in precipitates was evaluated by Bradford colorimetric method (Table 2). The results showed that precipitates of systems with H_2O_2 contained more protein than the same systems without H_2O_2. Therefore in the samples containing H_2O_2 the significant part of protein molecules does not leave MNPs surface when centrifuged while in the samples "A2-MNP-0" and "T1-MNP-0" the most of protein molecules leaves the surface. This indicates the stability of adsorption layer formed in the presence of free radical generation initiator and proves cross-links formation.

TABLE 2 The amount of protein in precipitates after centrifugation of the samples "A2-MNP-0", "A2-MNP-1", "T1-MNP-0", and "T1-MNP-1" of 3.2 mL in volume

Sample name	Amount of protein in precipitates, mg
"A2-MNP-0"	0.05
"A2-MNP-1"	0.45
"T1-MNP-0"	0.15
"T1-MNP-1"	1.05

Analysis of content of supernatant liquids obtained after ultracentrifugation of reaction systems containing MNPs and BSA which differed by H_2O_2 and ascorbic acid presence ("A2-MNP-0", "A2-MNP-1" and "A2-MNP-1-acid") allows evaluating the scale of free radical processes in the presence of H_2O_2. As it was mentioned above in the presence of ascorbic acid free radicals generate not only on MNPs surface but also in solution. So both molecules on the surface and free molecules in solution can undergo free radical modification in this case. From Figure 10 we can see that the IR-spectrum of "A2-MNP-1-acid" differs from the spectra of "A2-MNP-0" and "A2-MNP-1", while *the spectra of* "A2-MNP-0" and "A2-MNP-1" *almost* have no differences. The IR-spectra differ in the region of 1200–800 cm^{-1}. The changes in this area are explained by free radical oxidation of amino acid residues of methionine, tryptophane, histidine, cysteine, and phenylalanine. These residues are sulfur-containing and cyclic ones which are the most sensitive to free radical oxidation [13,24]. The absence of differences in "A2-MNP-0" and "A2-MNP-1" proves that cross-linking of protein molecules in the presence of H$_2$O$_2$ is selective and takes place only on MNPs surfaces.

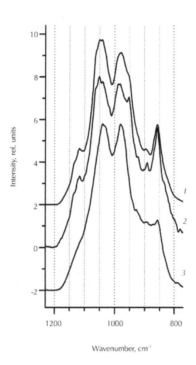

FIGURE 10   The IR-spectra of supernatant solutions obtained after centrifugation of the samples "A2-MNP-0" (*1*), "A2-MNP-1" (*2*), and "A2-MNP-1-acid" (*3*).

When proteins are used as components of coating on MNPs for biology and medicine their functional activity retaining is very important. Proteins fixed on MNPs can lose their activity as a result of adsorption on MNPs or free radical modification which is cross-linking and oxidation but it was shown that they do not lose it. Estimation of enzyme activity of thrombin cross-linked on MNPs surface was performed by comparison of the rates of conversion of FG to fibrin under the influence of thrombin contained in reaction mixtures "T1-MNP-0", "T1-MNP-1", and "T1-0-0" (Figure 11). The curves for the samples containing thrombin and MNPs which differ by the presence of H$_2$O$_2$ had no fundamental differences that illustrates preservation of enzyme activity of thrombin during free radical cross-linking on MNPs surface. Fibrin gel was formed during ~15 min in both cases. Rayleigh light scattering intensity was low when "T1-0-0" was used and small fibrin particles were formed in this case. The reason of this phenomenon is autolysis (self-digestion) of thrombin. Enzyme activity of thrombin, one of serine proteinases, decreases spontaneously in solution [25]. So the proteins can keep their activity longer when adsorbed on MNPs. This way, the method of free radical cross-linking of proteins seems promising for enzyme immobilization.

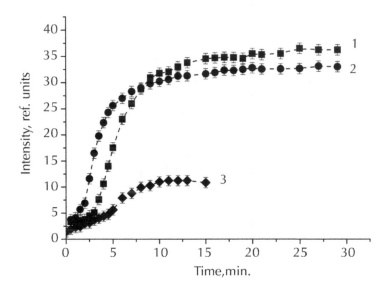

FIGURE 11   Kinetics curves of growth of Rayleigh light scattering intensity in the process of fibrin gel formation in the presence of "T1-MNP-0" (*1*), "T1-MNP-1", (*2*) and "T1-0-0" (*3*).

CONCLUSION

The novel method of fixation of proteins on MNPs proposed in the work was successfully realized on the example of albumin and thrombin. The blood plasma proteins are characterized by a high biocompatibility and allow decreasing toxicity of nanoparticles administered into organism. The method is based on the ability of proteins to form interchain covalent bonds under the action of free radicals. The reaction mixture for stable coatings obtaining should consist on protein solution, nanoparticles containing metals of variable valence (for example, Fe, Cu, and Cr) and water-soluble initiator of free radicals generation. In this work albumin and thrombin were used for coating being formed on magnetite nanoparticles. Hydrogen peroxide served as initiator. By the set of physical (ESR-spectroscopy, ferromagnetic resonance, dynamic and Rayleigh light scattering, IR-spectroscopy) and biochemical methods it was proved that the coatings obtained are stable and formed on individual nanoparticles because free radical processes are localized strictly in the adsorption layer. The free radical linking of thrombin on the surface of nanoparticles has been shown to almost completely keep native properties of the protein molecules. Since the method provides enzyme activity and formation of thin stable protein layers on individual nanopaticles it can be successfully used for various biomedical goals concerning a smart delivery of therapeutic products and biologically active substances (including enzymes). It reveals principally novel technologies of one-step creation of biocompatible magnetically targeted nanosystems with multiprotein polyfunctional coatings which meet all the requirements and contain both biovectors and therapeutic products (Figure 12).

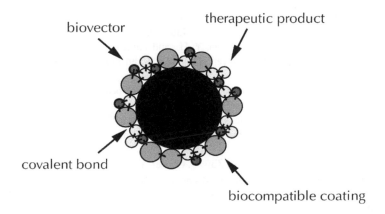

FIGURE 12 The scheme of magnetically targeted nanosystem for a smart delivery of therapeutic products based on the free radical protein cross-linking.

KEYWORDS

- **Bovine serum albumin**
- **Magnetic nanoparticles**
- **Nanosystems**
- **Novel method**
- **Spin labels technique**

REFERENCES

1. Gupta, A. K. and Gupta, M. Synthesis and surface engineering of iron oxide nanoparticles for biomedical applications. Biomaterials, 26, 3995–4021 (2005).
2. Vatta, L. L., Sanderson, D. R., and Koch, K. R. Magnetic nanoparticles: Properties and potential applications. Pure Appl. Chem., 78, 1793–1801 (2006).
3. Lu, A. H., Salabas, F., Schűth, E. L. Magnetic Nanoparticles: Synthesis, Protection, Functionalization, and Application. Angew. Chem. Int. Ed., 46, 1222–1244 (2007).

4. Laurent, S., Forge, D., Port, M., Roch, A., Robic, C., Elst, L. V., and Muller, R. N. Magnetic Iron Oxide Nanoparticles: Synthesis, Stabilization, Vectorization, Physicochemical Characterizations, and Biological Applications. Chem. Rev., 108, 2064–2110 (2008).
5. Pershina, A. G., Sazonov, A. E., and Milto, I. V., Application of magnetic nanoparticles in biomedicine. Bulletin of Siberian Medicine [in Russian], 2, 70–78 (2008).
6. Trahms, L. Biomedical Applications of Magnetic Nanoparticles. Lect. Notes Phys., 763, 327–358 (2009).
7. Bychkova, A. V., Sorokina, O. N., Rosenfeld, M. A., and Kovarski, A. L. Multifunctional biocompatible coatings on magnetic nanoparticles. Uspekhi Khimii (Russian journal), in press (2012).
8. Koneracka, M., Kopcansky, P., Antalik, M., Timko, M., Ramchand, C. N., Lobo, D., Mehta, R. V., and Upadhyay, R. V. Immobilization of proteins and enzymes to fine magnetic particles. J. Magn. Magn. Mater., 201, 427–430 (1999).
9. Xu, L., Kim, M. J., Kim, K. D., Choa, Y. H., and Kim, H. T. Surface modified Fe3O4 nanoparticles as a protein delivery vehicle. Colloids and Surfaces A: Physicochem. Eng. Aspects, 350, 8–12 (2009).
10. Peng, Z. G., Hidajat, K., and Uddin, M. S. Adsorption of bovine serum albumin on nanosized magnetic particles. Journal of Colloid and Interface Science, 271, 277–283 (2004).
11. Šafařík, I., Ptáčková, L., Koneracká, M., Šafaříková, M., Timko, M., and Kopčanský, P. Determination of selected xenobiotics with ferrofluid-modified trypsin. Biotechnology Letters, 24, 355–358 (2002).
12. Li, F. Q., Su, H., Wang, J., Liu, J. Y., Zhu, Q. G., Fei, Y. B., Pan, Y. H., and Hu, J. H. Preparation and characterization of sodium ferulate entrapped bovine serum albumin nanoparticles for liver targeting. International Journal of Pharmaceutics, 349, 274–282 (2008).
13. Stadtman, E. R. and Levine, R. L. Free radical-mediated oxidation of free amino acids and amino acid residues in protein. Amino Acids, 25, 207–218 (2003).
14. Bychkova, A. V., Sorokina, O. N., Shapiro, A. B., Tikhonov, A. P., and Kovarski, A. L. Spin Labels in the Investigation of Macromolecules Adsorption on Magnetic Nanoparticles. The Open Colloid Science Journal, 2, 15–19 (2009).
15. Abragam, A. The Principles of Nuclear Magnetism, Oxford University Press, New York (1961).
16. Noginova, N., Chen, F., Weaver, T., Giannelis, E. P., Bourlinos, A. B., and Atsarkin, V. A. Magnetic resonance in nanoparticles: between ferro- and paramagnetism. J. Phys.: Cond. Matter, 19, 246208–246222 (2007).
17. Sorokina, O. N., Kovarski, A. L., and Bychkova, A. V. Application of paramagnetic sensors technique for the investigation of the systems containing magnetic particles. In Progress in Nanoparticles research, C. T. Frisiras (Ed.), Nova Science Publishers, New York, 91–102 (2008).
18. Bychkova, A. V., Sorokina, O. N., Kovarskii, A. L., Shapiro, A. B., Leonova, V. B., and Rozenfel'd, M. A. Interaction of fibrinogen with magnetite nanoparticles. Biophysics (Russian journal), 55, 4, 544–549 (2010).
19. Bychkova, A. V., Sorokina, O. N., Kovarski, A. L., Shapiro, A. B., and Rosenfeld, The Investigation of Polyethyleneimine Adsorption on Magnetite Nanoparticles by Spin Labels Technique. Nanosci. Nanotechnol. Lett. (ESR in Small Systems), 3, 591–593 (2011).
20. Bychkova, A. V., Rosenfeld, M. A., Leonova, V. B., Lomakin, S. M., Sorokina, O. N., and Kovarski, A. L. Surface modification of magnetite nanoparticles with serum albumin in dispersions by free radical cross-linking method. Russian colloid journal (2012).

21. Bradford, M. M. A rapid and sensitive method for the quantitation of microgram quanti-
    ties of protein utilizing the principle of protein-dye binding. Anal. Biochem., 72, 248–254
    (1976).
22. Antsiferova, L. I., Vasserman, A. M, Ivanova, A. N., Lifshits, V. A., and Nazemets, N. S.
    Atlas of Electron Paramagnetic Resonance Spectra of Spin Labels and Probes [in Rus-
    sian], Nauka, Moscow (1977).
23. Dolotov, S. V. and Roldughin, V. I. Simulation of ESR spectra of metal nanoparticle ag-
    gregates. Russian colloid journal, 69, 9–12 (2007).
24. Smith, C. E., Stack, M. S., and Johnson, D. A. Ozone effects on inhibitors of human neu-
    trophil proteinases. Arch. Biochem. Biophys., 253, 146–155 (1987).
25. Blomback, B. Fibrinogen and fibrin – proteins with complex roles in hemostasis and
    thrombosis. Thromb. Res., 83, 1–75 (1996).

CHAPTER 21

ANTIMICROBIAL PACKAGING FOR FOOD APPLICATIONS

S. REMYA, C. O. MOHAN, C. N. RAVISHANKAR, R. BADONIA,
and T. K. SRINIVASA GOPAL

CONTENTS

21.1   INTRODUCTION

In recent years, food quality and safety have become major concerns across the border in the food industry. Active packaging (AP) is becoming increasingly important to overcome these concerns. Of the major AP technologies like oxygen scavenging, moisture absorption and control, carbon dioxide emitter and absorber, ethanol generator, antimicrobial and antioxidant packaging, and antimicrobial packaging is of great importance. Antimicrobial packaging has attracted much attention from the industry as an emerging technology for extending the shelf life and food safety of minimally processed and preservative free products. Use of antimicrobial packaging is useful in controlling the microbial growth and to target specific microbial pathogens to provide higher safety to the products. The characteristics of some antimicrobial packaging system, manufacturing of antimicrobial films, factors influencing the AM films, and future applications of antimicrobial packaging are discussed in this chapter.

Packaging is a discipline of post-production process where the product is enclosed in a container for many purposes like protection, transportation, distribution, storage, retailing, and end-use. Nowadays, packaging has assumed increasingly important role in the whole food chain '*from farm to fork*'. As an example, many fish and shellfishes are harvested from the aquatic environment and packed appropriately in live form for transportation to distant places. Many fresh agricultural products like fruits and vegetables are picked in the field and packed directly into consumer packages. The advantage of this is product is touched only once before it reaches the consumers. Another example is ready-to-eat food and snack products which are packed in microwaveable trays which allow consumers to prepare the food immediately and even serve as an eating dish. Perishable nature of most food products lead to the development of wide variety of preservation techniques. Till today most of the preservation processes like asceptic, baking, canning and sous-vide methods are largely depend on effective packaging. Preservation methods like freezing and drying will be ineffective without proper protective packaging after processing, which controls the exposure of light, oxygen, water vapor, microbes, and other contaminants. Food packaging has developed strongly during recent decades, mainly due to increased demands on product safety, shelf- life extension, cost-efficiency, environmental issues, and consumer convenience.

Food packaging must perform several tasks and satisfy many demands and requirements. Basic packaging requirements include good physical properties, good marketing properties, reasonable price, technical feasibility (for example suitability for automatic packaging machines, sealing ability, and so on), suitability for food contact, low environmental stress, and suitability for recycling and refilling [1]. A package must satisfy each of these requirements both effectively and economically. Furthermore, packaging has a more significant role in the preservation of food and in ensuring the safety of food in order to avoid wastage, food poisoning and to reduce allergies. To ensure a longer shelf-life, packaging must play an active role in processing, preservation and in retaining the quality of food products. The development of modified atmosphere packaging (MAP) over three decades ago was one of the first examples showing that some product/package interactions may have a positive effect

[22,47]. A more recent and advanced class of food packaging systems is known as "active packaging" [52].

The AP is an innovative concept that can be defined as a type of packaging that changes the condition of the packaging and maintains these conditions throughout the storage period to extend shelf-life or to improve safety or sensory properties while maintaining the quality of packaged food [1,52,58]. The AP performs some desired role other than providing an inert barrier between the product and external conditions [58] and combines advances in food technology, bio-technology, packaging and material science, in an effort to comply with consumer demands for "fresh like" products. The AP techniques for preservation and improving quality and safety of foods can be divided into three categories: (i) absorbing systems, (ii) releasing systems, and (iii) other specialty system for temperature, ultraviolet light and microwave control systems [64]. Major AP techniques are concerned with substances that absorb oxygen, ethylene, moisture, carbon dioxide, flavors/odors and those which release carbon dioxide, antimicrobial agents, antioxidants, and flavors. Among these antimicrobial packaging is assuming greater importance for packing wide variety of food products.

21.2 ANTIMICROBIAL PACKAGING

Antimicrobial packaging is a type of active packaging, where the package is designed to release active agents to inhibit the growth of microorganisms inside the package. Contamination of foods by a small number of hazardous microorganisms is generally undetectable and people may consume these without recognition. The subsequent consumption of contaminated foods can cause food poisoning and other food-borne diseases. Adequate hygiene is very important in food processing in order to prevent or minimize contamination by microorganisms. All raw materials handling, machinery and the surrounding environment must therefore be in sanitary condition. Manufacturing in completely germ-free conditions is very expensive and all the food industries cannot afford this. It is therefore accepted that most manufactured foods contain microorganisms to some level. As a result these foods in the market are to be purchased and consumed within a limited time while the number of microorganisms is still at an adequate level for consumption. Raw materials can carry natural microorganisms and during transport they may be contaminated with microorganisms from the surrounding environments such as air, water, soil, and sewage. Moreover, they may be polluted by diseased plants and animals, including pests. During processing, they may be contaminated by humans, processing equipment, and packaging materials. Microorganisms from a variety of sources can contaminate the surface of solid foods and penetrate into the bulk of the food. In liquid foods, the contaminating microorganisms can spread relatively easily through the foods whereas in solid foods the contamination generally remains on the surface of the food. When food is contaminated, the entire population of contaminating microorganisms will not grow uniformly and will spoil the food. There will be a competition among the contaminating microorganisms and finally the fastest growing genera will dominate. For example, fresh fish and shellfish is likely contaminated by approximately 25 genera of bacteria in sub-classes of Gram positive rods and cocci, Gram negative rods and coccobacilli and few genera of moulds and yeasts. Only few genera of microorganisms like *Pseudomonas* spp and *Shewanella* spp dominate and subsequently spoil the products. Like fishes, fresh meat is likely contaminated by approx. 30 genera of bacteria including 12 genera of moulds and 6 genera of yeasts. Species of dominant microorganisms are altered in different foods and at different storage conditions. Knowledge of dominant microorganisms is necessary to choose the right antimicrobial agents to inhibit the growth of such microorganisms.

In most foods, the surface growth of microorganisms is the major cause of food spoilage. Antimicrobial agents are often mixed directly into foods to control microbial growth and to extend shelf life. The vast majority of these AM agents, however, are synthetic materials that have the following disadvantages like they are distributed in the bulk of the food at relatively large quantities and therefore may impart an off-flavor and consumers are concerned about the possible side-effects of synthetic additives. To overcome these problems, AM packaging systems are helpful. Antimicrobial packaging differs from that of addition of chemical preservatives directly into the food matrix, where an excess amount of these synthetic additives is believed to be of concern. For the consumer, it seems safer when active agents are indirectly integrated in the food package and released into the food product thereafter. Antimicrobial packaging systems are able to kill or inhibit spoilage and pathogenic microorganisms that can potentially contaminate food products. The inhibition of microbial activity is achieved by slow release of AM agents from the packaging system onto the food surface. When a packaging system acquires AM activity, the packaging system limits or prevents microbial growth by extending the lag period and reducing the growth rate or decrease live counts of microorganisms. The goals of an AM system are safety assurance, quality maintenance and shelf-life extension [1].

The AM packaging materials can also be used for self sterilizing. If the packaging materials have self-sterilizing ability because of their own antimicrobial activity, they may eliminate chemical sterilization of packages using peroxide and simplify the aseptic packaging process. The self-sterilizing materials could be widely applied for clinical uses in food packaging, hospitals, biological lab-ware, biotechnology equipment, and biomedical devices. Such AM packaging materials greatly reduce the potential for recontamination of processed products and simplify the treatment of materials in order to eliminate product contamination. AM polymers might also be used to cover surfaces of food processing equipment so that they self-sanitize during use.

21.3   ANTIMICROBIAL AGENTS

Antimicrobial agents can be either natural or synthetic compound that can be used in a variety of applications in the food, pharmaceutical, and cosmetic industries. The AM agents suitable for incorporating in packaging film includes organic acids, fungicides, bacteriocins, proteins, enzymes, polymers, inorganic gases, alcohols, antibiotics, and metal substituted zeolite. With the increase in consumer awareness for food safety and health standards, there is a general concern for use of chemical preservatives in food chain. In response to this, bio-preservatives and naturally derived AM additives are becoming more important as they represent a perceived lower risk to consumers. More extensive attempts are being made in the search for alternative AM compounds based on plant extracts. For example, the AM effect of essential oils and their active constituents against many foods borne pathogenic bacteria including *Salmonella enterica, Campylobactor jejuni, Staphylococcus aureus, Escherichia coli, Listeria monocytogenes*, and *Vibrio parahaemolyticus* have been studied extensively. The use of natural extracts is often preferred due to less complex regulation processes and consumer preference when compared to chemical AM agents Plant extracts such as grapefruit seed, cinnamon, horseradish, and cloves have been added to packaging system to demonstrate effectiveness against spoilage and pathogenic bacteria. The use of natural AM compounds is not only important in the control of human and plant diseases of microbial origin but also in preservation and packaging food products. Table 1 lists some typical natural and synthetic AM agents that are used in food packaging and few of them are discussed.

TABLE 1   Classes of Antimicrobial agents for food contact applications

Class	Examples
Acid Anhydride	Benzoic anhydride, Sorbic anhydride
Alcohol	Ethanol
Ammonium Compound	Silicon quaternary ammonium salt
Antibiotic	Natamycin
Antimicrobial peptides	Leucocin, Sakacin, Enterocin
Antioxidant Phenolic	Butylated hydroxyanisole (BHA), Butylated hydroxytoluene (BHT), Tertiary butylhydroquinone (TBHQ), Grape seed extract, pomegranate peel and seed extracts
Bacteriocin	Bavaricin, Lacticin, Nisin, Pediocin
Chelator	Citric acid, EDTA, Lactoferrin, Polyphosphate
Enzyme	Chitinase, Ethanol oxidase, Glucose oxidase, Glucosidase, Lysozyme, Lactoperoxidase, Hydrolases
Fatty Acid	Lauric acid, Palmitoleic acid
Fungicide	Benomyl, Imazalil, Sulfur dioxide

TABLE 1   *(Continued)*

Metal	Copper, Silver
Natural Phenol	Catechin, Hydroquinones
Organic Acid	Acetic acid, Benzoic acid, Citric acid, Lactic acid, Propionic acid, Sorbic acid, Tartaric acid
Organic Acid Salt	Potassium sorbate, Sodium benzoate, Acetic, propionic acid, Benzoic, sorbic acid, Calcium sorbate, Benzoic anhydride, Propionic acid, Propyl paraben
Paraben	Ethyl, methyl and propyl paraben
Plant-Volatile Component	Allyl isothiocyanate, Cinnamaldehyde, Eugenol, Terpineol, Thymol
Polysaccharide	Chitosan, carragenan

Lysozyme is a food grade enzyme (EC 3.2.1.17), abundantly present in a number of secretions, such as tears (except bovine tears). This is present in cytoplasmic granules of the polymorphonuclear neutrophils (PMN) (except for bovine neutrophils) and released through the mucosal secretions (such as tears and saliva). They can also be found in high concentration in egg white. This enzyme shows antimicrobial activity mainly on Gram positive bacteria mainly attacking peptidoglycans in the cell wall by hydrolyzing the bond that connects *N*-acetyl muramic acid with the fourth carbon atom of *N*-acetylglucosamine [14]. Due to its high stability over a wide range of pH and temperature, lysozyme might be used to extend the shelf-life of various food types [30].

Ethylene-diamine-tetra-acetic Acid (EDTA) play the main role as the chelating agent for lysozyme to inhibit the growth of gram negative bacteria by forming coordination compounds with most monovalent, divalent, trivalent and tetravalent metal ions, such as silver (Ag^+), calcium (Ca^{2+}), manganese (Mn^{2+}), copper (Cu^{2+}), iron (Fe^{3+}), or zirconium (Zr^{4+}). The EDTA at elevated concentrations is toxic to bacteria due to chelation of metals in the outer membrane.

Lauric acid or dodecanoic acid is a colorless, needle-like crystal which is vital in the construction of cellular membranes. Coconut and coconut milk are good source of lauric acid. Lauric acid has the potential to be used as antimicrobial agent because it is locally available abundantly. Besides, it is inexpensive, non-toxic and safe to handle and has a long shelf-life. It is also suitable for incorporating in to packaging film.

The hexamethylene-tetramine (HMT) is a white crystalline powder with slight amine odor. The use HMT as an AM packaging agent is mainly due to the formation of formaldehyde when the film comes into contact with an acidic medium. The ions of silver and copper, quaternary ammonium salts, and natural compounds such as Hinokitiol are generally considered safe AM agents. Silver-substituted zeolite (Ag-zeolite) is the most common agent with which plastics are impregnated. It retards a range of metabolic enzymes and has a uniquely broad microbial spectrum. Triclosan belongs to a large class of antimicrobials containing a phenolic ring as its chemical backbone. Triclosan at 0.5 to 1.0% w/w exhibits antimicrobial activity against *S. aureus*, *L. monocytogenes*, *E. coli* O157:H7, *S. enteritidis*, and *Brocothrix thermosphacta* in agar diffusion assay.

Another compound that exhibits antimicrobial effects is ethanol. The use of alcohol to prolong shelf life is a well-known method in food preservation. Ethanol is commonly used as surface disinfectant. The effect of ethanol depends on its concentration. At higher concentrations (60–70% v/v), ethanol denatures the proteins of the protoplasts of vegetative cells of microorganisms. Even at relatively low concentrations (4–12%), ethanol is effective in controlling growth of several moulds and bacteria. Ethanol can be applied on the surface of food by spraying prior to packaging. Another option is to incorporate the sachets generating ethanol vapor into the food package. In this, sachets contain food grade ethanol absorbed or encapsulated in a carrier material. A slow or rapid release of ethanol from the carrier material to the package headspace is regulated by the permeability of the sachet material to water vapor. The ethanol in the carrier material is exchanged with the water absorbed by the carrier material. A major disadvantage of ethanol vapor is its absorption by the food product. In some cases the ethanol concentration in the product might cause regulatory problems. If the product is heated prior to consumption the accumulated ethanol may evaporate. Another drawback is the cost of the sachets, which limits their use to products with higher profit margins. Ethanol vapor generators are widely used in many countries for high moisture bakery goods, fish products, and cheese.

Another most effective antimicrobial agent is the nano-sized antimicrobial agents. These have the advantage of high surface area-to-volume ratio and enhanced surface reactivity compared to their high size counterparts. The most commonly used nano-antimicrobial agents are silver nano particles, metal oxide nanoparticles, chitosan nanostructure, nanoclays, and so on. Silver exhibits broad spectrum antimicrobial activities against both Gram negative and Gram positive bacteria, fungi, protozoa, and certain viruses. It also has high temperature stability and low volatility which is advantageous in incorporating into films [38]. Nanoparticles of silver favors the interaction with microbial cells due to their larger surface area to volume ratio due to which they are more effective than larger silver particles [4]. This was well demonstrated by Damm et al [16] who found out that silver nanoparticles incorporated at low concentration of 0.06 wt% Ag with polyamide was 100% effective in killing *E coli* compared to only 80% killing by silver microcomposite containing 1.9 wt % of Ag incorporated polyamide film. The mechanism of the antimicrobial activity of silver nanoparticles is gradual release of silver ions resulting in inhibition of ATP production and DNA replication, direct damage to microbial cell membranes, generation of reactive oxygen species (ROS) which damages the cell and interaction of silver ions with thiol groups in protein which inactivates bacterial enzymes [15,23,38].

Titanium oxide (TiO_2) is another proven antimicrobial agent. The antimicrobial effect of TiO_2 is attributed to the production of active radicals when the photocatalyst TiO_2 is exposed to the UV illumination. These active radicals oxidize C–H bonds leading to degradation of the organic molecules [29,61], generation of hydroxyl radicals (OH) and ROS on the TiO_2 surface which results in the oxidation of polyunsaturated phospholipids of cell membranes of microorganisms. As a consequence, the microorganism is inactivated [13,37]. TiO_2 photocatalyst has been used to inactivate a wide spectrum of microorganisms [19,42]. First report on the antimicrobial effect of TiO_2 was reported on *E coli* [44] followed by *Pseudomonas aeruginosa* and *Enterococcus faecalis* [32,36], *Lactobacillus helveticus* [40], *Legionella pneumophila* [12], *Clostridium perfringens* and Coliphages [27], coliform bacteria [51], and food-borne pathogenic bacteria such as *Salmonella enterica Choleraesuis, Vibrio parahaemolyticus and Listeria monocytogenes* [33]. The use of TiO_2 for human food, drugs, cosmetics, and use in the food industry has been approved by the American Food and Drug Administration (FDA) as a non-toxic material. At present, considerable attraction is also given to the self-disinfecting property of TiO_2 for meeting hygienic design requirements in food processing and packaging surfaces [11].

Apart from synthetic chemicals, several natural compounds have commonly used as antimicrobial agents to incorporate in the packaging film. These include, organic acids and nisin [21,55], neem [31], chitosan, and essential oils [3,25,41,46,48]. Among these different substances, essential oils have an important role in food industry to use as antimicrobial agents in biodegradable films.

Various chemicals dip treatment such as cetylpyridinium chloride, chlorine dioxide, potassium sorbate, sodium lactate and lactic acid, Sodium and potassium lactates, sodium diacetate, sodium chloride (NaCl) and sodium nitrite (NaNO2), or potassium nitrate (KNO3), Sodium acetate, citrate, and lactate have been used in various food systems including fish products to improve the safety and to extend the shelf life. Among these sodium salts of the low molecular weight organic acids, such as acetic, lactic and citric, have been used extensively to control microbial growth, improve sensory attributes, and extend the shelf life of various food systems. In addition to their suppressing effect on the growth of food spoilage bacteria by reducing the water activity, organic salts of sodium acetate, lactate and citrate possess antibacterial activities against various food-borne pathogens. These salts are widely available, economical and generally recognized-as-safe (GRAS) by the U.S. Code of Federal Regulations, 21 CFR. Chemical preservative treatments coupled with the advanced packaging technologies and refrigerated storage will enhance the storage life of food products.

Essential oils are defined as any volatile oil(s) that have strong aromatic components which gives distinctive odor, flavor, or scent to a plant. These are the by-products of plant metabolism and are commonly referred to as volatile plant secondary metabolites. Normally, essential oils are found in glandular hairs or secretory cavities of plant-cell wall and are present as droplets of fluid in the leaves, stems, bark, flowers, roots, and/or fruits in different plants. The aromatic characteristics of essential oils provide many functions for plants like attracting or repelling insects, protecting themselves from heat or cold and utilizing chemical constituents in the oil as defence materials. Many of the essential oils have other uses as food additives, flavorings, and components of cosmetics, soaps, perfumes, plastics, and as resins. Typically these oils are liquid at room temperature and get easily transformed from liquid to gaseous state at room or slightly higher temperature without undergoing decomposition. The amount of essential oil found in most plants is 1–2%, but can contain amounts ranging from 0.01 to 10%.

Essential oils possess antibacterial, antioxidant, antiviral, and anti-mycotic properties [10], due to their active phenolic compounds, that is carvacrol, thymol [10,18,39]. Essential oils such as gingerol, cinnamon oil, anise oil and clove oil are well known inhibitors of mold, yeast, and bacteria growth. There have been a number of reports on a distinct property of substances in essential oil which inhibit the growth of mold and yeast [6,43].

Ginger, *Zingiber officinale,* is a monocotyledonous herbaceous perennial belonging to the Zingiberaceae family, which is characterized by a pale-yellow pungent aromatic rhizome that is the important part of this spice. It contains oleoresin and essential oils [62]. Ginger is used worldwide as a food ingredient and medicine. It has long been used to treat many gastrointestinal disorders and is often promoted as an effective antiemetic [7,8]. Gingerols, the pungent principles in the rhizome of ginger, were reported to have antiemetic, analgesic, antipyretic, anti-inflammatory, chemopreventive, and antioxidant properties [20,53,54]. Essential oils are considered as safe additive for food applications as they control foodborne pathogenic and spoilage bacteria [10]. Volatile components exhibit higher antimicrobial activity [26,35]. It is also observed that incorporation of active components of essential oils into edible films by vapor phase is very efficient which require much smaller concentrations to inhibit different microorganisms. Carvacrol is a phenolic compound extracted from oregano and thyme oil which has inhibitory effect on the growth of various microorganisms [35,45]. Studies on the effect of essential oils on different food products such as fish [34], meat [56], and against different microorganisms such as *E. coli* and *Bacillus* [9], *Campylobacter jejuni, Escherichia coli, Listeria monocytogenes,* and *Salmonella enterica* [24] are also reported.

Nisin is a common antimicrobial agent used as a food preservative and as a food additive. It has E number E 234. It is an inhibitory polycyclic peptide with 34 amino acid residues. Nisin is produced by fermentation using the bacterium *Lactococcus lactis.* Commercially it is obtained from natural substrates including milk and is not chemically synthesized. It is used in processed cheese production to extend shelf life by suppressing gram-positive spoilage and pathogenic bacteria. It is also used in fish and meat products to prevent the *Clostridium botulinum* growth and toxin production.

Chitosan is a linear polysaccharides prepared by cleavage of N-acetyl groups of the chitin. Apart from various other applications, chitosan is highly regarded for its antimicrobial properties. Chitosan has been reported as an antibacterial agent against a wide variety of microorganisms both gram positive and gram negative bacteria, fungi, and so on [60]. Compared to chitin, chitosan exhibits greater antimicrobial activity due to the greater number of free amino groups, which respond for the antimicrobial activity upon protonation [50]. Two main mechanisms have been suggested as the cause of the inhibition of microbial cells by chitosan. The interaction with anionic groups on the cell surface, due to its polycationic nature, causes the formation of an impermeable layer around the cell, which prevents the transport of essential solutes. The second mechanism involves the inhibition of the RNA and protein synthesis by permeation into the cell nucleus. Other mechanisms have also been proposed. Chitosan may inhibit microbial growth by acting as a chelating agent rendering metals, trace elements, or essential nutrients unavailable for the organism to grow at the normal rate. However, the most accepted antimicrobial mechanism for chitosan is related to interactions of the cationic chitosan with the anionic cell membranes, increasing membrane permeability and eventually resulting in rupture and leakage of the intracellular material. To avoid the use of bulk chitosan, chitosan nanoparticles (CSNP) are preferred as antimicrobial compounds. These CSNP are usually prepared by interaction of oppositely charged macromolecules. The mechanism of CSNP formation is based on electrostatic interaction between amine groups of chitosan and the negatively charged groups of a polyanion. Tripolyphosphate (TPP) is often used because it is non-toxic, multivalent, and able to form gels by ionic interactions. The interactions can be controlled by the charge density of TPP and chitosan, which depends on the pH [63]. Both the chitosan and its nanoparticles are ineffective at pH <6, probably because of the absence of protonated amino groups [49]. The CSNP present higher antibacterial activity than bulk chitosan because of their higher surface area and charge density, providing a higher interaction with the cationic surfaces of bacterial cells. The CSNP and silver loaded CSNP found more effective than bulk chitosan as their minimum inhibitory concentrations against *S. aureus* were 50 and 500 times (respectively) lower than that of bulk chitosan [2]. The higher antimicrobial activity of silver loaded nanoparticle is attributed to membrane disruption through the generation of intracellular ROS [57]. The electrospun nanofibrous mats developed from quaternized chitosan- organic rectorite modified by cetyltrimethyl ammonium bromide (OREC) -PVOH enhanced their antimicrobial activity [17]. These mats were found more effective against Gram-positive than Gram-negative bacteria as reported by Wang et al. [59].

Grapefruit seed extract (GFSE), grape seed extract and pomegranate peel extract also exhibits antimicrobial effect on aerobic bacteria and yeast. The GFSE at 1% w/w shows AM activity against E. *coli*, *S. aureus*, and *Bacillus subtilis*. Some of the commercially available antimicrobial packaging for food applications is given in Table 2.

TABLE 2 Commercial AM packaging available for food applications (Adopted from Han, [28])

AM Compounds	Trade Name(s)	Producer	Packaging Type
Silver zeolite	Aglon	Aglon Technologies	Paper, milk containers
	Novaron	Toagosei Co	Plastic
Triclosan	Microban	Microban	Deli-wrap, reheatable Containers

TABLE 2   *(Continued)*

Allylisothiocya-nate	WasaOuro	Lintec Corp.	Labels, sheets
		Dry Company	Sachets
Chlorine dioxide	Microsphere	Bernard Tech Inc.	Bags, coatings, labels
Carbon dioxide	Freshpax	Multisorb Tech.	Sachets
	Verifrais	Sarl Codimer	Sachets
Ethanol vapour	Ethicap, Ne-gamold,	Freund	Sachets
	Fretek	Nippon Kayaku	Sachets (Japan)
	Oitech		
Glucose oxidase	Bioka	Bioka Ltd	Sachets (Finland)

21.4   MANUFACTURING AM FILMS

There are four main methods used to produce AM packaging films: (i) use of polymers that have AM properties, (ii) immobilisation of AM agents in polymers by chemical bonding, (iii) incorporation of the AM agents into the polymer matrix during processing without chemical bonding like absorption or impregnation, and (iv) coating of AM agents onto a polymer surface [5].

21.4.1   ANTIMICROBIAL POLYMERS

There are number of polymers available that have an inhibitory effect against the growth of some particular microorganisms. The most common polymer material which acts itself as an active substance is Chitosan from shellfish sources which shows a broader antimicrobial activity against *L. monocytogenes, L. innocua, E. coli, V. vulnificus, S. typhimurium, S. enteritidis, Shigella* spp., *Pseudomonas* spp., *S. aureus*, and so on.

21.4.2   IMMOBILISATION

This involves binding AM enzymes to the polymer, to produces non-migratory bioactive polymers. Unlike other types of AM packaging, the AM enzymes do not migrate into the food but remain active on the packaging surface. This advantage makes this type of packaging desirable for the modern market, where customer requirements are high and safety regulations are strict. The use of this packaging is restricted for liquid food only.

21.4.3   INCORPORATION

This involves the integration of AM agents into the polymer matrix in which the molecules of AM additives are physically bound in the structure of the polymer, and are released during the storage of the product. There are several types of incorporation mechanisms including absorption and impregnation. Organic acids like benzoic and sorbic acids are incorporated by absorption method. Certain antimicrobial agents can be incorporated by impregnation method by mixed with polymer pellets and melting together before the AM films are extruded from the polymer blend

21.4.4   COATING

There are several coating methods for producing AM polymeric films like immersing, spraying, and solvent casting. Nisin can be coated either by immersing or by spraying method. Chitosan mixed with various organic acids can be coated by casting method by pouring the chitosan and organic acid mixture in the mould.

Because of the environmental concerns and technological problems associated with the manufacturing of plastic films, natural biopolymers are becoming popular and are attracting great interest from different sectors. Most of these materials are edible and their film formation occurs under mild conditions. These include zein, methylcellulose, hydroxypropyl methylcellulose, carrageenan, alginate, and whey proteins. Edible AM materials produced by incorporating Lysozyme and Nisin in whey protein isolate (WPI) films are effective in inhibiting *B. thermosphacta*. The incorporation of EDTA in WPI films improves the inhibitory effect on *L. monocytogenes*.

21.5   ANTIMICROBIAL EFFICIENCY OF FILM

The antimicrobial efficiency of the film can be tested using standard strains of spoilage and pathogenic bacteria. For this, the film is cut into circular discs with known mass and sterilized and used for testing both qualitative and quantitative efficiency. For the qualitative test, the circular disc of the film is placed over the inoculated medium with known number of particular bacteria. It was then incubated at desired conditions and the clearance zone around the film disc is measured to assess the efficiency of the antimicrobial film against particular bacteria. For assessing the efficiency of antimicrobial film quantitatively, the film is cut into known area and placed on the sterile bottle and to this the known quantity of overnight cultured bacteria is inoculated and incubated at desired condition. The reduction in the number of bacteria over different time intervals was determined by culturing using serial dilution and the percentage reduction due to antimicrobial film is quantified.

21.6   FACTORS CONSIDERED IN THE MANUFACTURING OF ANTIMICROBIAL PACKAGING

The selection of both packaging material and antimicrobial agent is very important in developing AM packaging. The incorporation of AM agent may change the inherent physico-mechanical properties of the packaging film. To obtain the optimum advantages certain factors to be considered while manufacturing AM packaging.

21.6.1 PROCESS CONDITIONS AND RESIDUAL ANTIMICROBIAL ACTIVITY

Process conditions in the manufacture of packaging film affects the effectiveness of the AM agents. If the AM agents are applied by impregnation method, it may deteriorate during film fabrication, storage and distribution. Further if the extrusion method is used the conditions such as high temperature, pressure and shearing forces may affect the chemical stability of the AM agents. Other operations like lamination, printing, and drying of the film during manufacture may also affect the AM activity. Certain volatile AM compounds may be lost during the preparation and storage. This can be minimized using master batches of the AM agents in the resin for preparation of AM packages [28].

21.6.2 CHARACTERISTICS OF ANTIMICROBIAL SUBSTANCES AND FOODS

Foods with different chemical constituents are stored under different environmental conditions to alter the pattern of microbial growth. In ordinary air pack, aerobic microorganisms utilize headspace O_2 for their growth where as vacuum packed conditions favor the growth of anaerobic microflora. Effectiveness of AM agents alters with the package conditions like head space gas composition, humidity, water activity (a_w), and pH of the food. The pH of a product affects the growth rate of target microorganisms and changes the degree of ionization of the most active chemicals, as well as the activity of the AM agents [28]. Certain AM agents like benzoic anhydride and sorbic acid will be more effective at low pH values. The diffusion of potassium sorbate through polysaccharide films increases with a_w.

21.6.3 CHEMICAL INTERACTION OF ADDITIVES WITH FILM MATRIX

The knowledge of polarity and molecular weight of the additives are important while incorporating the additives into a packaging film. The AM agents with high molecular weight and low polarity are more compatible with the low density polyethylene (LDPE) as it is non-polar in nature. Apart from this, the diffusion of AM agents in the polymer is also affected by the ionic strength and solubility.

21.6.4 STORAGE TEMPERATURE

The storage temperature may also affect the activity of AM packages. The protective action of AM films deteriorates at higher temperatures, due to high diffusion rates in the polymer. The diffusion rate of the AM agent and its concentration in the film must be sufficient to remain effective throughout the shelf life of the product. If the packaging film is stored at refrigerated temperature instead of room temperature, the quantity used for incorporating into the packaging film may be reduced.

21.6.5   PROPERTIES OF PACKAGING MATERIALS

Addition of antimicrobial agents may affect the physical properties of packaging materials and process ability. The major physical properties which are altered with the incorporation of the AM agents are thickness, tensile strength, transparency, heat-seal ability, water vapor permeability, oxygen permeability, heat seal strength, swelling ability, morphology, and ability to withstand varied temperature conditions.

21.6.6   COST

As the AM packaging is still in the developmental stages, there are no reports on the cost of the AM films. But compared to the basic packaging materials, the AM packaging are expected to be slightly expensive and are most suitable for high value food products.

21.6.7   FOOD CONTACT APPROVAL

The AM agents to be used in the manufacture of AM packages should be approved by the competent authority. These AM agents should not be toxic to humans and they should be tested for their safety before using in the AM packages. The approval of AM agents is not uniform in all the countries. For example the use of allyl isothionate is not approved in U.S.A. whereas in Japan it is approved if this is extracted from a natural source. The use of Ag-zeolite is approved by the FDA for food contact applications where as its approval is not clear in EU. The use of Trichlosan is also not approved by EU directives. Overall migration limit from the material into the food or food simulant is set at 60 ppm, which is not compatible with the AP particularly when the system is designed to release active ingredients into the foods. A new approach in food packaging regulations is needed to safeguard the use of active packaging.

21.7   FUTURE

With the increasing consumer demand for fresh, convenient and safe food products, the innovative packaging options, particularly AM packaging for preservation, transportation and storage have bright future. There are plenty of antimicrobial compounds of natural origin are present in environment. Research has to focus on identifying and extracting these natural compounds and characterize them for their safety and antimicrobial activity in compatible with the packaging system. As there are a wide variety of microorganisms to be taken care of in food systems, different combinations of antimicrobial agents can be incorporated in a single packaging system to increase their efficacy. The migration of the added antimicrobial agents into food is the much debated topic. Further detailed research is needed to assess the safety of incorporated antimicrobial agents in the packaging material. An additional challenge is in the area of odor/flavor transfer by natural plant extracts to packaged food products. Thus, research is needed to determine whether natural extracts could act as both an antimicrobial agent and as an odor/flavor enhancer.

KEYWORDS

- **Antimicrobial agents**
- **Antimicrobial efficiency**
- **Antimicrobial packaging**
- **Chitosan**
- **Packaging**

REFERENCE

1. Ahvenainen, R. Active and intelligent packaging: An introduction. In: Novel Food Packaging Techniques, R. Ahvenainen (Ed.), Woodhead Publishing Ltd., Cambridge, UK, pp 5–21 (2003).
2. Ali, S.W., Rajendran, S., and Joshi. M. Synthesis and characterization of chitosan and silver loaded chitosan nanoparticles for bioactive polyester. Carbohydrate Polymers, 83(2), 438–446 (2011).
3. Altiok, D., Altiok, E., and Tihminlioglu. F. Physical, antibacterial and antioxidant properties of chitosan films incorporated with thyme oil for potential wound healing applications. Journal of Materials Science: Materials in Medicine, 21, 2227–2236 (2010).
4. An, J., Zhang, M., Wang, S., and Tang, J. Physical, chemical and microbiological changes in stored green asparagus spears as affected by coating of silver nanoparticles-PVP. LWT—Food Science and Technology, 41(6), 1100–1107 (2008).
5. Appendini, P. and Hotchkiss, J. H. Review of Antimicrobial Food Packaging. Innovative Food Science & Emerging Technology, 3, 113–126 (2002).
6. Arora, D. S. and Kaur, J. Antimicrobial activity of spices. International Journal of Antimicrobial Agent, 12, 257–262 (1999).
7. Bhattarai, S., Tran, V. H., and Duke, C. C. The stability of [6] gingerol and shogaol in aqueous solutions. Journal of Pharmacology Science, 90(10), 1658–1664 (2001).
8. Bhattarai, S., Tran, V. H., and Duke, C. C. Stability of [6]-gingerol and [6]-shogaol in simulated gastric and intestinal fluids. Journal of Pharmacy & Biomedical Anal, 45(4), 648–653(2007).
9. Burt, S. A., Van der Zee, R., Koets, A. P., de Graaff, A. M., van Knapen, F., and Gaastra, W. Carvacrol induces heat shock protein 60 and inhibits synthesis of flagellin in Escherichia coli O157:H7. Applied and Environmental Microbiology, 73, 4484–4490 (2007).
10. Burt. S. Essential oils: their antibacterial properties and potential applications in foods—a review. International Journal of Food Microbiology, 94 (3), 223–253. (2004).
11. Chawengkijwanich, C. and Hayata. Y. Development of TiO2 powder–coated food packaging film and its ability to inactivate Escherichia coli in vitro and in actual tests. International Journal of Food Microbiology, 123, 288–292(2008).
12. Cheng, Y. W., Chan, R. C. Y., and Wong, P. K. Disinfection of Legionella pneumophila by photocatalytic oxidation. Water Research, 41, 842–852 (2007).
13. Cho, M., Chung, H., Choi, W., and Yoon, J. Linear correlation between inactivation of E. coli and OH radical concentration in TiO2 photocatalytic disinfection. Water Research, 38(4), 1069–1077 (2004).

14. Conte, A., Buonocore, G. G., Sinigaglia, M., Lopez, L.C., Favia, P., and d'Agostino, R. Antimicrobial activity of immobilized lysozyme on plasma-treated polyethylene films. Journal of Food Protection, 71, 119–125 (2008).

15. Dallas, P., Sharma, V. K., and Zboril. R. Silver polymeric nanocomposites as advanced antimicrobial agents: classification, synthetic paths, applications, and perspectives. Advances in Colloid and Interface Science, 166(1–2), 119–135 (2011).

16. Damm, C. Münsted, H., and Rösch. A. The antimicrobial efficacy of polyamide 6/silvernano- and microcomposites. Materials Chemistry and Physics, 108, 61–66 (2008).

17. Deng, H., Lin, P., Xin, S., Huang, R., Li, W., and Du. Y. Quaternized chitosan-layered silicate intercalated composites based nanofibrous mats and their antibacterial activity. Carbohydrate Polymers, 89, 307–313 (2012).

18. Dormana, H. J. D., Peltoketo, A., Hiltunen, R., and Tikkanen, M. J. Characterisation of the antioxidant properties of de-odourised aqueous extacts from selected lamiaceae herbs. Food Chemistry, 83(2), 255–262 (2003).

19. Duffy, E. F., Touati, F. A., Kehoe, S. C., McLoughlin, O. A., Gill, L. W., and Gernjak. W. A novel TiO2-assisted solar photocatalytic batch-process disinfection reactor for the treatment of biological and chemical contaminants in domestic drinking water in developing countries. Solar Energy, 77(5), 649–655 (2004).

20. Dugasani, S., Pichika, M. R., Nadarajah, V. D., Balijepalli, M. K., Tandra, S., and Korlakunta, J. N. Comparative antioxidant and anti-inflammatory effects of [6]-gingerol, [8]-gingerol, [10]-gingerol and [6]-shogaol. Journal of Ethnopharm, 127(2), 515–520 (2010).

21. Eswaranandam, S., Hettiarachchy, N. S., and Johnson, M. G. Antimicrobial activity of citric, lactic, malic, or tartaric acids and nisin-incorporated soy protein film against Listeria monocytogenes, Escherichia coli 0157:H7, and Salmonella gaminara. Journal of Food Science, 69, 79–84 (2004).

22. Farber, J. M. Microbiological Aspects of Modified Atmosphere Packaging Technology- A Review. Journal of Food Protection, 54, 58–70 (1991).

23. Feng, Q. L., Wu, J., Chen, G. Q., Cui, F. Z., Kim, T. N., and Kim, J. O. A mechanistic study of the antibacterial effect of silver ions on Escherichia coli and Staphylococcus aureus. Journal of Biomedical Materials Research, 52(4), 662–668 (2000).

24. Friedman, M., Henika, P. R., and Mandrell, R. E. Bactericidal activities of plant essential oils and some of their isolated constituents against Campylobacter jejuni, Escherichia coli, Listeria monocytogenes, and Salmonella enteric. Journal of Food Protection, 65, 1545–1560 (2002).

25. Gómez-Estaca, J., López de Lacey, A., López-Caballero, M. E., Gómez-Guillén, M. C., and Montero, P. Biodegradable gelatin–chitosan films incorporated with essential oils as antimicrobial agents for fish preservation. Food Microbiology, 27, 889–896 (2010).

26. Goñi, P., López, P., Sánchez, C., Gómez-Lus, R., Becerril, R., and Nerín, C. Antimicrobial activity in the vapour phase of a combination of cinnamon and clove essential oils. Food Chemistry, 116, 982–989 (2009).

27. Guimaraes, J. R. and Barretto, A. S. Photocatalytic inactivation of Clostridium perfringens and Coliphages in water. Brazilian Journal of Chemical of Engineering, 20(4), 403–411 (2003).

28. Han, J. H. Antimicrobial Food Packaging. Food Technology, 54, 56–65 (2000).

29. Hoffmann, M. R., Martin, S. T., Choi, W., and Bahnemann, D. W. Environmental applications of semiconductor photocatalysis. Chemical Reviews, 95(1), 69–96 (1995).

30. Hughey, V. L. and Johnson, E. A. Antimicrobial activity of lysozyme against bacteria involved in food spoilage and food-borne disease. Applied and Environmental Microbiology, 53, 2165–2170 (1987).

31. Jagannath, J. H., Radhika, M., Nanjappa, C., Murali, H. S., and Bawa. A. S. Antimicrobial, mechanical, barrier, and thermal properties of starch–casein based, neem (Melia azardirachta) extract containing film. Journal of Applied Polymer Science, 101, 3948–3954 (2006).

32. Jeffery, B., Peppler, M., Lima, R. S., and McDonald, A. Bactericidal effects of HVOF-sprayed nanostructured TiO2 on Pseudomonas aeruginosa. Journal of Thermal Spray Technology, 19, 344–349 (2009).

33. Kim, B., Kim, D., Cho, D., and Cho, S. Bactericidal effect of TiO2 photocatalyst on selected food-borne pathogenic bacteria. Chemosphere, 52(1), 277–281 (2003).

34. Kim, J. M., Marshall, M. R., Cornell, J. A., Preston, J. F., and Wei, C. I. Antibacterial activity of carvacrol, citral, and geraniol against Salmonella Typhimurium in culture medium and on fish cubes. Journal of Food Science, 60, 1364–1374 (1995).

35. Kloucek, P., Smid, J., Frankova, A., Kokoska, L., Valterova, I., and Pavela, R. Fast screening method for assessment of antimicrobial activity of essential oils in vapor phase. Food Research International, 47(2), 161–165 (2011).

36. Kubacka, A., Ferrer, M., Cerrada, M. L., Serrano, C., Sanchez-Chaves, M., and Fernandez-Garcia, M. Boosting TiO2-anatase antimicrobial activity: polymer-oxide thin films. Applied Catalysis B: Environmental, 89, 441–447 (2009).

37. Kuhn, K. P., Chaberny, I. F., Massholder, K., Stickler, M., Benz, V. W., and Sonntag, H. G. Disinfection of surfaces by photocatalytic oxidation with titanium dioxide and UVA light. Chemosphere, 53(1), 71–77 (2003).

38. Kumar, R. and Münstedt, H. Silver ion release from antimicrobial polyamide/silver composites. Biomaterials, 26, 2081–2088 (2005).

39. Lee, S. J., Umano, K., Shibamoto, T., and Lee, K. G. Identification of volatile components in basil (Ocimum basilicum L.) and thyme leaves (Thymus vulgaris L.) and their antioxidant properties. Food Chemistry, 91, 131–137 (2005).

40. Liu, H. L. and Yang. T. C. K. Photocatalytic inactivation of Escherichia coli and Lactobacillus helveticus by ZnO and TiO2 activated with ultraviolet light. Process Biochemistry, 39, 475–481 (2003).

41. Maizura, M., Fazilah, A., Norziah, M. H., and Karim. A. A. Antibacterial activity and mechanical properties of partially hydrolyzed sago starch–alginate edible film containing lemongrass oil. Journal of Food Science, 72, C324–C330 (2007).

42. Maneerat, C. and Hayata, Y. 2006. Antifungal activity of TiO2 photocatalysis against Penicillium expansum in vitro and in fruit tests. International Journal of Food Microbiology, 107(2), 99–103.

43. Matan, N., Rimkeeree, H., Mawson, A. J., Chompreeda, P., Haruthaithanasan, V., and Parker, M. Antimicrobial activity of cinnamon and clove oils under modified atmosphere conditions. International Journal of Food Microbiology, 107(2), 180–185 (2006).

44. Matsunaga, T., Tomada, R., Nakajima, T., and Wake, H. Photochemical sterilization of microbial cells by semiconductor powders. FEMS Microbiology Letters, 29(1–2), 211–214 (1985).

45. Nostro, A., Marino, A., Blanco, A. R., Cellini, L., Di Giulio, M., and Pizzimenti. F. In vitro activity of carvacrol against staphylococcal preformed biofilm by liquid and vapour contact. Journal of Medical Microbiology, 58, 791–797 (2009).

46. Ojagh, S. M., Rezaei, M., Razavi, S. H., and Hosseini, S. M. H. Development and evaluation of a novel biodegradable film made from chitosan and cinnamon essential oil with low affinity toward water. Food Chemistry, 122, 161–166 (2010).

47. Parry, R. T. Introduction. In: Principles and applications of MAP of foods. R. T. Parry (Ed.), Blackie Academic and Professional, New York, USA, pp 1–18 (1993).

48. Pranoto, Y., Salokhe, V. M., and Rakshit, S. K. Physical and antibacterial properties of alginate-based edible film incorporated with garlic oil. Food Research International, 38, 267–272 (2005).

49. Qi, L. F., Xu, Z. R., Jiang, X., Hu, C., and Zou, X. Preparation and antibacterial activity of chitosan nanoparticles. Carbohydrate Research, 339, 2693–2700 (2004).

50. Rabea, E. I., Badawy, M. E., Stevens, C. V., Smagghe, G., and Steurbaut, W. Chitosan as antimicrobial agent: applications and mode of action. Biomacromolecules, 4, 1457–1465 (2003).

51. Rahmani, A. R., Samadi, M. T., and Enayati Moafagh, A. Investigation of photocatalytic degradation of phenol by UV/TiO2 process in aquatic solutions. Journal of Health Research, 8(2), 55–60 (2008).

52. Rooney, M. L. Active packaging in polymer films. In: Active Food Packaging, M. L. Rooney (Ed.), Blackie Academic and Professional, London, pp 74–110 (1995).

53. Shukla, Y. and Singh. M. Cancer preventive properties of ginger: A brief review, Food & Chemical Toxicology, 45(5), 683–690 (2007).

54. Singh, G., Kapoor, I. P. S., Pratibha, S., Heluani, C. S. D., Lampasona, M. P. D., and Catalan, C. A. N. Chemistry, antioxidant and antimicrobial investigations on essential oil and oleoresins of Zingiber officinale. Food & Chemical Toxicology 46(10), 3295–3302 (2008).

55. Sivarooban, T., Hettiarachchy, N. S., and Johnson, M. G. Physical and antimicrobial properties of grape seed extract, nisin, and EDTA incorporated soy protein edible films. Food Research International, 41, 781–785 (2008).

56. Skandamis, P. N. and Nychas, G. J. E. Preservation of fresh meat with active and modified atmosphere packaging conditions. International Journal of Food Microbiology, 79, 35–45 (2002).

57. Su, H. L., Chou, C. C., Hung, D. J., Lin, S. H., Pao, I. C., and Lin, J. H. The disruption of bacterial membrane integrity through ROS generation induced by nano hybrids of silver and clay. Biomaterials, 30, 5979–5987 (2009).

58. Vermeiren, L., Devlieghere, F., Van Beest, M., De Kruijf, N., and Debevere, J. Developments in the active packaging of foods. Trends Food Science & Technology, 10(3), 77–86 (1999).

59. Wang, X., Du, Y., Luo, J., Yang, J., Wang, W., and Kennedy, J. F. A novel biopolymer/rectorite nanocomposite with antimicrobial activity. Carbohydrate Polymers, 77, 449–456 (2009).

60. Wu, T., Zivanovic, S., Draughon, F. A., Conway, W. S., and Sams, C. E. Physicochemical properties and bioactivity of fungal chitin and chitosan. Journal of Agricultural and Food Chemistry, 53(10), 3888–3894 (2005).

61. Yu, J. G., Yu, H. G., Cheng, B., Zhao, X. J., Yu, J. C., and Ho, W. K. The effect of calcination temperature on the surface microstructure and photocatalytic activity of TiO2 thin films prepared by liquid phase deposition. Journal of Physical Chemistry B, 107(50), 13871–13879 (2003).

62. Zarate, R. and Yeoman. M. M. Change in the amounts of [6] gingerol and derivatives during a culture cycle of ginger, Zingiber officinale. Plant Science, 121(1), 115–122 (1996).

63. Zhao, L. M., Shi, L. E., Zhang, Z. L., Chen, J. M., Shi, D. D., and Yang, J. Preparation and application of chitosan nanoparticles and nanofibers. Brazilian Journal of Chemical Engineering, 28: 353–362 (2011).
64. Han, J. H. Antimicrobial food packaging. Novel food packaging techniques, R. Ahvenainen (Ed.), Woodhead, CRC Press, Cambridge, Boca Raton, Florida, 47, 590 (2003).

INDEX